NUREG-1575, Supp. 1
EPA 402-R-09-001
DOE/HS-0004

I0488106

Multi-Agency Radiation Survey and Assessment of Materials and Equipment Manual (MARSAME)

Department of Defense
Department of Energy
Environmental Protection Agency
Nuclear Regulatory Commission

January 2009

DISCLAIMER

This supplement was developed by four agencies of the United States Government. Neither the United States Government nor any agency or branch thereof, or any of their employees, makes any warranty, expressed or implied, or assumes any legal liability of responsibility for any third party's use, or the results of such use, of any information, apparatus, product or process disclosed in this supplement, or represents that its use by such third party would not infringe on privately owned rights.

References within this supplement to any specific commercial product, process, or service by trade name, trademark, or manufacturer does not constitute an endorsement or recommendation by the United States Government.

ABSTRACT

The *Multi-Agency Radiation Survey and Assessment of Materials and Equipment* manual (MARSAME) is a supplement to the *Multi-Agency Radiation Survey and Site Investigation Manual* (MARSSIM) providing information on planning, conducting, evaluating, and documenting radiological disposition surveys for the assessment of materials and equipment. MARSAME is a multi-agency consensus document that was developed collaboratively by four Federal agencies having authority and control over radioactive materials: Department of Defense (DOD), Department of Energy (DOE), Environmental Protection Agency (EPA), and Nuclear Regulatory Commission (NRC). The objective of MARSAME is to provide a multi-agency approach for planning, performing, and assessing disposition surveys of materials and equipment, while at the same time encouraging an effective use of resources.

CONTENTS

LIST OF FIGURES

LIST OF TABLES

ACKNOWLEDGEMENTS

The *Multi-Agency Radiation Survey and Site Investigation Manual* (MARSSIM) and the *Multi-Agency Radiation Survey and Assessment of Materials and Equipment* manual (MARSAME) supplement came about as a result of individuals—at the management level—within the Environmental Protection Agency (EPA), Nuclear Regulatory Commission (NRC), Department of Energy (DOE), and Department of Defense (DOD) who recognized the necessity for a standardized guidance document for investigating radioactively contaminated sites. The creation of MARSSIM and MARSAME was facilitated by the cooperation of subject matter specialists from these agencies with management's support and a willingness to work smoothly together toward reaching the common goal of creating a workable and user-friendly guidance manual. Special appreciation is extended to Robert A. Meck of the NRC and Anthony Wolbarst of EPA for developing the concept of a multi-agency workgroup and bringing together representatives from the participating agencies.

MARSAME could not have been possible without the technical workgroup members who contributed their time, talent, and efforts to develop this consensus guidance document:

CAPT Colleen F. Petullo, U.S. Public Health Service, EPA, Chair

DOD David P. Alberth (Army)
    Dennis Chambers, CHP (Army, Retired)
    Gerald Falo, Ph.D., CHP (Army)
    Steven Doremus, Ph.D. (Navy)
    CAPT Vincent DeInnocentiis (Navy)
    Ramachandra Bhat, Ph.D., CHP (Air Force)
    Lt Col Craig Bias, Ph.D., CHP (Air Force)
    Lt Col Daniel Caputo, Ph.D. (Air Force Reserve)

EPA Kathryn Snead
    Nidal Azzam
    Lindsey Bender
    Vicki Lloyd
    Eugene Jablonowski

DOE W. Alexander Williams, Ph.D.
    Emile Boulos
    Harold T. Peterson, Jr., CHP (Retired)
    Amanda Anderson
    Wayne Glines, CHP

NRC Robert A. Meck, Ph.D.
    George E. Powers, Ph.D.
    Joseph DeCicco, CHP
    Anthony Huffert, CHP

DHS Carl V. Gogolak, Ph.D. (Retired)

Special mention is extended to the Federal Agency contractors for their assistance in developing the MARSAME supplement:

Scott Hay (Cabrera Services, Inc.)
Carl V. Gogolak (Environmental Management Support, Inc.)
Nicholas Berliner (Cabrera Services, Inc.)
Robert Coleman (Oak Ridge National Laboratory)
Deborah Schneider (S. Cohen & Associates, Inc.)
Kerri Wachter (S. Cohen & Associates, Inc.)

A special thank you is extended to Mary Clark (EPA), Schatzi Fitz-James (EPA), Paul Giardina (EPA), Bonnie Gitlin (EPA), Sally Hamlin (EPA), David Kappelman (EPA), Sophie Kastner

(EPA), Joseph LaFornara (EPA), Juan Reyes (EPA), Colby Stanton (EPA), Dennisses Valdes (EPA), Jean-Claude Dehmel (NRC), MAJ David Pugh, CHP (Air Force), Brian Renaghan (Air Force), Andrew Wallo III (DOE), Ethel Jacob (DHS), Kevin Miller (DHS), Peter Shebell (DHS), Jenny Goodman (NJ Bureau of Environmental Radiation), Nancy Stanley (NJ Bureau of Environmental Radiation), and Eric Abelquist (Oak Ridge Institute for Science and Education).

The Workgroup would also like to thank EPA's Science Advisory Board Radiation Advisory Committee for their consultations and peer review supporting development of the MARSAME supplement:

Chair

Bernd Kahn, Ph.D., Georgia Institute of Technology
Jill Lipoti, Ph.D., New Jersey Department of Environmental Protection (Past Chair)

Members

Thomas B. Borak, Ph.D., Colorado State University
Antone L. Brooks, Ph.D., Washington State University Tri-Cities
Faith G. Davis, Ph.D., University of Illinois at Chicago
Brian Dodd, Ph.D., Consultant
Shirley A. Fry, Ph.D., Consultant
William C. Griffith, Ph.D., University of Washington
Jonathan M. Links, Ph.D., Johns Hopkins University
Bruce A. Napier, Pacific Northwest National Laboratory
Daniel O. Stram, Ph.D., University of Southern California
Richard J. Vetter, Ph.D., Mayo Clinic

SAB Consultants

Bruce W. Church, BWC Enterprises, Inc.
Kenneth Duvall, Environmental Scientist/Consultant
Janet A. Johnson, Ph.D., Consultant
Paul J. Merges, Ph.D., Environment & Radiation Specialists, Inc.

Science Advisory Board Staff

K. Jack Kooyoomjian, Ph.D., Designated Federal Officer, EPA

The Workgroup acknowledges the interest of the NRC's Advisory Committee on Nuclear Waste and Materials in the development of MARSAME.

ACRONYMS AND ABBREVIATIONS

AL   action level
ALARA as low as reasonably achievable
ANSI   American National Standards Institute
ASTM   American Society for Testing and Materials

BKGD   background

CERCLA Comprehensive Environmental Response Compensation and Liability Act
CFR   Code of Federal Regulations
cpm   counts per minute
cps   counts per second
CSM   conceptual site model
CSU   combined standard uncertainty
CZT   cadmium zinc telluride

DAC   derived air concentration
DCGL   derived concentration guideline level
DL   discrimination limit
DOD   Department of Defense
DOE   Department of Energy
DOT   Department of Transportation
dpm   disintegrations per minute
DQA   data quality assessment
DQO   data quality objective

EMC   elevated measurement comparison
EPA   Environmental Protection Agency
EPRI   Electric Power Research Institute
EU   European Union
EZ   exclusion zone

FIDLER field instrument for the detection of low-energy radiation
FRER   fluence rate to exposure rate

GM   Geiger Mueller

HASP   health and safety plan
HEU   high-enriched uranium
HPGe   high-purity germanium
HPS   Health Physics Society
HSA   Historical Site Assessment
HPSR   Health Physics Society Report
HWP   hazard work permit

IA   initial assessment
IAEA   International Atomic Energy Agency
IEEE   Institute of Electrical & Electronics Engineers
ISGS   in situ gamma spectroscopy

ISO   International Organization for Standardization

JSA   job safety analysis

LBGR   lower bound of the gray region
LEU   low-enriched uranium
LSA   low specific activity
LSC   liquid scintillation cocktail

M&E   materials and equipment
MARLAP   Multi-Agency Radiological Laboratory Analytical Protocols manual
MARSAME   Multi-Agency Radiation Survey and Assessment of Materials and Equipment
                manual
MARSSIM   Multi-Agency Radiation Survey and Site Investigation Manual
MCA   multi-channel analyzer
MDC   minimum detectable concentration
MDCR   minimum detectable count rate
$MDCR_{surveyor}$ MDCR by a less than ideal surveyor
MDER   minimum detectable exposure rate
MQC   minimum quantifiable concentration
MQO   measurement quality objective

NARM   naturally occurring and accelerator-produced radioactive material
NCRP   National Council on Radiation Protection and Measurements
NIST   National Institute of Science and Technology
NJBER   New Jersey Bureau of Environmental Radiation
NORM   naturally occurring radioactive material
NRC   Nuclear Regulatory Commission
NUREG   Nuclear Regulatory Commission technical report prepared by NRC staff
NUREG/CR   Nuclear Regulatory Commission technical report prepared by NRC contractor

ORISE   Oak Ridge Institute for Science and Education
OSHA   Occupational Safety and Health Administration
OSWER   EPA Office of Solid Waste and Emergency Response

PCB   polychlorinated biphenyl
pH   hydrogen ion concentration (acidity or basicity)
PIC   pressurized ion chamber
PPE   personal protective equipment
PVC   polyvinylchloride

QA   quality assurance
QAPP   quality assurance project plan
QC   quality control

RCA   radiological control area
RCRA   Resource Conservation and Recovery Act
RCSU   relative combined standard uncertainty
RDR   relative detector response
RESRAD   RESidual RADioactivity computer code (exposure pathway model)
ROC   radionuclide of concern
RTG   Radioisotopic Thermoelectric Generator

RWP   radiation work permit

SCO   surface-contaminated object
SI   International System of Units (Système International d'Unités)
SOP   standard operating procedure

TEDE   total effective dose equivalent
TENORM   technologically enhanced naturally occurring radioactive material
TRU   transuranic

UBGR   upper bound of the gray region
UCL   upper confidence limit
UMTRCA   Uranium Mill Tailings Radiation Control Act
UNSCEAR   United Nations Scientific Committee on the Effects of Atomic Radiation
USEPA   United States Environmental Protection Agency
U.S.   United States
WRS   Wilcoxon Rank Sum

SYMBOLS, NOMENCLATURE, AND NOTATIONS

$<$	less than
$>$	greater than
$\leq$	less than or equal to
$\geq$	greater than or equal to
$\circ$	degrees (angle or temperature)
%	percent
$1-\beta$	statistical power of a hypothesis test
α	Type I decision-error rate
α_Q	quantile test ($\alpha_Q = \alpha/2$)
a	half-width of a rectangular or triangular probability distribution
A	area
A	overall sensitivity of a measurement
Ac	actinium (isotope listed: ^{228}Ac)
AL_i	action level value an individual radionuclide ($i = 1, 2, ..., n$)
$AL_{meas,mod}$	modified action level for the radionuclide being measured when it is used as a surrogate for other radionuclide(s)
AL_{meas}	action level for the radionuclide being measured
AL_{infer}	action level for the inferred radionuclide (in surrogate measurements)
Am	americium (isotope listed: ^{241}Am)
β	Type II decision-error rate
b	background count rate
b_i	the average number of counts in the background interval (scanning)
Be	beryllium (isotope listed: ^{7}Be)
Bi	bismuth (isotopes listed: ^{210}Bi, ^{212}Bi, ^{214}Bi)
Bq	becquerel
C	carbon (isotope listed: ^{14}C)
C	radionuclide concentration or activity
Ci	curie
C_i	concentration value an individual radionuclide ($i = 1, 2, ..., n$)
c_i	sensitivity coefficient
$c_i\mu(x_i)$	component of the uncertainty in y due to x_i
C_{infer}/C_{meas}	ratio of amount of the inferred radionuclide to that of the measured surrogate radionuclide
°C	degrees Celsius
cm	centimeter
cm^2	square centimeter
cm^3	cubic centimeter
Cd	cadmium (isotope listed: ^{109}Cd)
Co	cobalt (isotopes listed: ^{57}Co, ^{60}Co)
Cs	cesium (isotope listed: ^{137}Cs)
CsI(Tl)	cesium iodide (thallium activated)
Δ	shift (width of the gray region, UBGR–LBGR)

Δ/σ	relative shift
d	parameter in the Stapleton Equation for the critical net signal
d'	detectability index (scanning)
ε_i	instrument efficiency
ε_s	surface efficiency for surveyed media
eV	electron-volt
E_γ	energy of a gamma photon of concern in kiloelectron-volts (keV)
E_i	energy of a photon of interest
°F	degrees Fahrenheit
f_i	relative fraction of activity contributed by radionuclide i to the total
ft	foot (feet)
ft^3	cubic foot (feet)
Fe	iron (isotope listed: ^{55}Fe)
g	gram
GBq	gigabecquerel (1×10^9 becquerels)
GG_{AL}	gross gamma action level
h	hour
H	hydrogen (isotope listed: ^{3}H [tritium])
H_0	null hypothesis
H_1	alternative hypothesis
i	observation time interval length (scanning)
I	iodine (isotopes listed: ^{123}I, ^{125}I, ^{131}I)
in	inch
Ir	iridium (isotope listed: ^{192}Ir)
k	coverage factor for the expanded uncertainty, U
K	potassium (isotope listed: ^{40}K)
kBq	kilobecquerel (1×10^3 becquerels)
keV	kiloelectron-volt (1×10^3 electron-volts)
kg	kilogram
k_Q	multiple of the standard deviation defining y_Q, usually chosen to be 10
L	grid size spacing
L	liter
lb	pound
μ	micro (10^{-6})
μ	theoretical mean of a population distribution
$(\mu_{en}/\rho)_{air}$	mass energy absorption coefficient in air centimeters squared per gram (cm^2/g)
μR	microroentgen (1×10^{-6} roentgen)
m	number of reference measurements (WRS test or Quantile test)
m	meter
m^2	square meter
MeV	megaelectron-volt (1×10^6 electron-volt)
mrem	millirem (1×10^{-3} rem)

mSv	milliseivert (1×10^{-3} Sv)
n	number of survey unit measurements (WRS test or Quantile test)
N	sample size, i.e. number of data points (or samples) for the Sign test
n_{EA}	survey unit area divided by the maximum area corresponding to the area factor, which yields the number of measurements needed so the scan MDC is adequate
Na	sodium (isotope listed: ^{22}Na)
NaI(Tl)	sodium iodide (thallium activated)
Ni	nickel (isotope listed: ^{63}Ni)
Np	neptunium (isotope listed: ^{237}Np)
ξ_B	non-Poisson variance component of the background count rate correction
p	coverage probability for expanded uncertainty, also used for efficiency of a less than ideal surveyor (scanning)
P	probability of interaction between radiation and a detector
Pa	protactinium (isotopes listed: ^{234}Pa, ^{234m}Pa)
PA	probe area
Pb	lead (isotopes listed: ^{212}Pb, ^{214}Pb)
PC	personal computer
pCi	picocurie (1×10^{-12} curies)
Pm	promethium (isotope listed: ^{147}Pm)
Po	polonium (isotopes listed: ^{210}Po, ^{212}Po, ^{214}Po, ^{216}Po)
Pu	plutonium (isotopes listed: ^{238}P, ^{239}Pu, ^{240}Pu, ^{241}Pu)
q	critical value for statistical tests (Table A.3, Table A.4)
ρ	density
$\rho(X_i,X_j)$	correlation coefficient for two input quantities, X_i and X_j
R	ratio
R	roentgen (exposure rate)
Ra	radium (isotopes listed: ^{224}Ra, ^{226}Ra, ^{228}Ra)
R_B	mean background count rate
R_I	mean interference count rate
Rn	radon (isotopes listed: ^{220}Rn, ^{222}Rn)
$r(x_i,x_j)$	correlation coefficient for two input estimates, x_i and x_j
σ	theoretical total standard deviation of the population distribution being sampled
σ_M	theoretical measurement standard deviation of the population distribution being sampled, estimated by the combined standard uncertainty of the measurement
σ_M^2	theoretical measurement variance of the population distribution being sampled
σ_{MR}	required measurement method standard deviation (upper limit)
σ_s	theoretical sampling standard deviation of the population distribution being sampled
σ_S^2	theoretical sampling variance of the population distribution being sampled
$\sigma(\bar{R}_I)$	standard deviation of the measured interference count rate
$\sigma(y\|Y=y_Q)$	variance of the estimator y given the true concentration Y equals y_Q
$\sigma(X_i, X_j)$	covariance for two input quantities, X_i and X_j

$S+$	Sign test statistic
$s(x)$	sample standard deviation of the input estimate, x_i
S_C	critical value of the net instrument signal
S_D	mean value of the net signal that gives a specified probability, $1-\beta$, of yielding an observed signal greater than its critical value S_C
s_i	minimum detectable number of net source counts in the observation interval (scanning)
$s_{i,surveyor}$	minimum detectable number of net source counts in the observation interval by a less than ideal surveyor (scanning)
Sr	strontium (isotope listed: ^{90}Sr)
Sv	seivert
Tc	techicium (isotopes listed: ^{99}Tc, ^{99m}Tc)
Th	thorium (isotopes listed: ^{228}Th: ^{230}Th, ^{232}Th, ^{234}Th)
Tl	thalium (isotopes listed: ^{201}Tl, ^{208}Tl)
t_B	count time for the background
t_S	count time for the source
U	expanded uncertainty
U	uranium (isotopes listed: ^{234}U, ^{235}U, ^{238}U)
$u(x_i)$	standard uncertainty of the input estimate, x_i
$u(x_i)/\;\mid x_i\mid$	relative standard uncertainty of x_i
$u(x_i,x_j)$	covariance of two input estimates, x_i and x_j
$u_c(y)$	combined standard uncertainty of y
$u_c(y)/y$	relative combined standard uncertainty of the output quantity for a particular measurement
$u_c^2(y)$	combined variance of y
$u_i(y)$	component of the combined standard uncertainty, $u_c(y)$, generated by the standard uncertainty of the input estimate x_i, $u(x_i)$, multiplied by the sensitivity coefficient, c_i
u_M	measurement method uncertainty
u_{MR}	required measurement method uncertainty
φ_{MR}	required relative measurement method uncertainty
ϕ_A^2	relative variance of the measured sensitivity
$\varphi(x_i)$	relative standard uncertainty of a nonzero input estimate, x_i, for a particular measurement. $\varphi(x_i) = u(x_i)/x_i$
$\Phi(z)$	cumulative normal distribution function
W_r	sum of the ranks of the (adjusted) reference measurements (WRS test)
W_s	sum of the ranks of the (adjusted) sample measurements (WRS test)
WS	weighted instrument sensitivity
x	estimate of the input quantity, X
X_i	an input quantity
x_C	the critical value of the response variable, x
x_Q	minimum quantifiable value of the response variable, x
y	year

y	estimate of the output quantity for a particular measurement, Y
Y	output quantity, measurand
y_C	critical value of the concentration
y_D	minimum detectable concentration (MDC)
y_Q	minimum quantifiable concentration (MQC)
yd	yard
yd^3	cubic yard
Z	atomic number
$z_{1-\alpha}$	$(1 - \alpha)$-quantile of the standard normal distribution
$z_{1-\beta}$	$(1 - \beta)$-quantile of the standard normal distribution
ZnS(Ag)	zinc sulfide (silver activated)

CONVERSION FACTORS

To Convert From	To	Multiply "From" Quantity By	To Convert From	To	Multiply By
acre	hectare	0.405	meter (m)	inch	39.4
	sq. meter (m^2)	4,050		mile	0.000621
	sq. feet (ft^2)	43,600	sq. meter (m^2)	acre	0.000247
becquerel (Bq)	curie (Ci)	2.7×10^{-11}		hectare	0.0001
	dps	1		sq. feet (ft^2)	10.8
	pCi	27		sq. mile	3.86×10^{-7}
Bq/kg	pCi/g	0.027	m^3	liter	1,000
Bq/m^2	dpm/100 cm^2	0.60	mrem	mSv	0.01
Bq/m^3	Bq/L	0.001	mrem/y	mSv/y	0.01
	pCi/L	0.027	mSv	mrem	100
centimeter (cm)	inch	0.394	mSv/y	mrem/y	100
Ci	Bq	3.70×10^{10}	ounce (oz)	liter (L)	0.0296
	pCi	1×10^{12}	pCi	Bq	0.037
dps	dpm	60		dpm	2.22
	pCi	27	pCi/g	Bq/kg	37
dpm	dps	0.0167	pCi/L	Bq/m^3	37
	pCi	0.451	rad	Gy	0.01
dpm/100 cm^2	Bq/m^2	1.67	rem	mrem	1,000
gray (Gy)	rad	100		mSv	10
hectare	acre	2.47		Sv	0.01
liter (L)	cm^3	1000	seivert (Sv)	mrem	100,000
	m^3	0.001		mSv	1,000
	ounce (fluid)	33.8		rem	100

ROADMAP

Introduction to MARSAME

The *Multi-Agency Radiation Survey and Assessment of Materials and Equipment* manual (MARSAME) is a supplement to the *Multi-Agency Radiation Survey and Site Investigation Manual* (MARSSIM 2002). MARSAME provides technical information on approaches for planning, implementing, assessing, and documenting surveys to determine proper disposition of materials and equipment (M&E).

The technical information in MARSAME is based on the data life cycle, similar to MARSSIM. Survey planning is based on the data quality objectives (DQO) process and is discussed in MARSAME Chapters 2, 3, and 4. Implementation of the survey design is described in MARSAME Chapter 5, with discussions on selection of instruments and measurement techniques as well as handling and segregating the M&E. MARSAME also includes the concept of measurement quality objectives (MQOs) for selecting and evaluating instruments and measurement techniques from the *Multi-Agency Radiological Laboratory Analytical Protocols* manual (MARLAP 2004). Assessment of the survey results uses data quality assessment (DQA) and the application of statistical tests as described in MARSAME Chapter 6. In addition to the first six chapters, which present the MARSAME process, the MARSAME manual contains the statistical basis for the DQOs, MQOs, and survey designs (Chapter 7) and illustrative examples of the information and process presented in MARSAME (Chapter 8).

The scope of MARSSIM was limited to surfaces soils and building surfaces. The scope of MARSAME is M&E potentially affected by radioactivity, including metals, concrete, tools, equipment, piping, conduit, furniture and dispersible bulk materials such as trash, rubble, roofing materials, and sludge. The wide variety of M&E requires additional flexibility in the survey process, and this flexibility is incorporated into MARSAME.

The Goal of the Roadmap

The increased flexibility of MARSAME comes with increased complexity. The goal of the roadmap is to assist the MARSAME user in negotiating the information in MARSAME and determining where important decisions need to be made on a project-specific basis, as summarized in Roadmap Figure 1. Roadmap Figure 2 provides additional detail and illustrates how the data life cycle is applied to disposition surveys. (Shaded blocks within the figures depict significant decisions or milestones.)

This roadmap is not designed to be a stand-alone document, but to be used as a quick reference to MARSAME for users already familiar with the process of planning, implementing, and assessing surveys. Roadmap users will find flowcharts summarizing major decision points in the survey process combined with references to sections in MARSAME with more detailed information. The roadmap assumes a familiarity with MARSAME terminology. Section 1.2 of MARSAME discusses key terminology, and a complete set of definitions is provided in the glossary.

Initial Assessment

The initial assessment (IA) is the first step in the investigation of M&E, similar to the historical site assessment (HSA) in MARSSIM. The purpose of the IA is to collect and evaluate information about the M&E to support a categorization decision and support potential disposition of the M&E (e.g., release or interdiction). Project managers are encouraged to use the IA to evaluate M&E for other hazards (e.g., lead, PCBs, asbestos) that could increase the complexity of the disposition survey design or pose potential risks to workers during subsequent survey activities (Section 5.2), or to human health and the environment following subsequent disposition of the M&E.

Categorization

MARSAME uses the term categorization to describe the decision of whether M&E are impacted or non-impacted. Non-impacted is a term that applies to M&E where there is no reasonable potential to contain radionuclide concentration(s) or radioactivity above background. Impacted is a term that applies to M&E that are not classified as non-impacted. Roadmap Figure 3 shows the categorization process as part of the IA.

Standardized Survey Designs

Most operating radiological facilities maintain standard operating procedures (SOPs) as part of a quality system. In many cases these SOPs include instructions for conducting disposition surveys. The first step in evaluating an existing SOP is to determine whether there is adequate information available to design a disposition survey. If the existing information is inadequate to design a disposition survey, it is inadequate for determining if an existing survey design is adequate either. Roadmap Figure 4 addresses assessing the adequacy of existing information for designing disposition surveys. Roadmap Figure 5 shows how implementing an existing SOP that is applicable to the M&E being investigated takes the user from MARSAME Chapter 2 to MARSAME Chapter 6. If a project-specific survey design needs to be developed, Roadmap Figure 5 directs the user to the information in MARSAME Chapter 3.

In some cases, it may be possible to modify the M&E to match the assumptions used to develop the existing SOP, or modify the existing SOP to address the M&E being investigated. M&E may be modified by changing the physical attributes described in Table 2.1 or the radiological attributes described in Table 2.2. Modifications to the SOP are most often associated with MQOs such as the measurement detectability (Section 5.7) or measurement quantifiability (Section 5.8). Modifying the MQOs may result in small changes such as an increased count time (e.g., to account for an increase in measurement uncertainty or a decrease in counting efficiency) or larger changes such as selecting a different instrument (e.g., a gas-proportional detector instead of a Geiger-Mueller detector) or a different measurement technique (e.g., in situ measurements instead of scan measurements). Information on evaluating an existing survey design to determine if it will meet the DQOs for the M&E being investigated is provided in Section 3.10.

Develop a Decision Rule

MARSAME Chapter 3 focuses on developing a decision rule by identifying inputs to the decision (see Roadmap Figure 6, which depicts the various inputs to the decision). A decision rule is a theoretical "if...then..." statement that defines how the decision maker would choose among alternative actions. There are three parts to a decision rule:

- An action level that causes a decision maker to choose between the alternative actions (see Roadmap Figure 7 and Section 3.3),
- A parameter of interest that is important for making decisions about the target population (see Section 3.4), and
- Alternative actions that could result from the decision (Section 3.5).

Other inputs to the decision that are discussed in MARSAME Chapter 3 include selecting radionuclides or radiations of concern (Section 3.2), developing survey unit boundaries (see Roadmap Figure 8 and Section 3.6), inputs for selecting provisional measurement methods (Section 3.8), and identifying reference materials if necessary (Section 3.9).

Survey Design

Once a decision rule has been developed, a disposition survey can be designed for the impacted M&E being investigated. The disposition survey incorporates all of the available information to determine the quality and quantity of data required to support a disposition decision. Roadmap Figure 9 shows how a disposition survey design is developed.

MARSAME, like MARSSIM, provides information on using a null hypothesis that radionuclide concentrations or activity levels associated with the M&E exceed the action level (i.e., Scenario A). MARSAME also incorporates additional technical information from NUREG-1505 (NRC 1998a) and MARLAP for designing surveys using Scenario B where the null hypothesis is that the radionuclide concentrations or activity levels are less than the action level. The assignment of values to the lower bound of the gray region (LBGR) and upper bound of the gray region (UBGR), specification of decision error rates, and classification are all similar to information provided in MARSSIM.

MARSAME provides information on four types of survey designs:

- Scan-only survey designs (Section 4.4.1),
- In situ survey designs (Section 4.4.2),
- Survey designs that combine scans and static measurements (MARSSIM-type surveys, Section 4.4.3), and
- Method-based survey designs (Section 4.4.4).

A method-based survey design is a special type of scan-only, in situ, or MARSSIM-type survey design that incorporates a required measurement method or combination of measurement technique and instrumentation, so Roadmap Figure 9 only depicts the first three. It will still need to address all of the required components, such as number, type, location, and sensitivity of measurements.

Scan-only survey designs use scanning techniques to measure the M&E. In general, scan-only survey designs may be applied to all types of M&E, from small individual items to large quantities of materials to large, complex machines. Scan-only surveys range from hand-held instruments moving over the M&E to conveyorized systems that move the M&E past the detectors. Scan-only survey designs often require the least amount of resources to design and implement, and are easy to incorporate into SOPs or project-specific survey designs. In many cases it is not necessary to document the results of individual scanning measurements because it is easy to identify results that exceed some threshold corresponding to the action level. With the real-time feedback available during Class 1 scan-only surveys, the user can implement a "clean as you go" practice by segregating M&E that exceed the threshold for additional investigation. Drawbacks to scan-only surveys include increased measurement uncertainty because of variations in scan speed and source to detector distance making it difficult to detect or quantify radionuclides with action levels close to zero or background.

In situ surveys are characterized by limited numbers of static measurements with long count times (relative to scan-only surveys) to measure the M&E. In situ surveys generally are applicable to situations where scan-only surveys are determined to be unacceptable. For example, variations in source-to-detector distance, scan speed, and surface efficiency that are commonly associated with scanning measurements can often be effectively controlled using an in situ survey design. There are a wide variety of in situ measurement techniques available including box counters, portal monitors, in situ gamma spectroscopy systems, and direct measurements with hand-held instruments. The primary difference between an in situ survey and a MARSSIM-type survey is that an in situ survey measures 10–100% of an item (using one or several measurements) to determine the average radionuclide concentration for that item. A MARSSIM-type survey uses a statistically based number of measurements (that generally do not measure 10% of the item or group of items being surveyed) to calculate an average radionuclide concentration for that item or group of items.

MARSSIM-type survey designs combine a statistically based number of static measurements or samples (Roadmap Figure 10) to determine average radionuclide concentrations with scanning to identify localized areas of elevated activity (Roadmap Figure 11). MARSSIM-type surveys are designed using the information in MARSSIM. The process of identifying survey unit sizes, laying out systematic or random measurement grids, and calculating project- and item-specific area factors requires significantly greater effort during planning and implementation than either scan-only or in situ survey designs. In general, MARSSIM-type surveys of M&E are only performed on large, complicated M&E with a high inherent value after scan-only and in situ survey designs have been considered and rejected as inappropriate or unacceptable.

Measurement Quality Objectives

Measurement quality objectives (MQOs) are characteristics of a measurement method required to meet the objectives of the survey. Additional information on MQOs can be found in MARSAME Section 3.8, Section 5.5, and Section 7.3 as well as MARLAP Chapter 3. MQOs are an important concept that was not presented in MARSSIM, and should be an important factor when evaluating existing survey designs and SOPs for applicability to surveying M&E. MQOs for a project include, but are not limited to—

- The measurement method uncertainty at a specified concentration expressed as a standard deviation (Sections 3.8.1, 5.5, and 7.4);
- The measurement method's detection capability expressed as the minimum detectable concentration (MDC; see Sections 3.8.2, 5.7, and 7.5);
- The measurement method's quantification capability expressed as the minimum quantifiable concentration (MQC; see Sections 3.8.3, 5.8, 7.6, and 7.7);
- The measurement method's range, which defines the measurement method's ability to measure the radionuclide or radiation of concern over some specified range of concentration or activity (see Section 3.8.4 and Appendix D);
- The measurement method's specificity, which refers to the ability of the measurement method to measure the radionuclide or radiation of concern in the presence of interferences (Section 3.8.5); and
- The measurement method's ruggedness, which refers to the relative stability of measurement method performance for small variations in measurement method parameter values (see Section 3.8.6 and Appendix D).

Implement the Survey Design

The implementation phase of the data life cycle is when the activities described in the survey design are performed. Roadmap Figure 12 illustrates the process for implementing disposition surveys.

MARSAME, like MARSSIM, does not provide prescriptive guidance for implementing survey designs. Chapter 5 presents topics to be considered while implementing disposition surveys. This approach allows MARSAME users flexibility to use either existing or new and innovative techniques that meet the survey objectives.

Evaluate the Results

The assessment phase of the data life cycle involves evaluating the results of the survey as shown in Roadmap Figure 13. DQA is used to evaluate the survey results. DQA is a scientific and statistical evaluation that determines whether data are the type, quality, and quantity to support their intended use. When individual measurement results are not recorded, as allowed in some scan-only survey designs, the preliminary data review will be brief and based primarily on the results of quality control (QC) measurements. To increase the flexibility and general applicability of MARSAME, several evaluation methods have been incorporated in addition to the Sign test and Wilcoxon Rank Sum (WRS) test used in MARSSIM. Roadmap Figure 14 presents information on interpreting survey results for scan-only and in situ surveys. Roadmap Figure 15 presents information on interpreting survey results for MARSSIM-type surveys.

Summary

The roadmap presents a summary of the data life cycle as it applies to disposition surveys in MARSAME and identifies where information on important topics are located in MARSAME. Flow charts are provided to summarize major steps in the survey design process, again citing appropriate references in MARSAME.

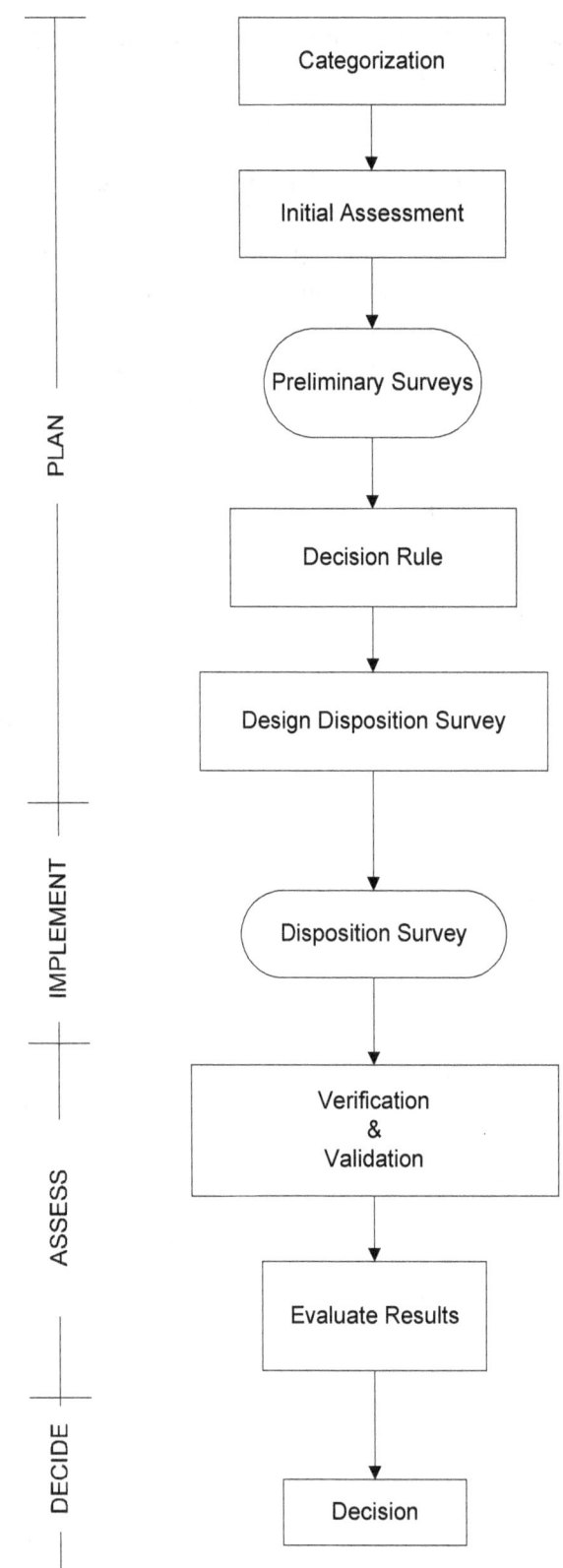

Roadmap Figure 1. Overview of MARSAME Process

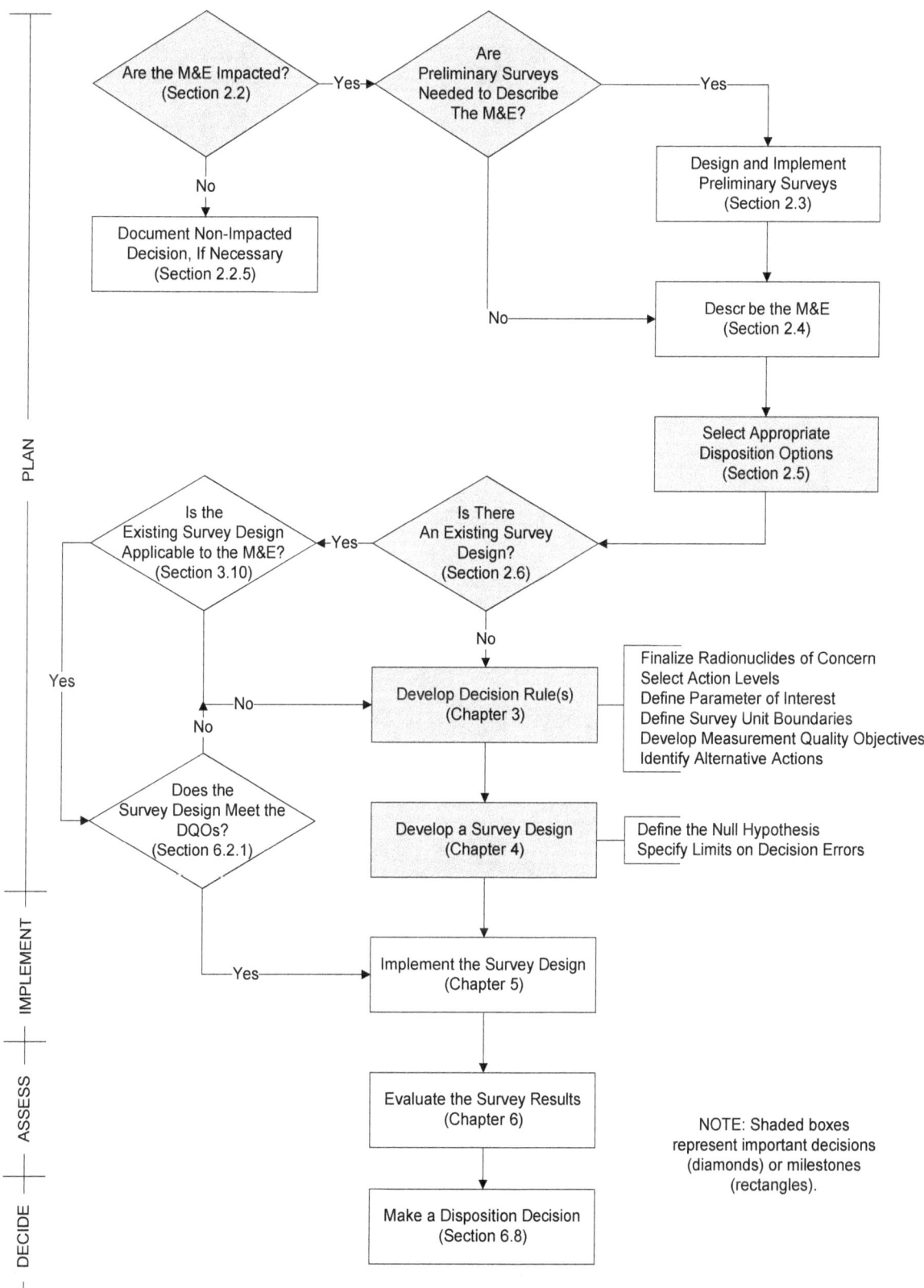

Roadmap Figure 2. The Data Life Cycle Applied to Disposition Surveys

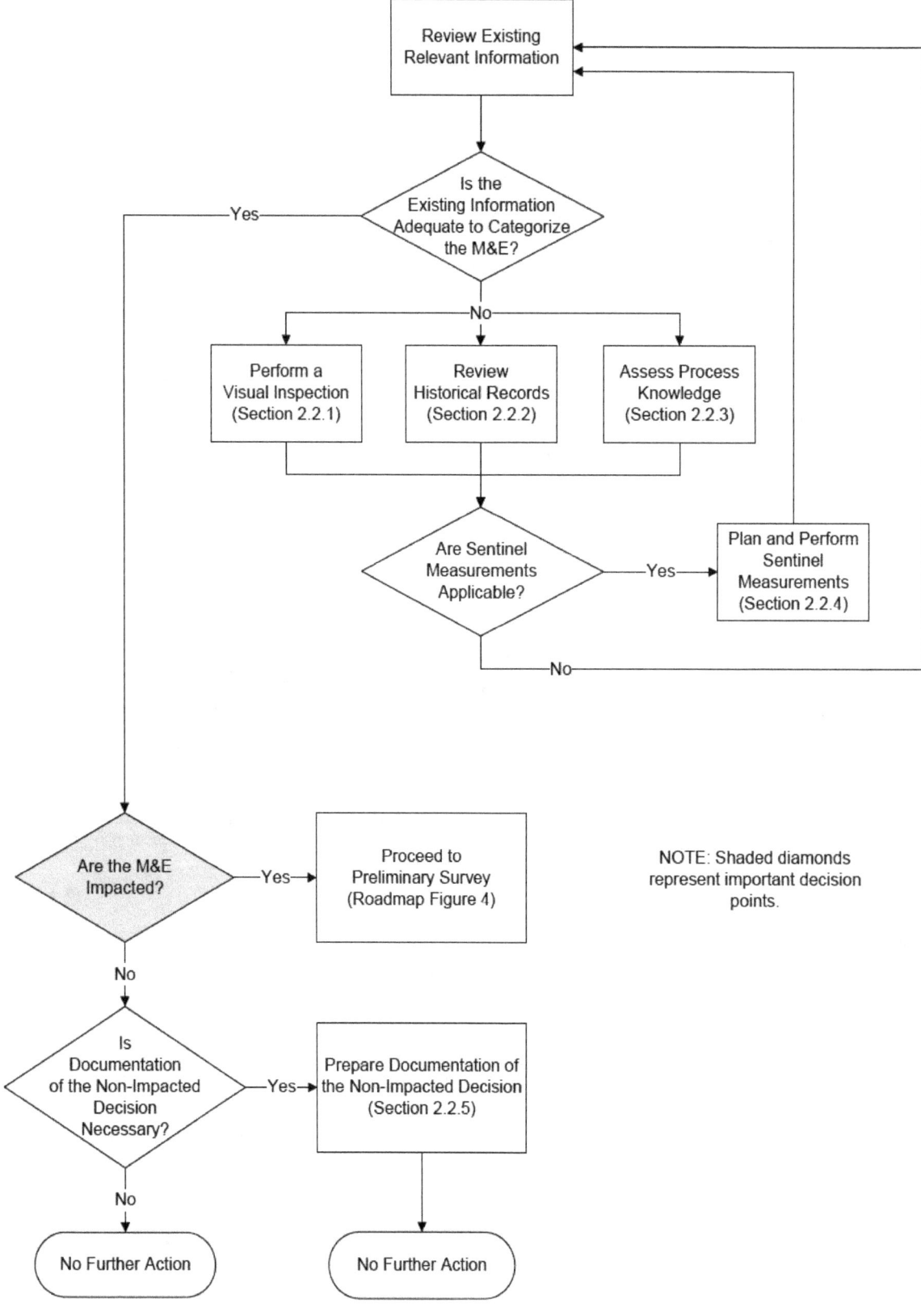

Roadmap Figure 3. The Categorization Process as Part of Initial Assessment

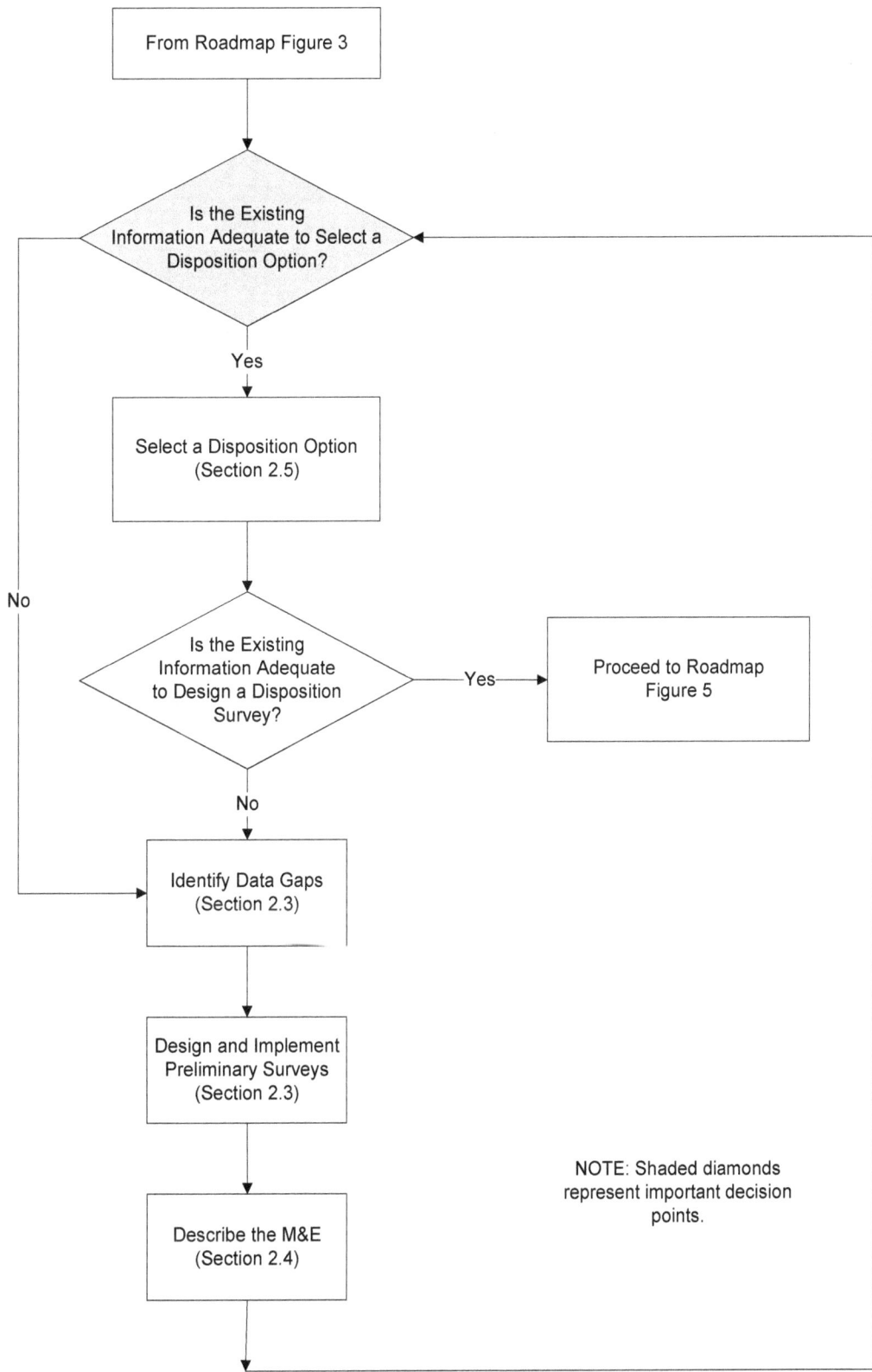

Roadmap Figure 4. Assessing Adequacy of Information for Designing

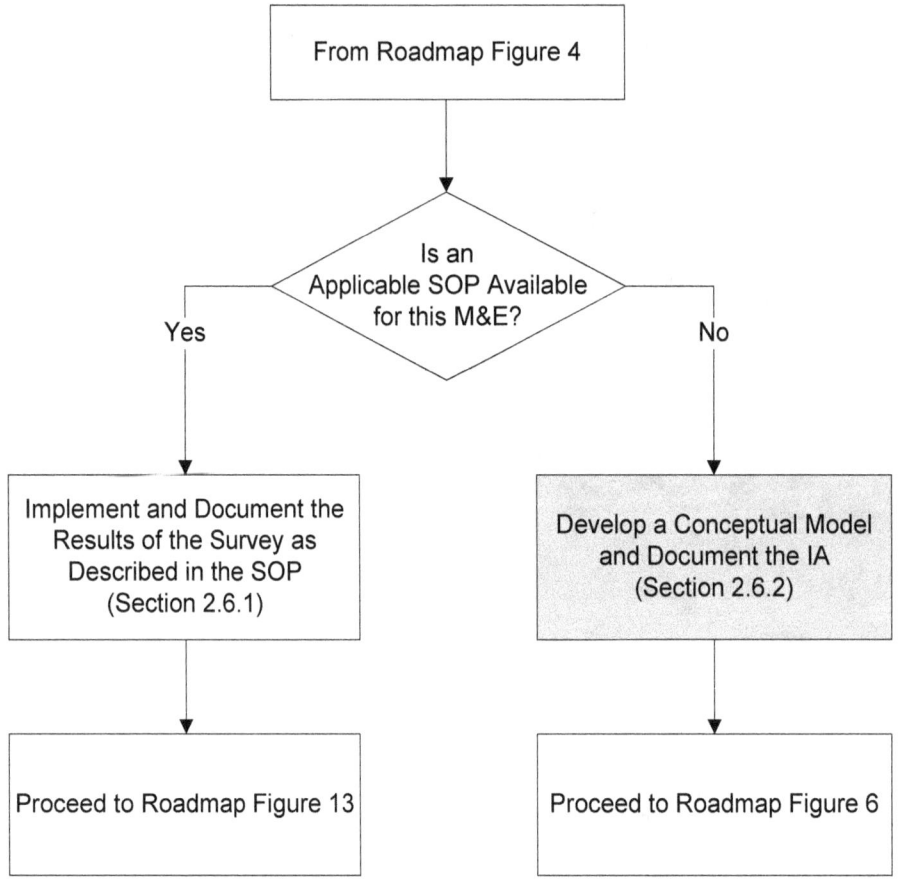

NOTE: Shaded box represents important milestone.

Roadmap Figure 5. Assessing the Applicability of Existing SOPs

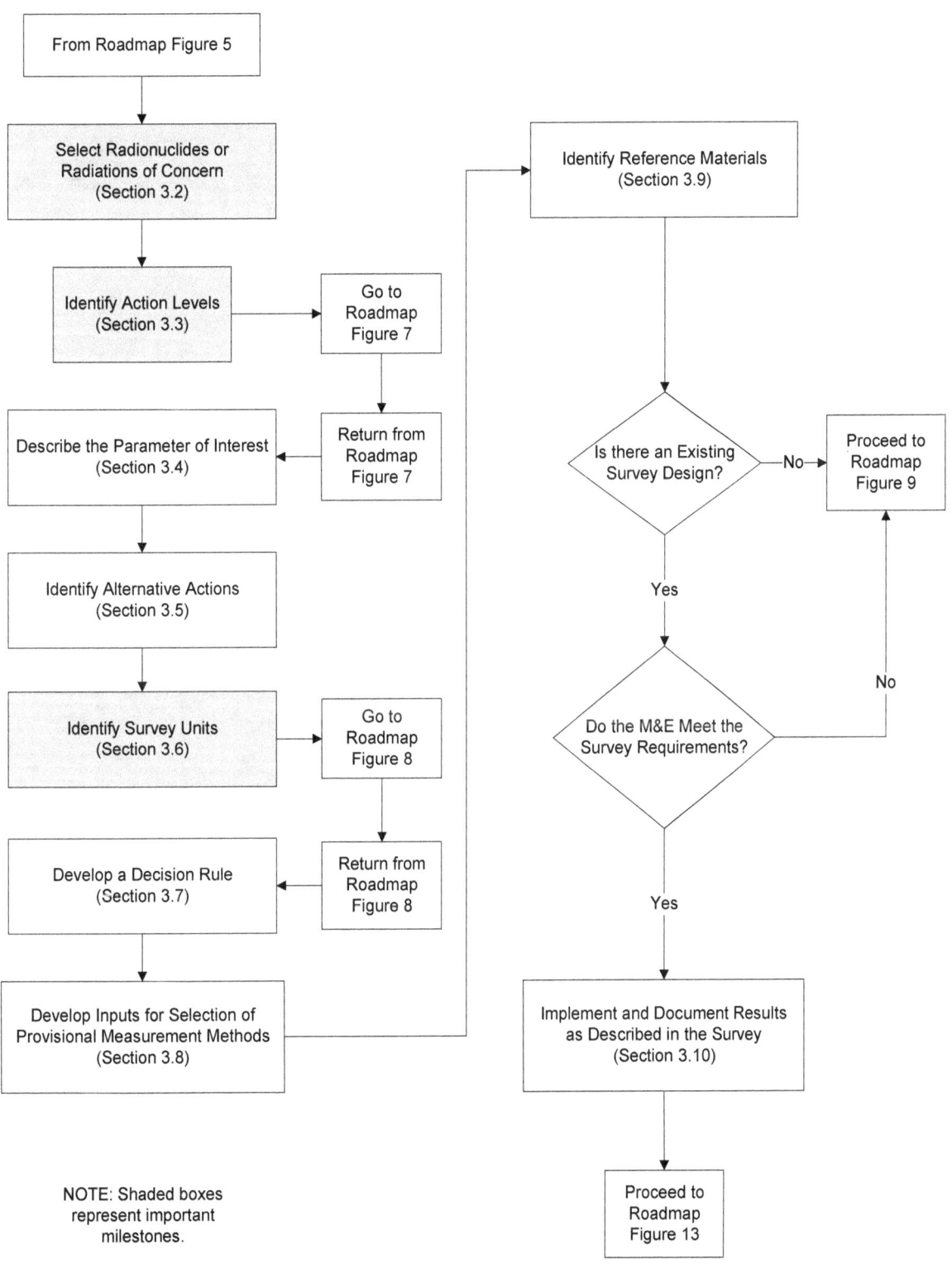

Roadmap Figure 6. Identify Inputs to the Decision

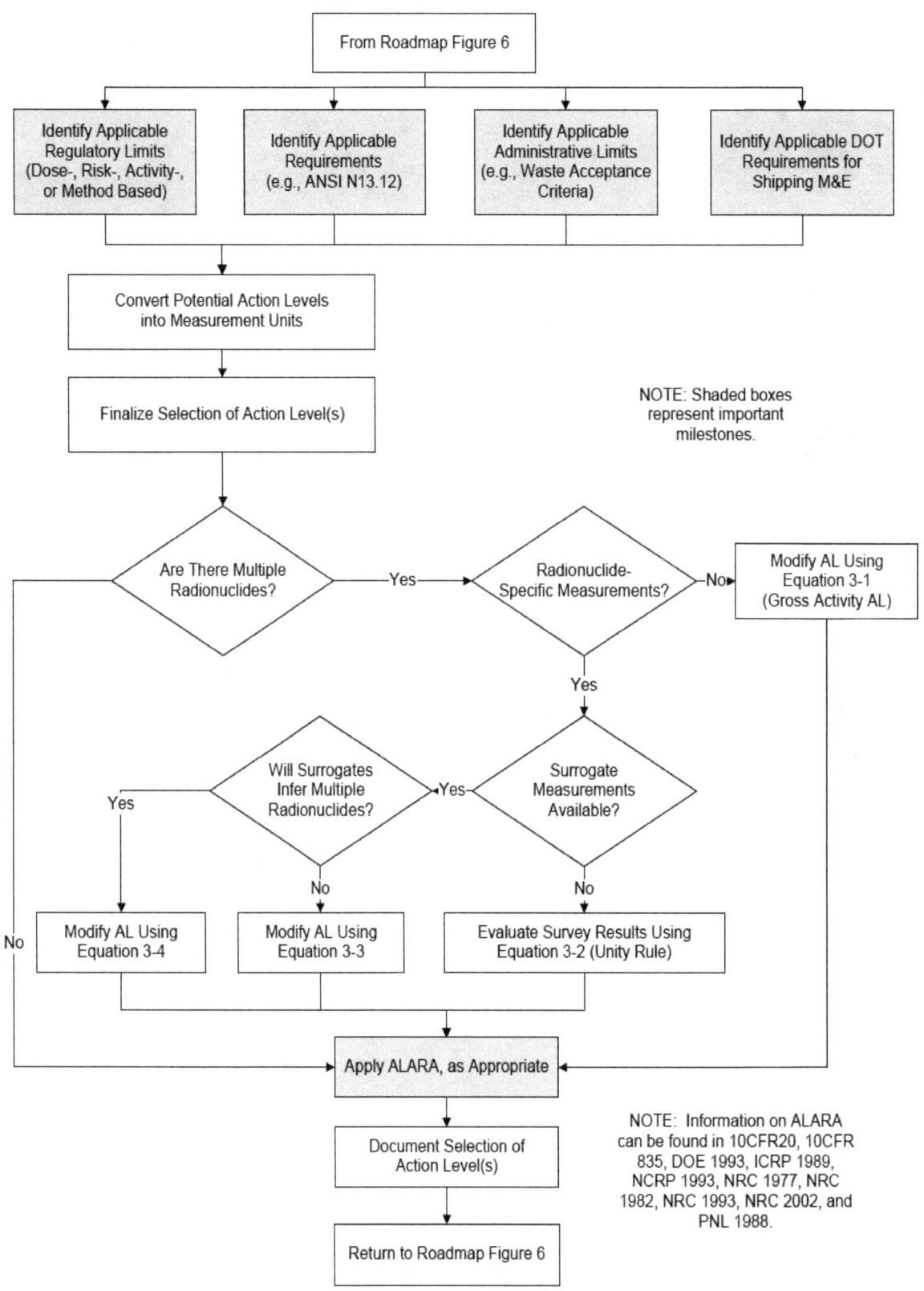

Roadmap Figure 7. Identify Action Levels

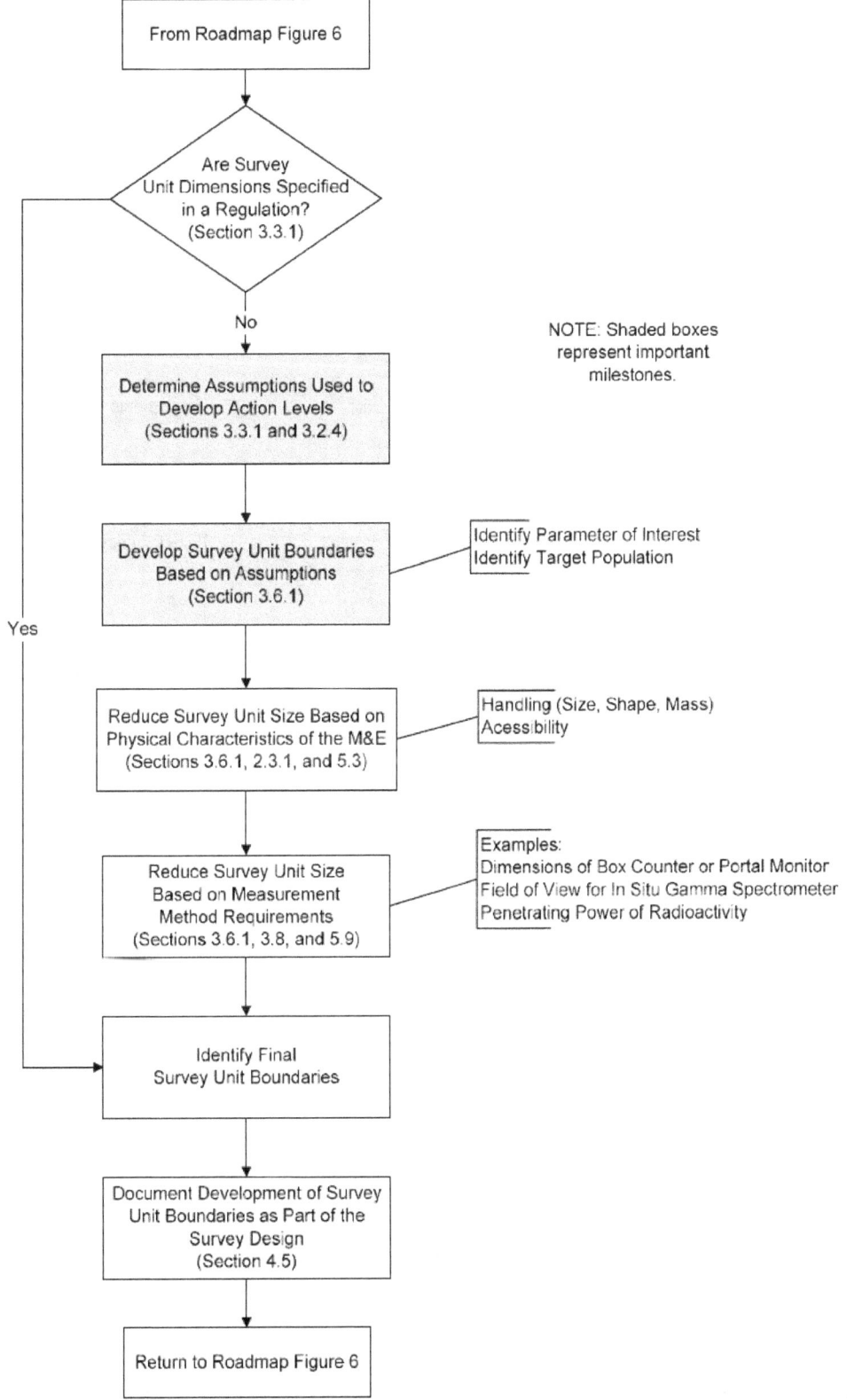

Roadmap Figure 8. Developing Survey Unit Boundaries (Apply to all Impacted M&E for each set of Action Levels Identified in Section 3.3)

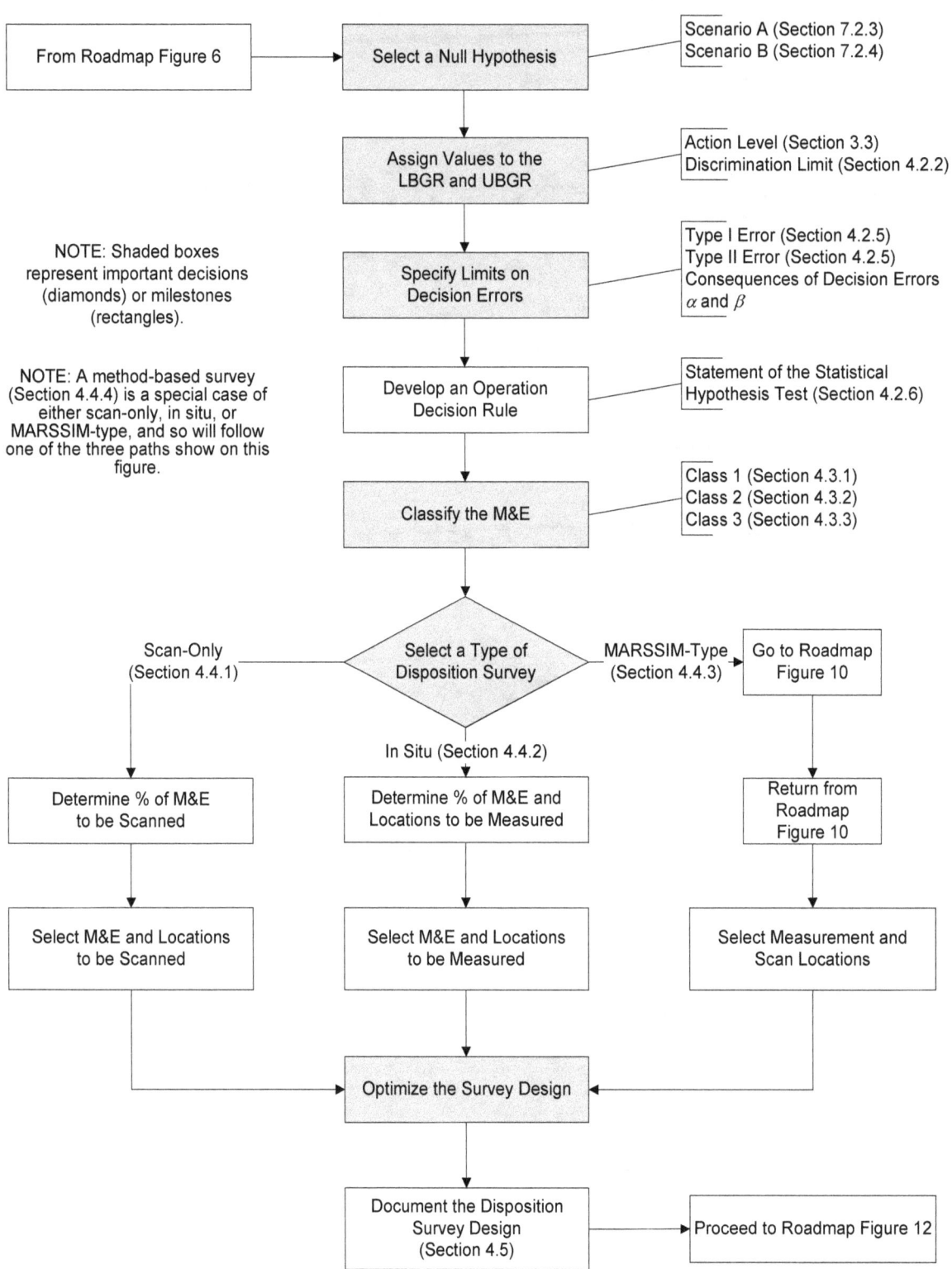

Roadmap Figure 9. Flow Diagram for Developing a Disposition Survey Design

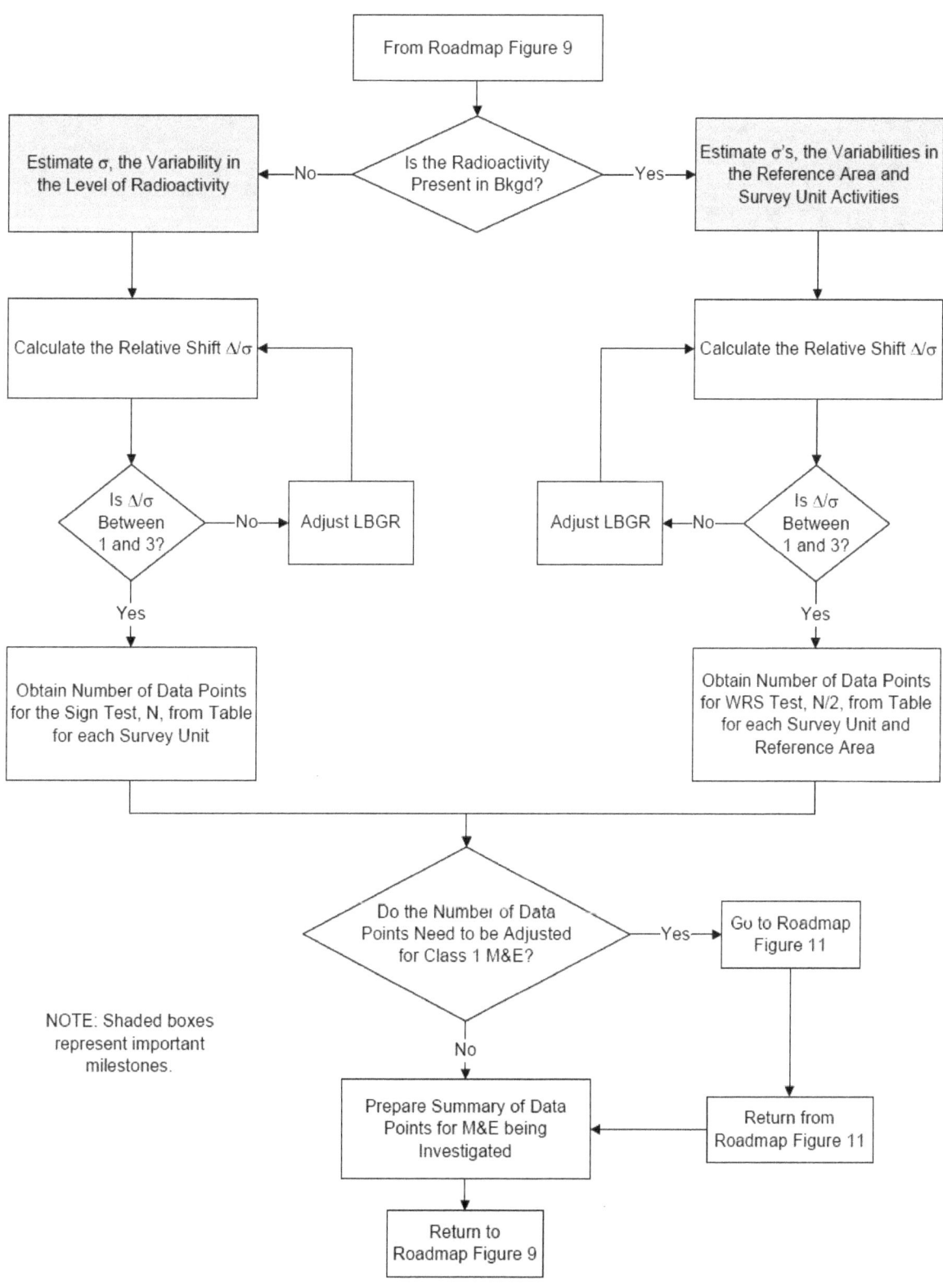

Roadmap Figure 10. Flow Diagram for Identifying the Number of Data Points for a MARSSIM-Type Disposition Survey

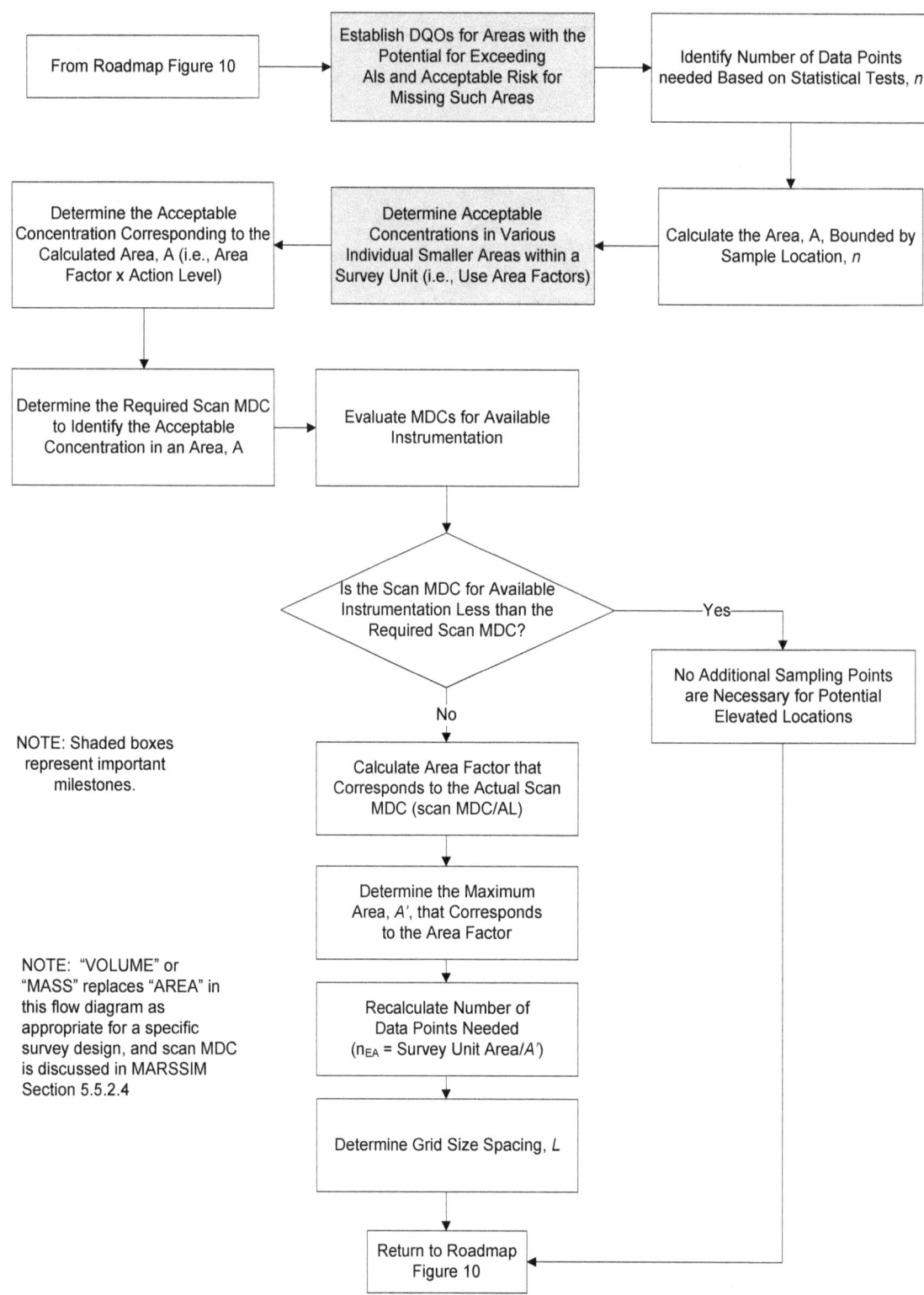

Roadmap Figure 11. Flow Diagram for Identifying Data Needs for Assessment of Potential Areas of Elevated Activity in Class 1 Survey Units for MARSSIM-Type Disposition Surveys

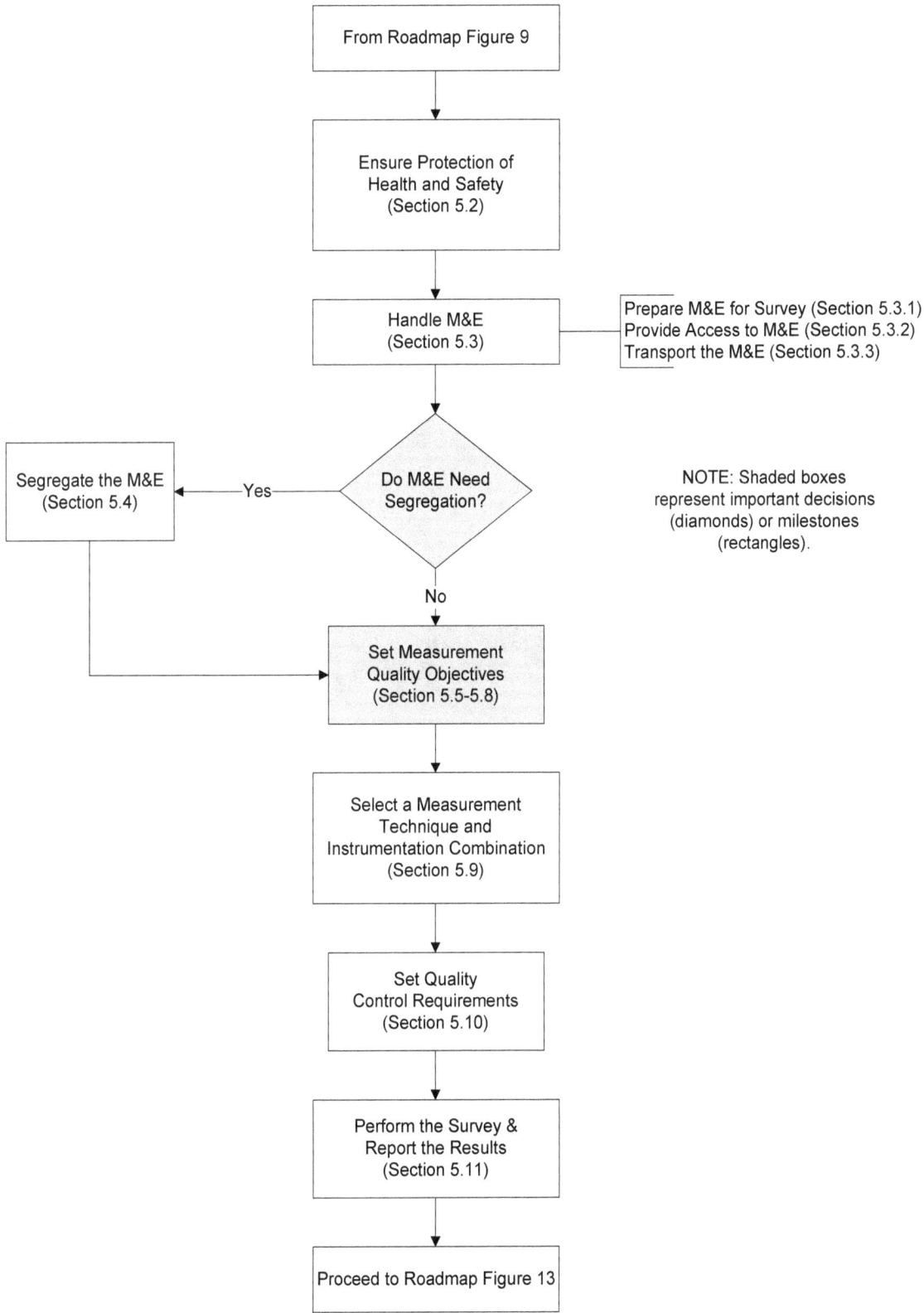

Roadmap Figure 12. Implementation of Disposition Surveys

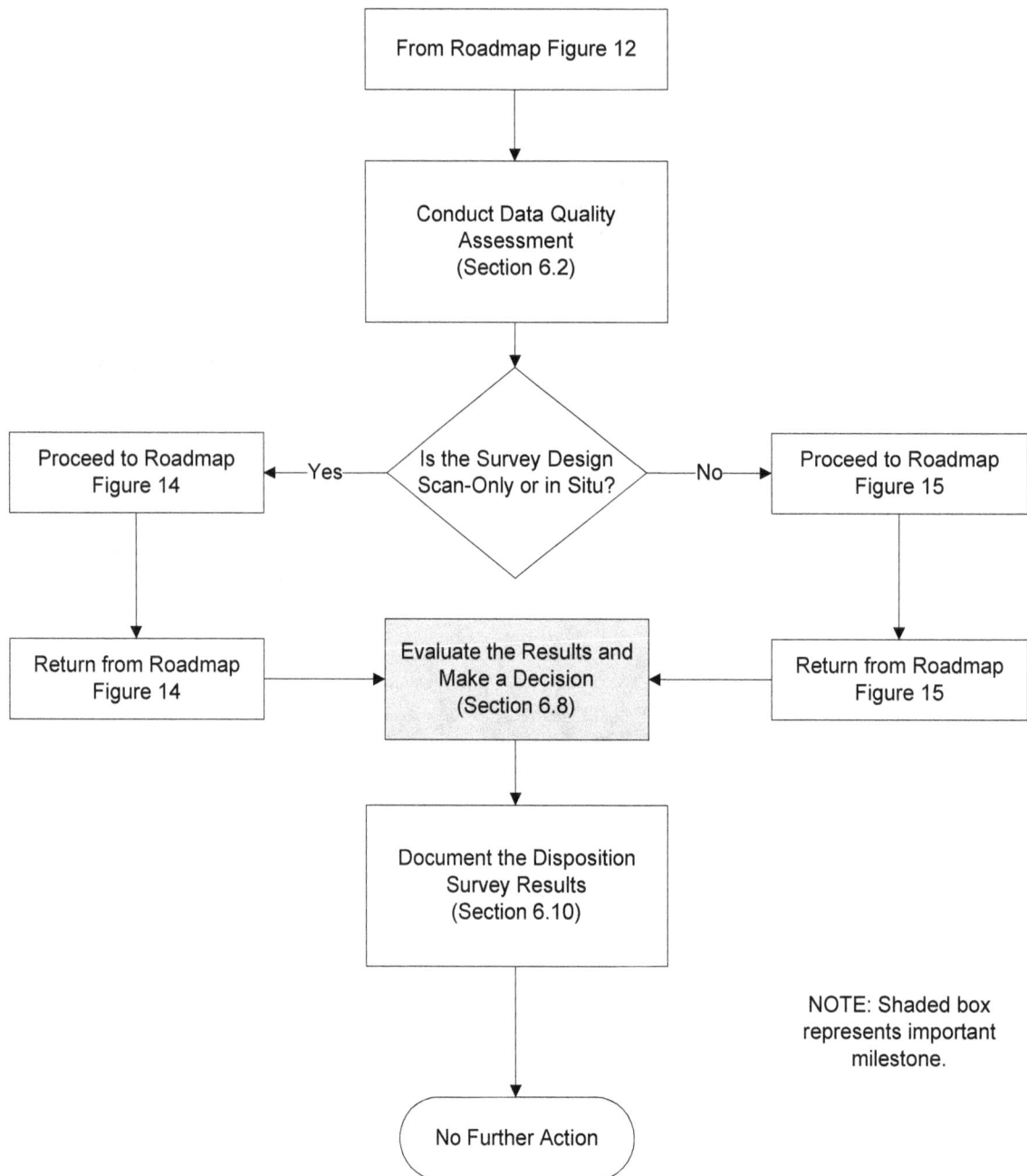

Roadmap Figure 13. Assess the Results of the Disposition Survey

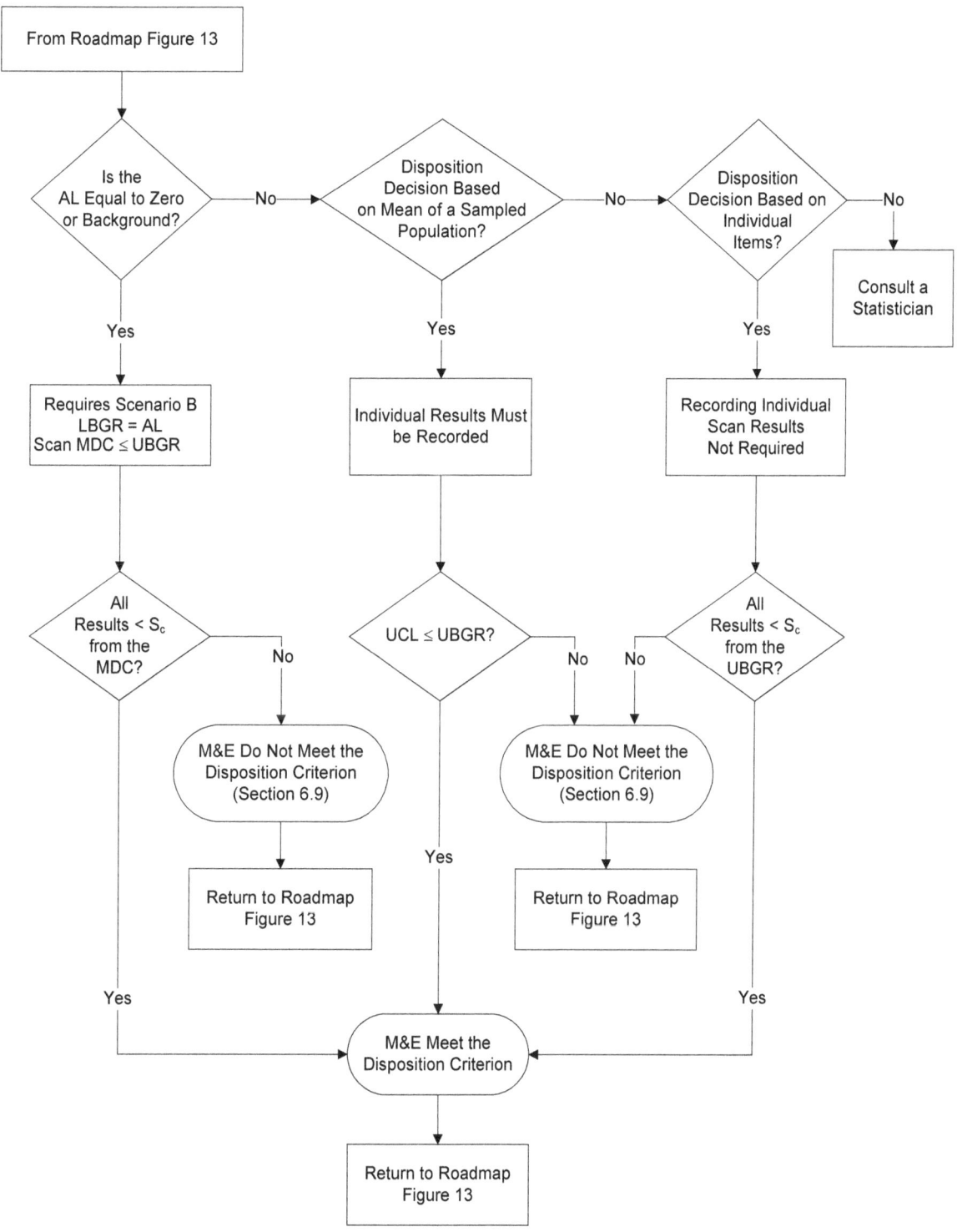

Roadmap Figure 14. Interpretation of Survey Results for Scan-Only and In Situ Surveys

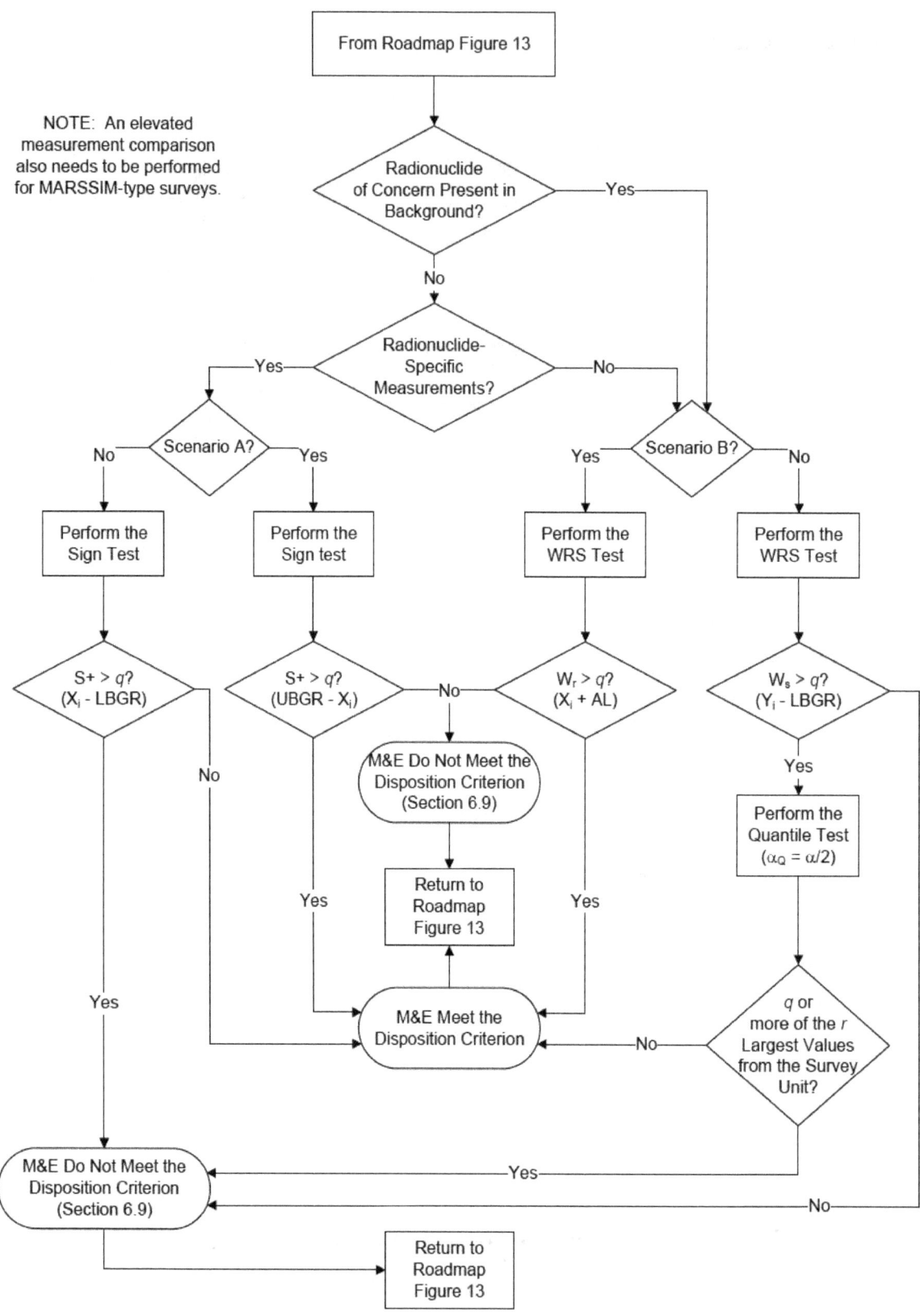

Roadmap Figure 15. Interpretation of Survey Results for MARSSIM-Type Surveys

1 INTRODUCTION AND OVERVIEW

1.1 Purpose and Scope of MARSAME

Large quantities of materials and equipment (M&E) potentially affected by radioactivity are present throughout the United States. The potential for residual radioactivity can come from use of source, byproduct, and special nuclear materials as well as naturally occurring radioactive material (NORM), naturally occurring and accelerator-produced radioactive materials (NARM) and technologically enhanced naturally occurring radioactive material (TENORM). This M&E may be commercial, research, education, or defense related. The M&E might be—

- Used or stored at sites and facilities licensed to handle radioactivity,
- Commercial products purposely containing radionuclides (e.g., smoke detectors),
- Commercial products incidentally containing radionuclides (e.g., phosphate fertilizers), or
- Associated with NARM and TENORM.

The owners of M&E potentially affected by radioactivity need to determine acceptable disposition options for M&E currently under their control. Industries or facilities sensitive to the presence of radioactivity need to evaluate the acceptability of M&E coming under their control. Regulatory agencies need to distinguish items in general commerce that are inherently radioactive from illicit trafficking of radioactive M&E.

This *Multi-Agency Radiation Survey and Assessment of Materials and Equipment* manual (MARSAME) is a supplement to the *Multi-Agency Radiation Survey and Site Investigation Manual* (MARSSIM). Like MARSSIM, MARSAME is a joint effort by the Department of Defense (DOD), Department of Energy (DOE), Environmental Protection Agency (EPA), and Nuclear Regulatory Commission (NRC). Information on MARSSIM can be found on the World Wide Web (MARSSIM 2002). MARSAME also incorporates information for measuring radioactivity from the *Multi-Agency Radiological Laboratory Analytical Protocols* manual (MARLAP 2004). MARSAME provides information on surveys where radiological control of M&E could be initiated, maintained, removed, or transferred (i.e., an M&E disposition) to another responsible party. In addition, MARSAME discusses the need for a graded approach to surveying M&E.

MARSAME provides technical information on approaches for planning, implementing, assessing, and documenting surveys to determine proper disposition of M&E. Release (including clearance) and interdiction are types of disposition options in MARSAME. Detailed descriptions of these disposition options are provided in Chapter 2.

Examples of M&E include metals, concrete, tools, equipment, piping, conduit, furniture, and dispersible bulk materials such as trash, rubble, roofing materials, and sludge. Liquids, gases, and solids stored in containers (e.g., drums of liquid, pressurized gas cylinders, containerized soil) are also included in the scope of this document.

Radionuclides or radioactivity on workers or members of the public are outside the scope of the document, as are liquid and gaseous effluent releases and real property (e.g., fixed buildings and structures, surface and subsurface soil remaining in place).

The purpose of this supplement is to provide information for the design and implementation of technically defensible surveys for disposition of M&E. MARSAME provides information on selecting and properly applying disposition survey strategies and selecting measurement methods. The data quality objectives (DQO) process is used for selecting the best disposition survey design based on the selected disposition option, action level, description of the M&E (e.g., size, accessibility, component materials), and description of the radioactivity (e.g., radionuclides, types of radiation, surficial versus volumetric activity). Detailed information on the DQO process can be found in EPA QA/G-4 (EPA 2006a), MARSSIM Appendix D, and MARLAP Appendix B.

This supplement describes a number of different approaches for performing technically defensible disposition surveys and provides information for optimizing survey designs. However, MARSAME does not represent the only acceptable approach to radiologically evaluate M&E. MARSAME describes a graded approach that the signatory agencies find acceptable and useful for most situations. The signatory agencies recognize that alternative approaches or modification of the MARSAME procedures may be appropriate or necessary for some situations. Nothing in MARSAME should be construed to prohibit the use of other appropriate procedures.

Disposition surveys may be performed as a single event or as part of a routine process. Single event disposition surveys are usually performed once in association with a specific project. Surveying a backhoe at the completion of a decommissioning project is one example of a single event disposition survey. Routine process disposition surveys are usually associated with ongoing tasks where similar surveys are performed repeatedly. One example of a routine process disposition survey would be a radiological survey of tools prior to removal from a controlled area at a nuclear facility. Both single event and routine process types of surveys are included in the scope of MARSAME.

The guidance in MARSAME is designed to incorporate existing survey methods whenever possible, while at the same time allowing the use of new and innovative survey techniques when appropriate. The use of previously established and accepted standard operating procedures (SOPs) as part of a standardized initial assessment (IA) is described in Section 2.6.1. The use of SOPs that document approved methods for performing disposition surveys along with assessing the results of these surveys can reduce the effort required to develop new survey designs, since the survey design effort was applied when the SOP was developed. MARSAME also allows consideration of innovative survey techniques through the modification of existing SOPs or development of new survey designs as described in Chapters 3, 4, and 5. Prior to implementation, existing SOPs should be evaluated to ensure they meet the survey design objectives.

MARSAME assumes the user has some historical knowledge of the M&E being investigated. The historical information is gathered during the initial assessment (IA) to determine acceptable disposition options (Chapter 2). The characteristics, history of prior use, and inherent radioactivity of the M&E are important when determining the appropriate disposition options.

The historical information is termed "process knowledge." The role of process knowledge (discussed in Chapter 2) is important in providing information on the nature and amount of radioactivity that might be expected on, or incorporated in, the M&E being investigated. If no historical information is available, information on the current status of the M&E can be determined using preliminary surveys (i.e., scoping, characterization, remedial action support) prior to designing a disposition survey.

The recommendations in this supplement may be applied to a broad range of regulations, including dose-, risk-, or radionuclide concentration-based regulations. The translation of a regulatory dose- or risk-based limit to a corresponding concentration level is not addressed in MARSAME. The terms dose, risk, and dose- or risk-based regulation are used throughout the supplement, but these terms are not intended to limit the applicability of this supplement. MARSAME can be applied to activity concentrations (e.g., Bq/m^2) without associated dose or risk values. MARSAME does not address the regulatory status of the M&E (e.g., NRC exempted or excluded materials).

MARSAME uses the word "should" as a recommendation. This is not to be interpreted as a requirement. The user need not assume that every recommendation in this supplement will be taken literally and applied to every project. Rather, it is expected the survey documentation will address how the recommendations will be applied on a project-specific basis.

1.2   Understanding Key MARSAME Terminology

In order to understand the information in MARSAME, the user should first become familiar with the scope of this supplement, the terminology, and the concepts in this document. As a supplement to MARSSIM, MARSAME uses terms generally consistent with MARSSIM. Some additional terms were developed for MARSAME, while other commonly used terms were adopted from other sources. This section explains some of the terms used in this supplement. The terms *impacted, non-impacted,* and *graded approach* are defined in MARSSIM. These terms are used consistently in MARSSIM and MARSAME. Unlike MARSSIM, which applies to land, structures, or buildings, MARSAME applies to M&E. The action taken may initiate, maintain, remove, or transfer radiological controls associated with the M&E. The decision to take action may be largely based on the results of a radiological survey designed to evaluate the disposition of the M&E, either through release or interdiction. Therefore, the terms *release criterion, derived concentration guideline level* (DCGL), and *final status survey* used in MARSSIM are replaced by the more generic terms *disposition criterion, action level,* and *disposition survey,* respectively, in MARSAME.

Disposition is the future use, fate, or final location for something (e.g., recycle, reuse, disposal). Disposition options range from release to interdiction:

- *Release* is a reduction in the level of radiological control, or a transfer of control to another party. Release includes clearance. Examples of release (other than clearance) include recycle, reuse, disposal as waste, or transfer of control of radioactive M&E from one authorized user to another.

- *Interdiction* is an increase in the level of radiological control or a decision not to accept control from another party. Examples of interdiction include identification of radioactive material that results in the initiation of radiological controls or identification of unauthorized movement of radioactive material.

Categorization is the act of determining whether M&E are impacted or non-impacted. This is a departure from MARSSIM where this decision was referred to as classification. This change was made to emphasize the difference between the decision of whether a survey is needed (i.e., impacted or non-impacted) and the determination of the appropriate level of survey effort (i.e., classification).

Classification is the act or result of separating impacted M&E or survey units into one of three designated classes: Class 1, Class 2, or Class 3. Classification is the process of determining the appropriate level of survey effort based on estimates of activity levels and comparison to action levels, where the activity estimates are provided by historical information, process knowledge, and preliminary surveys.

Measurable radioactivity is radioactivity that can be quantified using known or predicted relationships developed from historical information, process knowledge or preliminary measurements as long as the relationships are developed, verified, and validated as specified in the DQOs and measurement quality objectives (MQOs). Measurability is of primary importance in MARSAME.

Surficial radioactive material is radioactive material distributed on any of the surfaces of a solid object. Surficial radioactive material may be *removable* (by non-destructive means such as casual contact, wiping, brushing, or washing) or *fixed*. Surfaces may either be accessible or difficult-to-measure. Changes to the surface (e.g., paint, dirt, oxidation) may affect the measurability and the physical condition of surficial radioactive material.

Survey unit for M&E is the specific lot, amount, or piece of M&E on which measurements are made to support a disposition decision concerning the same specific lot, amount, or piece of M&E. The survey unit defines the spatial boundaries for the disposition decision and a separate decision is made for each survey unit, similar to MARSSIM. The survey unit boundaries also define the population for the parameter of interest.

Volumetric radioactive material is radioactive material that is distributed throughout or within the material or equipment being measured, as opposed to a surficial distribution. Volumetric radioactive material may be homogeneously (e.g., uniformly activated metal) or heterogeneously (e.g., activated reinforced concrete) distributed throughout the M&E. Volumetric radioactive material may be distributed throughout the M&E being measured or distributed in layers. Layers of volumetric radioactive material may start at the surface (e.g., porous surfaces penetrated by radioactive material) or under a layer of other material (e.g., activated rebar inside a concrete wall). By definition all radioactive liquids and gases in containers and all bulk quantities of radioactive material when measured as a whole are volumetric radioactive material.

The concept of whether radioactivity is measurable is the major factor in demonstrating compliance with an action level. MARSAME does not provide an exact definition for the transition between surficial and volumetric radioactive material. Rather, the assumptions used to quantify the radioactivity need to be clearly defined and identified so they can be compared to the DQOs and MQOs. Individual action levels may specify applicability to surficial or volumetric radioactivity. In these cases, the definition of surficial and volumetric radioactivity should be specified as part of the definition of the action level.[1]

Accessible area is an area that can be easily reached or obtained. In many cases an area must be physically accessible to perform a measurement. However, radioactivity may be measurable even if M&E are not physically accessible (e.g., energetic gamma rays may be quantified even after passing through a layer of shielding).

Difficult-to-measure radioactivity is radioactivity that is not measurable until the M&E to be surveyed is prepared. Preparation of M&E may be relatively simple (e.g., cleaning) or more complicated (e.g., disassembly or complete destruction). Given sufficient resources, all radioactivity can be made measurable; however, it is recognized that increased survey costs can outweigh the benefit of some dispositions.

Initial assessment (IA) is an investigation to collect existing information describing M&E and is similar to the Historical Site Assessment (HSA) described in MARSSIM. The IA provides initial categorization of M&E as impacted or non-impacted. In addition to the HSA activities described in MARSSIM, the IA may lead to grouping or segregating M&E with similar characteristics as well as designing and implementing preliminary surveys. The IA also identifies the expected disposition of the M&E (e.g., clearance, radiological control, recycle, reuse, disposal). The results of the IA provide most, if not all, information needed to design a disposition survey for impacted M&E. A graded approach is used to determine the level of effort applied during the IA.

Sentinel measurement is a biased measurement performed at a key location to provide information specific to the objectives of the IA (see Section 2.2.4). Sentinel measurements cannot be used as the only source of information to support a decision that M&E are non-impacted. The objective of performing sentinel measurements as part of the IA is to gather additional information to support a decision regarding further action, verify assumptions based on process knowledge, provide additional support to a finding of impacted or non-impacted for M&E, and to distinguish illicit or inadvertent transport of radioactive materials from items in general commerce that are inherently radioactive (e.g., fertilizers, phosphates, sand-blasting grit).

1.3   Use of MARSAME

MARSAME provides technical information describing a framework for planning, implementing, and assessing radiological surveys of M&E. MARSAME does not establish or supersede any regulatory or license requirements. Federal and State regulatory agencies may have requirements or guidance that differs from what is presented in MARSAME and may be implemented as

---

[1] This idea is consistent with the definition of a surface soil sample provided in the MARSSIM Glossary. A surface soil sample is a sample that reflects the modeling assumptions used to develop the DCGL for surface soil activity. The example in MARSSIM references 40 CFR 192, which defines surface soil as the first 15 cm of soil.

appropriate. Consequently, persons planning, implementing, and assessing disposition surveys should also obtain appropriate regulatory approval for the procedures that are in use to maintain regulatory compliance.

Potential users of this supplement are Federal, State, and local government officials having authority for control of radioactive M&E, their contractors, and other parties such as organizations with licensed authority to possess and use radioactive materials. This supplement to MARSSIM is intended for a technical audience having knowledge of radiation health physics and an understanding of statistics as well as experience with the practical applications of radiation protection. Understanding and applying the recommendations in this supplement requires knowledge of instrumentation and measurement methodologies as well as expertise in planning, approving, and implementing radiological surveys. Certain situations and projects may require consultation with more experienced or specialized personnel (e.g., a statistician).

MARSAME users with less professional experience than described above should still be able to apply the majority of guidance found in this supplement although obtaining technical support is recommended. The wide range of topics and subjects covered by MARSAME emphasizes the need for a well rounded planning team as described in the first step of the DQO process. While it may be difficult to identify a single person with all the required technical experience to design an appropriate survey, it is easier to assemble a small group of experts with the required range of knowledge. Consultation with the responsible regulatory agency is critical for the success of all disposition surveys, and even more so for MARSAME users with less professional experience. In addition, MARSAME provides information in Appendix B, Appendix C, and Appendix D that may be useful to users with less professional experience.

MARSAME recommends that a graded approach be applied to the disposition of M&E. Non-impacted M&E are removed from further consideration early in the process through categorization. Impacted M&E are classified based on the level of residual radioactivity so that a higher level of scrutiny can be applied to M&E with the highest potential for residual radioactivity. Finally, MARSAME includes practical considerations such as inherent value of the M&E and handling the M&E when evaluating options for disposition. The combination of these considerations results in a graded approach where an appropriate level of survey effort is applied to M&E to minimize the impacts of any decision errors.

1.4   Overview of MARSAME

The data life cycle is the foundation for the design, implementation, and assessment of surveys for disposition of M&E in this supplement. However, before commencing survey planning the user must select an appropriate disposition option. Multiple disposition options may exist. Consider all of the various disposition options and develop the most appropriate option for a given situation. Survey designs may then be planned using the DQO process, which is often iterative. The DQO process iterations may take place at different times during the disposition process, for example during the IA as well as during the disposition survey. The different survey designs are compared and the most resource-effective design that meets the survey objectives is selected for implementation. Following implementation of the selected survey design, the results are evaluated using data quality assessment (DQA). A technically defensible decision regarding disposition of the M&E can then be made.

Whenever practical, MARSAME recommends designing disposition surveys where one hundred percent of the M&E are measurable. This means that all radioactivity associated with the M&E has been measured and quantified (e.g., 100% scan with conventional instruments, measurement with a box counter, or measurement using in situ gamma spectroscopy), a known or accepted relationship was used to estimate concentrations for difficult to measure radionuclides using surrogate measurements,[2] or that a known or accepted relationship allows quantification of radioactivity in areas that were not measured. MARSAME employs the use of a graded approach to determine if a 100% measurable survey is practical and to ensure that a sensible, commensurate balance is achieved between resource expenditures and risk reduction.

MARSAME uses the data life cycle to design disposition surveys. The data life cycle is described in MARSSIM Section 2.3, and consists of four phases:

- Planning phase (MARSAME Chapters 2, 3, and 4; MARSSIM Chapters 3, 4, and 5),
- Implementation phase (MARSAME Chapter 5; MARSSIM Chapters 6 and 7),
- Assessment phase (MARSAME Chapter 6; MARSSIM Chapter 8), and
- Decision-making phase (MARSAME Chapter 6; MARSSIM Chapter 8).

A brief description of each of the phases and how they apply to the disposition survey design process is provided in the following sections. Table 1.1 provides a simplified overview of the principal steps in designing a disposition survey and illustrates how the data life cycle can be used in an iterative fashion within the survey process. Figure 1.1 illustrates how the data life cycle is applied to disposition surveys.

Table 1.1 The Data Life Cycle Used to Support Disposition Survey Design

Disposition Survey Design Process	Data Life Cycle		MARSAME Processes
Categorization	Categorization Data Life Cycle	Plan Implement Assess Decide	Provides information on collecting and assessing existing data (Section 2.2)
Preliminary Surveys	Preliminary Survey Data Life Cycle	Plan Implement Assess Decide	Discusses the purpose (i.e., filling data gaps) and general approach to performing preliminary surveys (Section 2.3)
Disposition Survey	Disposition Survey Data Life Cycle	Plan Implement Assess Decide	Provides detailed information for planning (Chapters 3 and 4), implementing (Chapter 5), and assessing (Chapter 6) disposition surveys

---

[2] The MARSSIM term "surrogate measurement" as used here is consistent with the MARLAP term "alternate radionuclide."

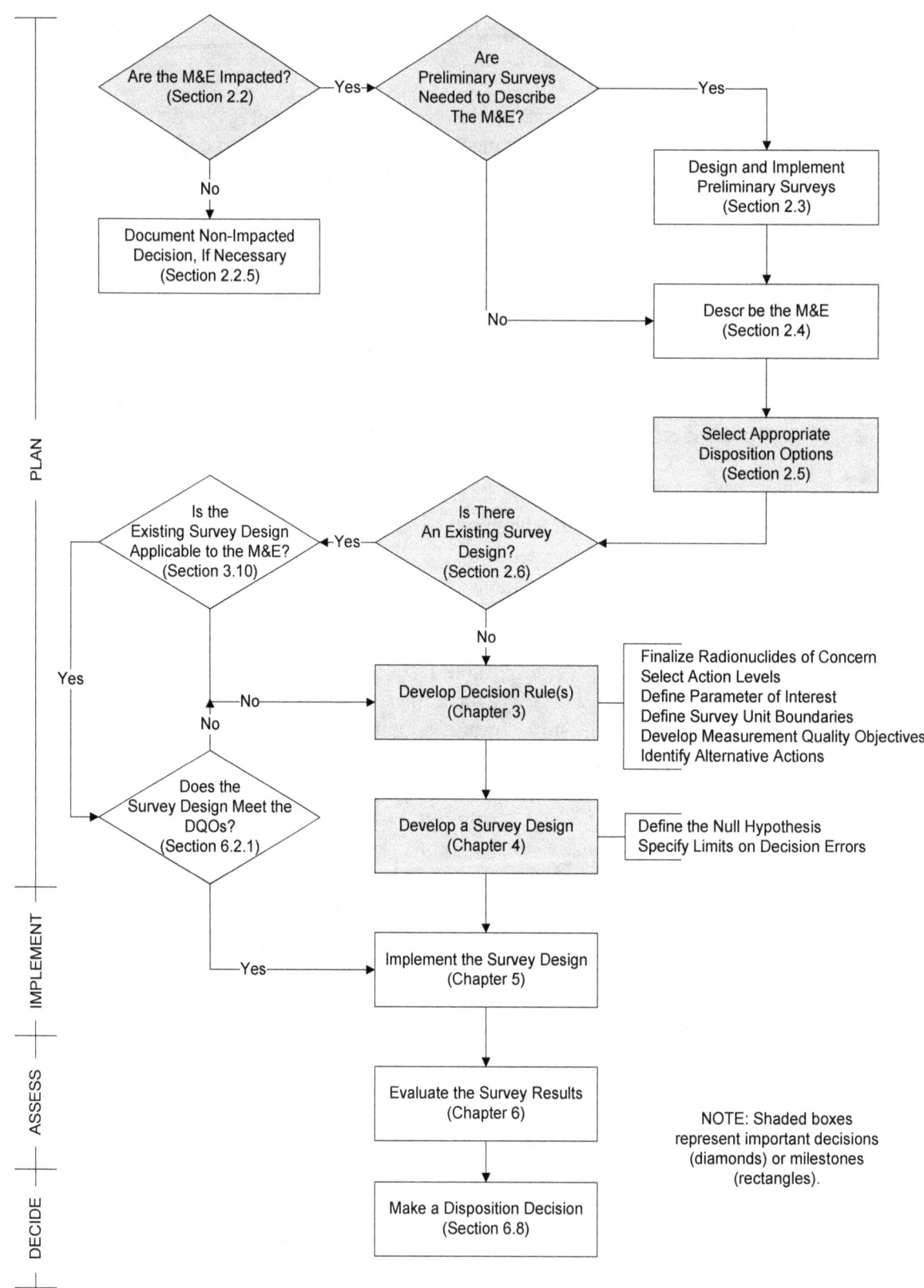

Figure 1.1 The Data Life Cycle Applied to Disposition Surveys

1.4.1 Planning Phase

The planning phase is where the survey design is developed and documented using the DQO process. The survey design documents the decision rule as well as the number, type, and location of measurements required to support the disposition decision. Soliciting input from regulatory agencies early in the planning phase helps ensure the disposition survey results will meet regulatory needs.

MARSAME processes begin with the historical evaluation of the M&E being investigated. This IA usually combines a review of process knowledge and historical records with a visual inspection of the M&E. The results of the IA are used to develop a conceptual model describing the physical characteristics of the M&E and providing information on the radioactivity potentially associated with the M&E. The physical description of the M&E should include information on the size, shape, complexity (e.g., can it be broken down or combined with other M&E), accessibility (e.g., can the surveyor physically access areas of concern to perform measurements), and inherent value (i.e., resources associated with reuse, recycle, repair, remediation, replacement, and disposal). Information on radioactivity should include the radionuclides of potential concern, the expected levels of radioactivity, the distribution of radioactivity (e.g., uniform or not), and the location of the radioactivity (i.e., surface or volume).

The IA may also include data collection in the form of sentinel measurements. The results of sentinel measurements can be used as the basis to reject assumptions based on process knowledge. However, sentinel measurements alone cannot be used to justify the categorization of M&E as non-impacted (see Section 2.2.4 for information on sentinel measurements).

There are two decisions associated with the IA. The first decision, called categorization, is whether or not the M&E are impacted. Non-impacted M&E do not require additional investigation, but may require documentation of the justification for the non-impacted decision. The second decision is to select an appropriate disposition option for impacted M&E at the end of the IA to provide direction for designing a disposition survey. Additional information may be required before a disposition survey can be designed. Preliminary surveys (e.g., scoping, characterization, and remedial action support surveys) may be performed as part of the IA to collect this additional information.

For single event surveys, the IA should focus on collecting the information necessary to develop a technically defensible disposition survey design. Information necessary to design a disposition survey includes a description of the M&E and the radioactivity potentially associated with the M&E. The results of the IA are carried forward and used to develop the survey design, which is usually documented in a project-specific work plan.

For routine process surveys, the IA should lead to an existing survey design from a standard operating procedure (SOP), if applicable, or develop a new survey design for documentation in an SOP. The SOP should clearly state the assumptions used to develop the survey design, along with a description of the M&E and radioactivity covered by the SOP. The selection process is based on evaluating the M&E to determine if the survey design in a specific SOP is applicable. Documentation of individual survey results may not be required as long as there are records showing that the SOP was approved, the instruments were working properly, and the personnel performing the survey were properly trained. Development of SOPs is usually accomplished using the same processes as those used to develop single event surveys. There may be regulatory or site-specific guidance that specifies documentation requirements for SOPs. Information on developing SOPs can be found in EPA QA/G-6 (EPA 2001).

Following the IA, it is necessary to develop a decision rule for the disposition of M&E being investigated. The decision rule is an "if...then..." statement consisting of three parts:

- Action level(s),
- Parameter of interest, and
- Alternative actions.

An example of a decision rule might be "If the average surficial activity concentration is less than a level specified by the regulator, then the M&E can be cleared, otherwise the M&E are not cleared." The parameter of interest is closely related to the description of the M&E, the description of the radioactivity, and the survey unit boundaries. The action level is influenced by the selection of a disposition option. The selected disposition option defines two alternative actions. A decision rule should be developed for each decision to be made concerning the M&E. For example, if the action level is stated in terms of total activity, generally only one decision rule is required. If, on the other hand, the action level provides limits for fixed, removable, and maximum levels of radioactivity, e.g., DOE Order 5400.5, Figure IV-1 (DOE 1993), then a decision rule is required to evaluate each action level. The measurement performance requirements, or MQOs, are also evaluated when developing a decision rule to ensure that an acceptable measurement technique is available to support the proposed survey design.

Once the decision rule(s) have been established, a survey design is developed. The survey design specifies the number and quality of measurements required to support a disposition decision recorded in the decision rule. MARSAME recommends applying a graded approach to designing disposition surveys (Section 4.4). The survey design, definitions of decision errors, and burden of proof are determined by the selection of a null hypothesis (Section 4.2).

The survey design should be documented in a quality document (e.g., QA Survey Plan, SOP) that has been reviewed and accepted by the appropriate authority (e.g., technical expert, management, or regulator). Survey designs that are often repeated may be documented in SOPs along with supporting records on instrument performance and personnel training. Other types of disposition surveys are usually documented in a project-specific work plan and survey results are presented in a disposition survey report (Sections 2.5 and 4.5). If the selected survey design is not technically or economically practical, the planning team can investigate additional disposition options if necessary (Sections 2.4 and 4.4).

1.4.2 Implementation Phase

To ensure flexibility and encourage the use of optimal measurement techniques for a specific project, MARSAME does not provide detailed information on specific implementation techniques. However, detailed descriptions of several measurement techniques are provided

(Chapter 5 and Appendix D). These descriptions serve as a template for information required to evaluate different measurement techniques. It is important to remember that the survey design is usually linked to a specific option for disposition of the M&E (Chapters 3 and 4).

During implementation, the descriptions of measurement techniques are compared to the MQOs defined during survey planning. A measurement method (i.e., combination of a measurement technique with an instrument; see Section 5.9) is selected based on its ability to meet the MQOs. The number and type of measurements specified in the documented survey design are performed at the locations specified in the survey design. If a measurement method is specified in the survey design, that method should generally be used during implementation. If the specified measurement method cannot be performed (e.g., the instrument is unavailable or the measurement method does not meet the MQOs), another measurement method should be selected based on the MQOs. The selection of the replacement measurement method should be documented in the survey design and survey report.

An action level may be established for implementing disposition surveys to support disposition decisions about individual objects or measurement locations. If this action level is in the same units as measurements performed in the field, then the surveyor can make final disposition decisions by directly comparing the measurement results to the action level as the measurements are performed. This may allow the surveyor to perform remediation as required and implement a "clean as you go" component to the survey design (Section 6.9). Clean as you go surveys may reduce the amount of M&E requiring additional consideration following completion of the disposition survey. This clean as you go approach to surveys is only applicable for Class 1 surveys (i.e., radionuclide concentrations or radiation levels exceed the action level and 100% of M&E are measured) where there is high confidence in the quality and accuracy of detection decisions.

Quality control (QC) data are collected and analyzed during implementation to provide an estimate of the uncertainty associated with the survey results. QC measurements are technical activities performed to measure the attributes and performance of a survey. A well-designed QC program increases efficiency and provides for early detection of problems. This can save time and money by averting rework and enables the user to make decisions more expeditiously (EPA 2002c).

1.4.3 Assessment Phase

The assessment phase begins with verification and validation of the survey results. Data verification is used to ensure the requirements documented in the survey design were implemented as prescribed. Data validation ensures the results of the data collection activities support the objectives of the survey (i.e., DQOs), or permit a determination that these objectives should be modified (MARSSIM Section 9.3 and MARSSIM Appendix N).

DQA determines if the collected data are of the right type, quality, and quantity to support their intended use. DQA helps complete the data life cycle by providing the assessment needed to determine that the planning objectives are achieved. DQA is described in detail in EPA QA/ G-9R (EPA 2006b), MARSSIM Section 8.2, and MARSSIM Appendix E.

The preliminary data review is performed to learn about the structure of the data (e.g., identifying patterns, relationships, or potential anomalies). Graphical techniques are used to help visualize the data. Calculation of basic statistical quantities is used to help describe the distribution of data.

The survey data are evaluated using a statistical test. A test statistic is calculated and compared to a critical value. The critical value divides the potential values of the test statistic into two regions. The critical region includes values for the test statistic where the null hypothesis is rejected. The null hypothesis is not rejected for values of the test statistic outside the critical region. Keep in mind that a statistical test could be as simple as comparing survey results directly to the critical value to ensure no radiation is detected, or may involve using a more complex statistical evaluation such as the Wilcoxon Rank Sum (WRS) test.

In some cases the assessment phase may be performed during survey implementation. For example, the "clean as you go" approach described in Section 6.9 requires that field data be assessed and a decision made concerning the M&E being measured. The M&E are "cleaned," or remediated, as necessary and another disposition survey performed.

1.4.4   Decision-Making Phase

Following the assessment phase, a decision is made regarding the disposition of the M&E. The decision rule defines the final decision. The statistical test or data comparison determines whether the parameter of interest exceeds the action level. Based on the outcome, a decision can be made regarding the alternative actions. If multiple decision rules are defined for a single disposition survey (e.g., a MARSSIM-type survey where the average activity is evaluated using a statistical test and small areas of elevated activity are evaluated using the elevated measurement comparison) any one decision that the action level has been exceeded should result in additional investigation.

In some cases the decision making phase may be performed during survey implementation. For example, the "clean as you go" approach described in Section 6.9 requires that field data be assessed and a decision made concerning the M&E being measured. The M&E are "cleaned," or remediated, as necessary and another disposition survey is performed.

1.5   Organization of MARSAME

The planning, implementation, and assessment of disposition surveys in MARSAME are based on the data life cycle. Each chapter in MARSAME provides information for specific steps in the process. The planning phase is discussed in Chapters 2, 3, and 4. The implementation phase is discussed in Chapter 5, and Chapter 6 discusses the assessment phase and decision-making phase.

Chapter 2 focuses on the IA. Information is provided on categorizing whether the M&E are impacted or non-impacted using existing data and sentinel measurements in Section 2.2. Information on designing and implementing preliminary surveys to provide the information needed to design a disposition survey is provided in Section 2.3. Discussions on describing the

M&E being surveyed are provided in Section 2.4. The selection of a disposition option and development of a conceptual model are discussed in Section 2.5. Information pertaining to documenting the results of the IA is provided in Section 2.6.

Chapter 3 provides information on developing a decision rule and discusses other inputs needed to design a disposition survey. Section 3.2 addresses selecting the radionuclides or radiations of concern which must be established before forming a decision rule. There are three parts to a decision rule–

- Action level(s), discussed in Section 3.3,
- Parameter of interest, discussed in Section 3.4, and
- Alternative actions, discussed in Section 3.5.

Section 3.7 brings these three components together to develop decision rule(s) that are used to design the disposition survey in Chapter 4. Survey units are discussed in Section 3.6, and inputs for selecting measurement methods are presented in Section 3.8. Section 3.9 identifies reference materials that can be used to estimate background radionuclide concentrations or radiation levels. The process for evaluating an existing survey design is described in Section 3.10.

Chapter 4 completes the planning phase with the development of a survey design. This chapter discusses the selection of a null hypothesis and setting tolerable limits on decision errors (Section 4.2), determines the level of survey effort for the disposition survey (Section 4.3), and determines the type, number, and location of measurements to support a disposition decision (Section 4.4). Information pertaining to disposition survey design documentation is provided in Section 4.5. The processes in Chapter 4 result in a documented survey design.

The implementation processes in Chapter 5 focus on selection of an appropriate measurement technique. Recommendations are provided on issues related to health and safety that may impact the implementation of disposition surveys (Section 5.2). Chapter 5 also provides information on process control and handling of potentially radioactive M&E (Section 5.3). The use of segregation to help improve the efficiency of measurements and detectability of radioactivity, and as a tool to limit the uncertainty is described in Section 5.4. Sections 5.5 through 5.8 discuss the establishment of measurement uncertainty, measurement detectability, and measurement quantifiability as MQOs to validate the measurement method's ability to meet the established performance objectives. Information is provided on several measurement techniques (Section 5.9) that can be used for comparison to the MQOs developed in Chapter 3. These descriptions can also be used during the planning phase to specify a measurement technique in the survey design. Recommendations related to QC are also provided to ensure that survey instruments are functioning properly, and the data meet defined performance limits specified during planning (Section 5.10). Information related to collecting and documenting survey data is discussed in Section 5.11.

Chapter 6 provides methods for the assessment and decision-making phases. Recommendations are provided for performing the preliminary data review, calculating statistical quantities, and preparing graphic representations that will assist the user in exploring the data (Section 6.2). Disposition decisions about individual items may be based on individual measurement results by comparing data to the upper bound of the gray region (Section 6.3). Information is also provided for calculating the upper confidence limit (Section 6.4). Details on performing recommended statistical tests are also included (Sections 6.5 through 6.7). This chapter also describes how to make a disposition decision based on the survey results (Section 6.8), what to do if the selected disposition option is not accepted (Section 6.9), and the documentation to support the decision (Section 6.10).

Chapter 7 discusses the general concepts of statistical survey design and hypothesis testing. Detailed discussions and calculations of MQOs, measurement uncertainty, minimum detectable concentrations (MDCs), and minimum quantifiable concentrations (MQCs) are provided in this chapter. Details and examples of each topic are provided. A detailed example of a scan MDC calculation is provided that is used to support the illustrative examples in Chapter 8.

Chapter 8 provides detailed illustrative examples implementing specific concepts found throughout MARSAME. The illustrative examples cover a range of material, equipment, radionuclides, and disposition options. Sections of these illustrative examples are used to illustrate specific concepts throughout the supplement.

MARSAME contains several appendices to provide additional information on specific topics. Appendix A provides copies of statistical tables needed to implement the information in MARSAME. Appendix B lists sources of environmental radiation such as natural background and fallout. A list of potential radionuclides grouped by industry or type of facility is provided in Appendix C. Appendix D provides detailed information on specific measurement systems unique to disposition surveys. Appendix E lists and describes some of the potential sources of action levels applicable to decisions regarding disposition of M&E.

1.6   Similarities and Differences Between MARSSIM and MARSAME

During the 1990s, there was a concerted effort to improve the planning, implementation, evaluation, and documentation of building surface and surface soil final radiological surveys for demonstrating compliance with standards. This effort included the preparation of NUREG-1505 (NRC 1998a) and NUREG-1507 (NRC 1998b) by the NRC and culminated in 1997 with the issuance of MARSSIM (MARSSIM 2002). MARSSIM was a joint effort by DOD, DOE, EPA, and NRC to develop a multi-agency approach for planning, performing, and assessing the ability of surveys to meet dose- or risk-based standards while at the same time encouraging effective use of resources. MARSSIM provided recommendations for developing appropriate final status survey designs using the DQO process to ensure survey results were of sufficient quality and quantity to support a final decision. MARSSIM (MARSSIM 2002), NUREG-1505 (NRC 1998a), and NUREG-1507 (NRC 1998b) replaced the previous approach for such surveys contained in NUREG/CR-5849 (NRC 1992).

This MARSAME supplement expands the scope of MARSSIM methods and processes to provide technical information supporting the disposition decision for M&E, specifically the design and implementation of disposition surveys, to ensure the disposition decision is technically defensible and optimized for efficiency. MARSSIM addressed the disposition of real property (e.g., buildings and land) where the only disposition options were unrestricted release, restricted release, or maintaining radiological controls. MARSAME addresses the disposition of non-real property (e.g., M&E) and includes additional options for future use including recycle or disposal as radioactive waste (see Section 2.5). Increasing radiological controls and interdiction are also included as potential disposition options. While several, or all, disposition alternatives may be acceptable for a specific project, optimizing the disposition survey design based on the selected disposition alternative is described in MARSAME.

MARSAME as a supplement to MARSSIM expands the scope of technically sound measurement processes and methods to include M&E. Table 1.2 summarizes the major similarities between MARSSIM and MARSAME, which result from application of a graded approach to support a technically defensible decision regarding disposition. Table 1.3 summarizes the major differences between MARSSIM and MARSAME, which result from the change from real to non-real property.

Table 1.2 Similarities Between MARSSIM and MARSAME

Parameter	MARSSIM	MARSAME
Graded Approach	Used to place greater survey effort on areas that have, or had, the highest potential for residual radioactivity.	Used to place greater survey effort on M&E that have, or had, the highest potential for residual radioactivity.
Data Quality Objectives (DQO) Process	Used to design technically defensible surveys to support decisions on disposition of real property.	Used to design technically defensible surveys to support decisions on disposition of non-real property (e.g., M&E).
Data Quality Assessment (DQA)	Used to evaluate survey results and support a decision of whether to release real property.	Used to evaluate survey results and support a disposition decision for non-real property.
Process Knowledge	Used during the Historical Site Assessment to support the determination of whether an area is impacted and provide information for designing subsequent surveys.	Used during the Initial Assessment to support the determination of whether M&E are impacted and provide information for designing subsequent surveys.
Classification	Determines the level of survey effort based on the potential amount of residual radioactivity present.	Determines the level of survey effort based on the potential amount of residual radioactivity present.
Flexibility	MARSSIM allows and encourages flexibility in the design and implementation of final status surveys for application to diverse site conditions.	MARSAME allows and encourages flexibility in the design and implementation of disposition surveys for application to diverse M&E.
Statistics	Used to develop a technically defensible survey design.	Used to develop a technically defensible survey design.
Scale of Decision Making	A separate release decision is made for every survey unit.	A separate release decision is made for every survey unit.
Inherent Radioactivity	Inherent radioactivity is site-specific and generally cannot be separated from ambient radiation.	Inherent radioactivity is specific to the M&E being investigated. Segregation of M&E based on inherent radioactivity can be used to reduce measurement variability.

Table 1.3 Differences Between MARSSIM and MARSAME

Parameter	MARSSIM	MARSAME
Scope	Surface soil and building surface surveys (i.e., real property).	Materials and equipment (i.e., non-real property).
Disposition Options	Restricted or unrestricted release, or fail to release.	Release survey (maintain, remove, or transfer radiological control; clearance for reuse, recycling, or disposal), or Interdiction survey (increase in the level of radiological control or a decision not to accept control from another party).
Categorization	Included as part of classification in MARSSIM.	Separates the decision to survey from determining level of survey effort.
Application of the Graded Approach	Classification and survey unit size result in varying levels of survey effort.	Multiple disposition options result in varying levels of survey effort.
Sentinel Measurements	Not described in MARSSIM.	Allows use of sentinel measurements during IA to check validity of certain process knowledge assumptions.
Documentation of Survey Designs	Assumes project-specific survey designs will be developed for individual sites.	In addition to project-specific survey design, allows SOPs for categories of M&E to provide standard approach to disposition surveys.
Preliminary Surveys	Scoping and characterization surveys regularly used to obtain information needed to design a final status survey.	Scoping and characterization surveys rarely used to obtain information needed to design a disposition survey. Historical information obtained during the IA is generally sufficient to design a disposition survey. If not, preliminary surveys may be used to provide the necessary information.
Ambient Radiation	Ambient radiation is site-specific and generally cannot be separated from inherent radioactivity.	Ambient radiation is selected based on location where disposition surveys are performed, and can be separated from inherent radioactivity.
Interdiction	Not addressed in MARSSIM.	Surveys may be performed to identify radioactive material resulting in an increase in the level of radiological control or deciding not to accept control from another party. For example, identifying radioactive materials and initiating radiological controls, or identifying unauthorized movement of radioactive material.

Table 1.3 Differences Between MARSSIM and MARSAME (Continued)

Parameter	MARSSIM	MARSAME
Null Hypothesis	MARSSIM recommends using the null hypothesis: "The activity in the survey unit exceeds the action level (Scenario A)." MARSSIM allows using the null hypothesis: "The activity in the survey unit is indistinguishable from background (Scenario B) with information from NUREG-1505 (NRC 1998a)."	User selects the appropriate null hypothesis: "The activity in the survey unit exceeds the action level (Scenario A)." or "The activity in the survey unit is indistinguishable from background (Scenario B)."
Scan-Only Surveys	Not addressed in MARSSIM	M&E may be dispositioned based on the results of scan-only surveys provided the scan measurements meet the MQOs for the survey.
In Situ Surveys	Not addressed in MARSSIM	M&E may be dispositioned based on the results of in situ surveys provided the in situ measurements meet the MQOs for the survey.

2   INITIAL ASSESSMENT OF MATERIALS AND EQUIPMENT

2.1   Introduction

The initial assessment (IA) is the first step in the investigation of materials and equipment (M&E), similar to the historical site assessment (HSA) described in the *Multi-Agency Radiation Survey and Site Investigation Manual* (MARSSIM 2002). The purpose of the IA is to collect and evaluate information about the M&E in order to determine if it is impacted or non-impacted (i.e., categorization). During the IA process, additional information is collected to identify and support potential disposition of impacted M&E (e.g., clearance, increased radiological controls, remediation, or disposal). Project managers are encouraged to use the IA to evaluate M&E for other hazards (e.g., lead, PCBs, asbestos) that could increase the complexity of the disposition survey design or pose potential risks to workers during subsequent survey activities (Section 5.2) or to human health or the environment following subsequent disposition of the M&E.

There are five major activities associated with the performance of the IA:

- Categorize the M&E as impacted or non-impacted based on visual inspection, historical records, process knowledge, and results of sentinel measurements (Section 2.2).
- Design and implement preliminary surveys to adequately describe the M&E and address data gaps based on a preliminary description of the M&E (Section 2.3).
- Describe the physical and radiological attributes of the M&E (Section 2.4).
- Select appropriate disposition option(s) and define alternative actions applicable to impacted M&E (Section 2.5).
- Document the results of the IA through the use of a standard operating procedure (SOP) or development of a conceptual model (Section 2.6).

For M&E that have been categorized as impacted, an existing survey design in the form of an SOP may be available for investigating the radiological status of the M&E. If an applicable SOP is available, the instructions in the SOP should be followed for implementing and assessing the results of the survey. The information on performing preliminary surveys (Section 2.3) can be used to determine whether an SOP is applicable to the M&E being investigated. The information on describing the M&E (Section 2.4) can be used to determine if preliminary surveys are necessary. The information on selecting a disposition option (Section 2.5) and documenting the results of the IA (Section 2.6) can be used for project-specific applications, or for developing a new SOP.

2.2   Categorize the M&E as Impacted or Non-Impacted

The first decision made when investigating M&E is whether they are impacted or non-impacted. M&E with no reasonable potential for containing radioactivity in excess of natural background, fallout levels, or inherent levels of radioactivity are non-impacted. Impacted M&E have a reasonable potential to contain radionuclide concentration(s) or radioactivity above background.

The decision of whether M&E are impacted or non-impacted is primarily based on existing information. Figure 2.1 depicts the categorization process.

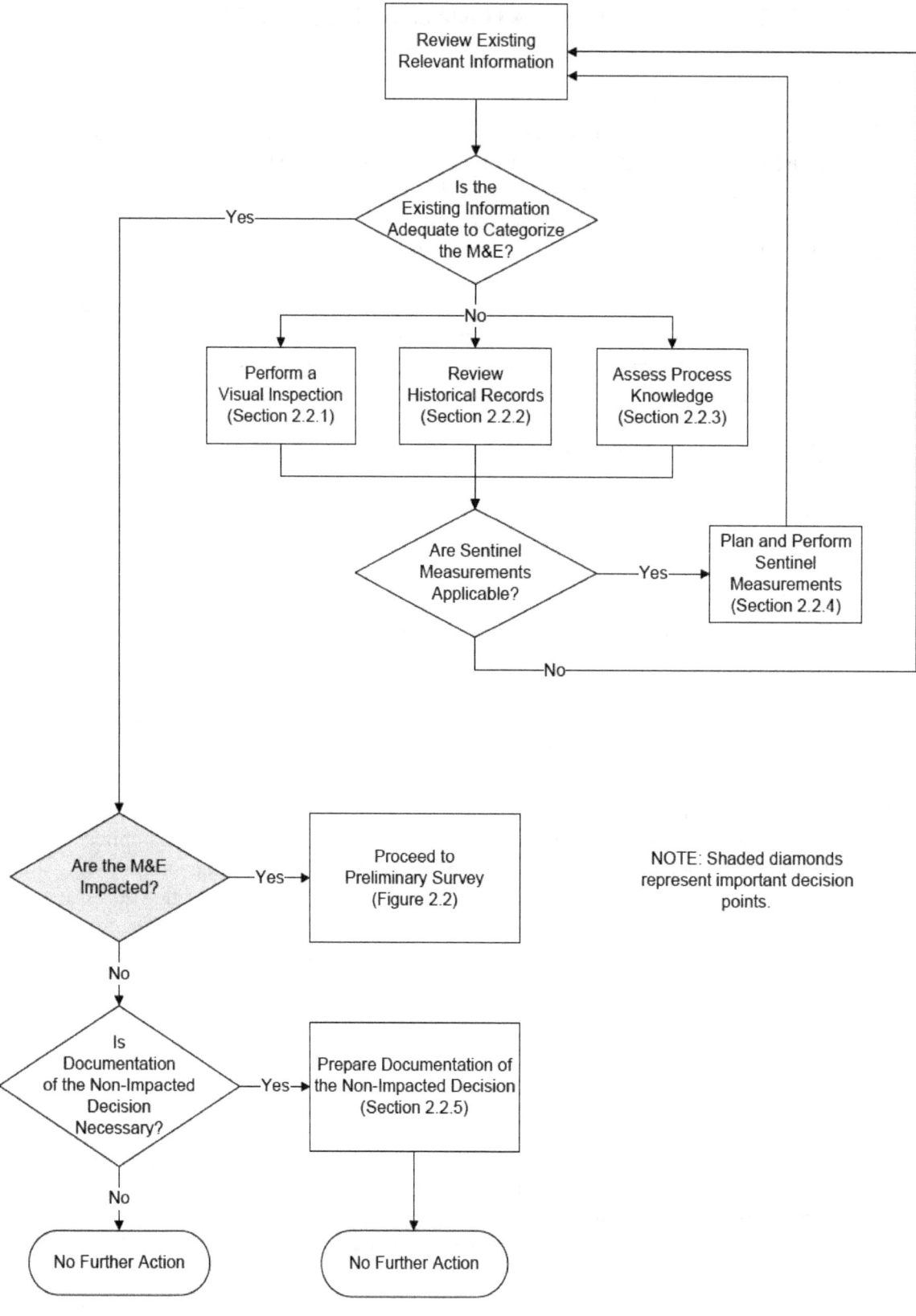

Figure 2.1 The Categorization Process as Part of Initial Assessment

If adequate information is readily available to support a categorization decision, the decision maker should decide if the M&E are impacted or non-impacted. A complex single unit or group of M&E may be divided into portions that are impacted and portions that are non-impacted. This is illustrated in the front loader example described in Section 8.3, where the bucket and tires may be impacted while the engine and cab interior are non-impacted. If additional information is required to support the categorization decision, visual inspection (Section 2.2.1), collection and review of historical records (Section 2.2.2), and assessment of process knowledge (Section 2.2.3) are the most common sources of additional existing information. Assumptions may be made regarding the use and interpretation of existing information. Data collection activities may be performed during the IA to specifically address questions about these assumptions. These data collection activities are called sentinel measurements and are discussed in Section 2.2.4.

Additional investigation is required to make technically defensible disposition decisions regarding impacted M&E. All impacted M&E must receive some level of additional investigation, even if the expected disposition is disposal as radioactive waste. For example, M&E shipped for disposal as radioactive waste must meet waste acceptance criteria at the disposal facility as well as Department of Transportation (DOT) requirements for transporting radioactive material. The results of any additional investigation must clearly demonstrate compliance with any applicable requirements, and be appropriately documented. Non-impacted M&E do not receive any additional radiological investigation.

2.2.1 Perform a Visual Inspection

The purpose of the visual inspection is to identify and document the physical characteristics of the M&E (e.g., size, kind of material, shape, condition) when this description is not readily available to support a categorization decision. The visual inspection may be performed during a site visit, or by reviewing photographs or videos of the M&E. Photographs and video also provide a means for documenting the results of the visual inspection. The visual inspection corresponds to the Site Reconnaissance presented in Section 3.5 of MARSSIM. Information will be used to support the following activities:

- Developing survey unit boundaries (Section 3.6).
- Defining the parameter of interest during the development of a decision rule for impacted M&E (Section 3.4).
- Verifying the requirements of an SOP are met before performing a routine survey (Section 4.5.1).
- Evaluating any health and safety concerns (Section 5.2).
- Developing handling protocols for implementation of the disposition survey (Section 5.3 and 5.4).

Prior to performing a visual inspection, the surveyor should review what is known about the M&E. If little or no information is available describing potential hazards associated with the M&E, care should be exercised in performing a visual inspection. Screening measurements for radiation, chemical, and other hazards, along with the use of personal protective equipment (e.g., gloves, coveralls, respirators), may be necessary depending on available information. Situations with known or expected risks (i.e., M&E that are radiologically or chemically impacted) may require preparation of a study plan or SOP anticipating activities to be performed and identifying specific information to be collected. Casual visual inspections of M&E with an unknown history are not recommended. Detailed visual inspections (e.g., disassembly of potentially impacted equipment to examine interior surfaces) should not be performed without proper precautions and are more appropriately performed by preliminary surveys (Section 2.3).

While the primary objective for performing a visual inspection is to collect information used to design a disposition survey, the information can be used for other purposes. Evaluation of health and safety concerns (Section 5.2) and development of handling protocols for implementation of the disposition survey (Section 5.3) are two examples where visual inspection information would be used.

2.2.2   Collect and Review Additional Historical Records

When information on the identity, concentration, and distribution of radioactivity are not readily available to support a categorization decision, historical records may provide this specific information. Information on the physical characteristics of the M&E (e.g., size, shape, condition) and the characteristics of the radioactivity (e.g., radionuclides of concern, expected concentrations) will be used to select a disposition option in Section 2.5 and describe initial survey unit boundaries in Section 3.6.1. The historical information is then used to define the action level, parameter of interest, and alternative actions during the development of a decision rule for impacted M&E (Section 3.7, EPA 2006a).

Types of historical records that provide useful information are described in MARSSIM Section 3.4.1, and may include—

- A facility or site radioactive materials license;
- Permits or other documents that authorize use of radioactive materials;
- Other permits and environmental program files;
- Operating records (e.g., previous surveys, waste disposal records, effluent releases);
- Corporate contract files (e.g., purchasing records, shipping records);
- A site or facility description (e.g., locations of M&E, site photographs); and
- Inspection reports, incident analyses, and compliance histories maintained by currently and formerly involved regulatory agencies.

Another source of historical information is interviews with current or previous employees. Interviews may be conducted early in the data collecting process or close to the end of the IA. Interviews conducted early in the IA cover general topics, and information gathered is used to guide subsequent data collection activities. Interviews conducted late in the IA allow the investigator to direct the investigation to specific areas that require additional information or clarification.

Once the historical records have been collected, they should be reviewed to identify information that supports the categorization decision. Historical information used to support the categorization decision should be evaluated using the data quality assessment (DQA) process (EPA 2006b). In particular, historical information should be examined carefully because—

- Previous data collection efforts may not be compatible with IA objectives,
- Previous data collection efforts may not be extensive enough to fully describe the M&E being investigated,
- Measurement techniques or protocols may not be known or compatible with IA objectives, or
- Conditions may have changed since the data were collected

Additional information on evaluating data can be found in the following documents–

- The Environmental Survey Manual Appendix A - Criteria for Data Evaluation (DOE 1987)
- Upgrading Environmental Radiation Data, Health Physics Committee Report HPSR-1 (EPA 1980)
- Guidance for Data Usability in Risk Assessment, Part A (EPA 1992a)
- Guidance for Data Usability in Risk Assessment, Part B (EPA 1992b)

Historical records describing impacted M&E may include additional information that can be used to support additional activities during the disposition process. For example, historical records may provide descriptions of the M&E that are sufficient to design a disposition survey (Chapter 4). On the other hand, the historical records can be used to identify data gaps that are addressed by performing preliminary surveys (Section 2.3).

2.2.3   Assess Process Knowledge

The characteristics, history of prior use, and inherent radioactivity are critical for evaluating the impacted status of M&E. This information is termed process knowledge. Process knowledge is obtained through a review of the operations conducted in facilities or areas where M&E may have been located and the processes where M&E were involved when this information is not readily available to support a categorization decision. This information is used to evaluate whether M&E—such as structural steel, ventilation ductwork, or process piping—had been in direct contact with radioactive materials or had been activated, which would lead to a decision the M&E are impacted. Descriptions of the physical attributes of the M&E (Section 2.4.1) and radiological attributes of the M&E (Section 2.4.2) can be obtained from process knowledge. In addition, process knowledge supports the selection of a disposition option (Section 2.5). The disposition option is then used to identify sources of action levels, a parameter of interest, and alternative actions during the development of a decision rule for impacted M&E (Section 3.7 of this supplement and EPA 2006a).

Process knowledge is obtained by researching the M&E and understanding the origin, use, and potential disposition. The level of detail required from process knowledge is project specific. The description of M&E could be simple, such as a set of hand tools being removed from a controlled area where the radiological conditions are well known. At the other extreme is a complex situation that requires knowledge of the manufacturing process, investigations of multiple processes that could impact the radiological conditions associated with the M&E, and understanding of recycle and reuse options that include movement of radionuclides through the environment. Sections 2.4.1 and 2.4.2 describe types of information that may be obtained from process knowledge and are necessary to support the development of a disposition survey.

In some cases, process knowledge of the equipment being investigated can be used to support categorization decisions. Consider a pump used to circulate demineralized make-up water. Maintenance records do not show the presence of radioactivity and operating records indicate no events where the pump could have been used with radioactivity. Radiological samples of the demineralized make-up water do not show the presence of radioactivity. Based on this process knowledge, the interior of the pump is categorized as non-impacted.

Historical records (Section 2.2.2) are one source of process knowledge. Historical records, including interviews, provide site- and project-specific information on historical use and radiological processes that may affect the M&E. Engineering and chemistry books and journals provide information on the origins (e.g., manufacturing) and potential disposition of the M&E. Industry documents and company records are also potential sources of process knowledge. Other sources of information on M&E should be considered during the IA, indicating how, where, and when the M&E were used in areas where they potentially could have been affected by radionuclides or activation. These sources of information include—

- Purchasing records showing when M&E were obtained,
- Maintenance records showing where and how they were used,
- Operating logs for systems that utilized or could have affected the M&E,
- Disposal records showing survey results for similar types of M&E indicating types, and Locations of radionuclides or radioactivity.

In some instances, process knowledge may not be available for the M&E being considered for release. For example, consider an outdoor material staging area for a nuclear facility where various pieces of surplus equipment and metal have accumulated over the years. The origin of these M&E is unknown. In this case, it is particularly important that preliminary surveys be performed on the M&E to determine if excess radioactivity is present and to finalize the list of radionuclides of concern.

Techniques used to protect equipment or prevent radioactivity from entering difficult-to-measure areas or penetrating porous surfaces can be used to support categorization decisions. Consider the following examples of protection and prevention techniques:

- Plan and coordinate all work to minimize exposure of equipment, tools, and vehicles to radioactivity.
- Evaluate materials, tools, and equipment for ease of decontamination and disassembly (that may be required for decontamination or release) prior to use.
- Use prefilters or have a separate source of outside air on the intake for internal combustion equipment subject to airborne radionuclides or radioactivity.
- Use a filtered inlet for high volume air handling equipment such as blowers, compressors, etc., to minimize the potential for internal contamination due to build up of low-level radioactivity.
- Do not bring electrically driven mobile equipment into controlled areas.
- Use protective sheathing/covers, strippable coatings, or protective caps to minimize the potential for surficial radionuclides or radioactivity.

- Cover and protect all openings on equipment, tools, or vehicles that may permit radioactivity to enter difficult-to-access or difficult-to-clean areas.
- Select technologies that minimize radiological airborne emissions, secondary wastes, and tool or equipment damage.

2.2.4 Perform Sentinel Measurements

Sentinel measurements are biased measurements performed at key locations to provide information specific to the objectives of the IA. The objective of performing sentinel measurements as part of the IA is to gather sufficient information to support a decision regarding further action (e.g., categorization). Sentinel measurements may also be used to verify assumptions based on existing information or obtain information on the current status of the M&E. Sentinel measurements are not a risk assessment, scoping survey, or study of the full extent of radionuclides or radioactivity associated with the M&E.

Sentinel measurements alone cannot be used to show that M&E are non-impacted. Positive results are definitive for determining that M&E are impacted. However, negative results provide only part of the evidence required for determining that the M&E are non-impacted. Since radioactivity in difficult-to-measure areas cannot be measured directly without accessing the area (e.g., disassembling equipment), sentinel measurements performed at access points to difficult-to-measure areas could be used to indicate that it is unlikely that radioactivity entered that area. For example, smears with elevated radioactivity, collected inside ductwork, can provide information to support categorization of the ventilation system as impacted. Because sentinel measurements are usually associated with difficult-to-measure areas, they are not generally applicable to dispersible bulk materials.

If protection and prevention techniques (described in Section 2.2.3) were applied to M&E used around radioactive material, sentinel measurements can be used in connection with process knowledge to support a decision of whether difficult-to-measure areas were impacted. For example, if prefilters are used to capture particulate airborne radioactivity of a specific size before the particulates enter difficult-to-measure areas, sentinel measurements can be made on the prefilters.

Sentinel measurement methods may involve any of the measurement techniques discussed in Section 5.9.1 combined with the instruments discussed in Section 5.9.2. Advantages and disadvantages of different combinations of measurement techniques and instrumentation are listed in Table 5.5 and discussed in Section 5.9.3. The selection of a measurement method for sentinel measurements should be made based on project-specific considerations using the DQO process.

It should be noted that access points are often modified to limit personnel radiation exposure to difficult-to-measure areas after use (e.g., capped, sealed, cleaned). Care should be taken to avoid performing sentinel measurements at modified access points to reduce the probability of making an incorrect decision about the status of the M&E. QA and QC should be considered during planning for collection of sentinel measurements. The measurement and subsequent evaluation of the results should be consistent with the assumptions used to define sentinel measurements.

2.2.5   Decide Whether M&E are Impacted

Once there is adequate information to support a categorization decision, the decision maker needs to decide whether the M&E are impacted or non-impacted. The categorization decision is based on four sources of information: visual inspection, historical records review, process knowledge, and the results of sentinel measurements.[1] If the results for any part of the categorization process indicate a reasonable potential for radionuclide concentrations or radioactivity above background, the decision is the M&E are impacted. For example, if the visual inspection, historical records, and process knowledge all indicate the M&E are non-impacted but the sentinel measurements indicate impacted, the M&E are impacted. Similarly, if the visual inspection and sentinel measurements indicate the M&E are non-impacted but the historical records and process knowledge indicate the M&E are impacted, the M&E are impacted. An important point is that sentinel measurements alone cannot be used to support a decision in declaring M&E as non-impacted.

In most cases, the categorization decision is obvious based on the available information. In cases where the decision is not obvious, the consequences of making a decision error usually result in a determination that the M&E are impacted. For example, the consequence of incorrectly categorizing M&E as impacted when they are not impacted includes performing a radiological survey. However, the consequence of incorrectly categorizing M&E as non-impacted when they are impacted could result in inadvertent exposure for members of the public, lack of confidence in other radiological decisions, and potential violation of regulatory requirements. The consequences of incorrectly categorizing M&E are also discussed in Section 4.3.4.

Collectively, this information should be used to develop survey strategies targeting different types of materials in recognition that a single survey method or procedure may not necessarily fit the technical requirements of all materials, given their diverse properties. For example, one procedure may be used to address only the routine releases of tools and equipment. On the other hand, a separate procedure may be developed to address infrequent releases of large amounts of bulk materials, such as concrete rubble. The approach suggested here is one of compartmentalizing the release activities into manageable and common functional elements with each one being optimized in the context of facility operations as to its effectiveness, while demonstrating compliance with applicable regulations. The development of standardized survey procedures for infrequent releases necessitates that the MARSAME user utilize processes in the remainder of this chapter and then move to Section 3.10 for evaluating and implementing standard operating procedures (SOPs).

If there is insufficient information available to design a disposition survey following categorization, preliminary surveys may be performed to obtain additional information describing the physical and radiological characteristics of the M&E (Section 2.4). These preliminary surveys facilitate the development of an effective and efficient disposition survey design.

If there are questions concerning the level of documentation for the categorization decision, consult the cognizant regulatory authority. The decision maker should consider the degree to

---

[1] Sentinel measurements are not required to support a categorization decision.

which documentation of the M&E categorization decision is necessary for M&E that are categorized as non-impacted, since no additional investigation is required. In most cases it is not necessary to document decisions that M&E are impacted since this decision will be documented later in the disposition process (e.g., documentation of the IA results in Section 2.6, documentation of the survey design in Section 4.5, and documentation of the disposition survey results in Section 6.6).

2.3   Design and Implement Preliminary Surveys

If there is insufficient information available to design a disposition survey following categorization, it may be necessary to perform preliminary surveys to obtain the required information. Preliminary surveys of M&E correspond to scoping and characterization surveys described in MARSSIM Sections 5.2 and 5.3.

Following a decision that the M&E being investigated are impacted, the decision maker should determine if an applicable standardized survey design is available, usually in the form of an SOP. If an SOP is available and applicable to the M&E being investigated, the instructions in the SOP should be implemented and the results of the survey evaluated as specified in the SOP (see Figure 2.2 and Section 2.6.1).

It may be necessary to evaluate the quantity and quality of data describing the M&E to determine if the existing data are adequate for implementing an existing SOP or developing a disposition survey design. If the data are adequate, no additional data collection is required. On the other hand, if there are data gaps that need to be addressed prior to completing a disposition survey design, preliminary surveys can be used to obtain the necessary data.

The purpose of performing preliminary surveys is to obtain information describing the physical and radiological characteristics of the M&E. The ultimate goal is to minimize heterogeneity in the subset of M&E being surveyed. Minimizing heterogeneity helps to control the measurement uncertainties (Section 5.6), and may be helpful in selecting a disposition option (Section 2.5). For example, if a subset of the M&E is identified as difficult-to-measure while the majority of the M&E is relatively easy to measure and is considered for release, minimizing heterogeneity of all the M&E by segregating the difficult-to-measure subset for potential disposal may simplify measurements and be cost-effective. See Section 5.4 for information on segregation of M&E to minimize heterogeneity during implementation of the disposition survey design.

In general, preliminary surveys are designed using professional judgment to address specific questions concerning the existing data. Once a data gap has been identified, a survey is designed and implemented to obtain the information required to fill that data gap. The results of the survey are evaluated to ensure the data gap has been adequately addressed and the results are documented. In some cases these surveys will be large and complicated, with written survey designs reviewed by stakeholders prior to implementation. In other cases, these will be simple surveys that quickly provide some small piece of information required to proceed with the disposition survey design. By necessity, there is no single approach that will address all types of preliminary surveys. However, the DQO process can be applied to successfully design a preliminary survey (EPA 2006a).

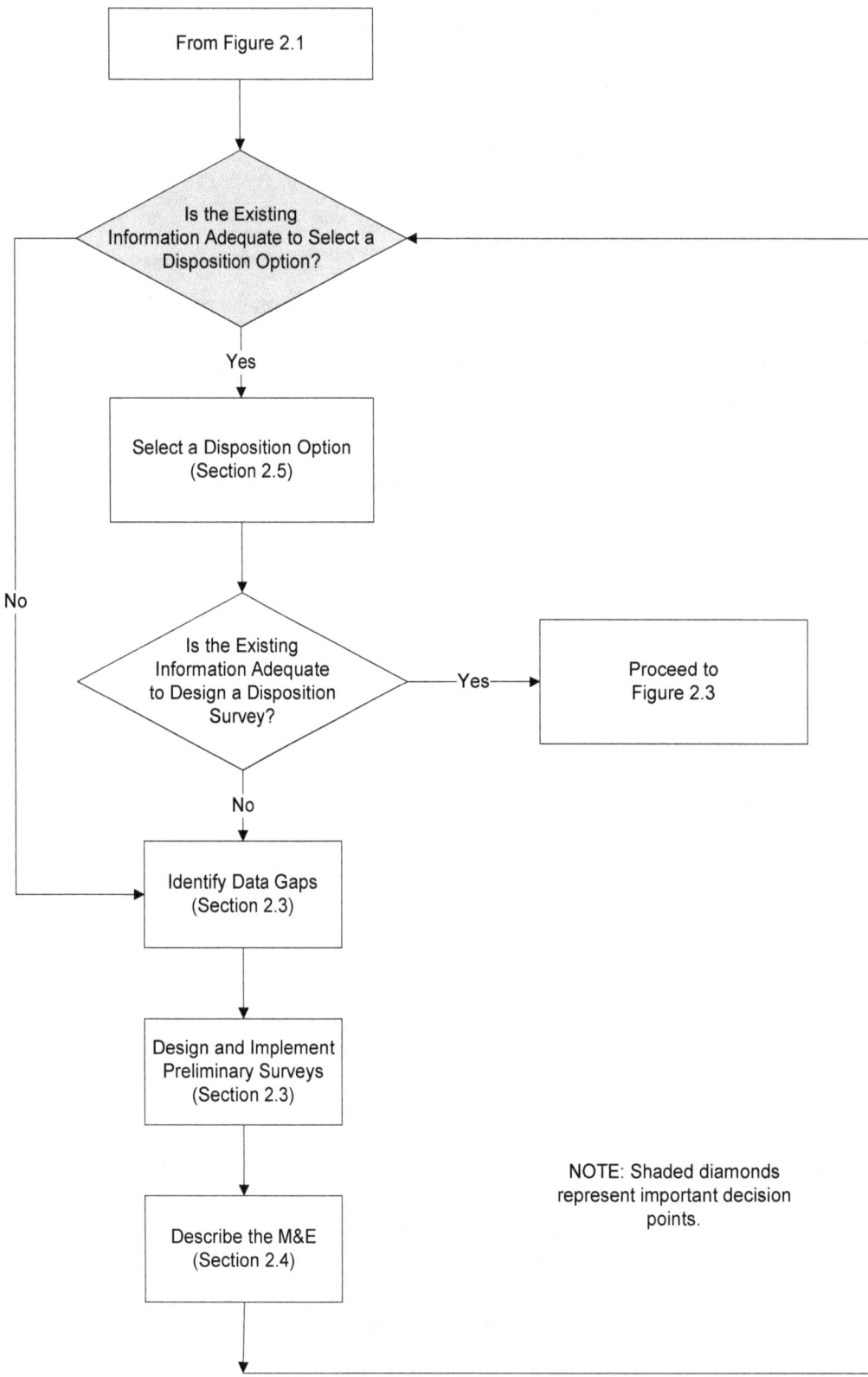

Figure 2.2 Assessing Adequacy of Information for Designing Disposition Surveys

The first step in designing a preliminary survey is to identify the data gaps to be addressed. Section 2.4.1 and Section 2.4.2 discuss the minimum information required to describe the M&E and design a disposition survey. Any of the required information that is not available or is not of sufficient quality represents a data gap. In addition, there may be project-specific information needed to complete the disposition survey design that could also represent potential data gaps. In order to complete the list of potential data gaps, it is recommended that the planning team work through the entire disposition survey planning process (Chapters 3 and 4). Whenever a data gap is identified, the planning team should make reasonably conservative assumptions or proceed with multiple survey designs based on a reasonable range of values to fill the data gap. Identifying a complete list of data gaps will help ensure the necessary additional information can be collected effectively and efficiently, with minimal waste of limited resources. If a separate preliminary survey is designed and implemented for every data gap as it is identified, there is an increased possibility of duplication of effort and increased demands on limited resources. As with all data collection activities, QA and QC should be considered during planning and evaluated during assessment of the results.

MARSAME uses an iterative planning process for designing surveys. Changes in the available information may result in multiple iterations of individual steps. Iteration may be necessary at any time that an assumption used to design a survey is shown to be false. For example, if a historical record is found that changes the description of the M&E from beta-gamma emitting radionuclides to include alpha emitting radionuclides, it is necessary to consider additional or different measurement techniques to account for the alpha radiation.

2.4   Describe the M&E

The M&E being investigated must be described with regards to its physical and radiological attributes in order to establish the information necessary to design a survey approach that can adequately measure the M&E. This description is intended to ensure that residual radioactivity associated with the M&E will not be missed by the disposition survey, the M&E is left in a usable condition, and that any data collected meet the objectives of the disposition survey.

2.4.1   Describe the Physical Attributes of the M&E

A description of the physical characteristics defining the investigated M&E is required to help the user develop a disposition survey design. The preliminary physical description is usually developed using some combination of the techniques presented in Section 2.2 (i.e., visual inspection, historical records, and process knowledge). The physical description of the M&E is used to help define survey unit boundaries (Section 3.6.1) and develop a decision rule (Section 3.7), which has a direct impact on the disposition survey design.

Table 2.1 lists the four attributes that should be addressed when describing the physical characteristics of the M&E being investigated (dimensions, complexity, accessibility, and inherent value). Questions related to the evaluation of the attributes are provided, along with a list of minimum information expected to be provided by the IA. The planning team should consider designing and implementing preliminary surveys (Section 2.3) to verify existing information and investigate data gaps identified during the initial steps of the IA. [2]

---

[2] The development of a planning team is discussed in MARSSIM Section 3.2.

Table 2.1 Physical Attributes Used to Describe M&E

Attribute	Minimum Information	Questions for Consideration
Dimensions	Size (Total Mass) Shape (Total Surface Area)	Are there issues with size and shape that affect how the M&E should be handled?
Complexity	M&E may require segregation to design a technically defensible disposition survey. M&E may be combined into similar groups and still allow a technically defensible disposition survey.	Are there situations where segregation (e.g., disassembly) could affect the usefulness of the M&E? Are there situations where segregation (e.g., disassembly) could result in the release of radioactivity or hazardous chemicals to non-impacted areas? Are there situations where engineering controls are required to prevent the release of radioactivity or hazardous chemicals to non-impacted areas? Are there component materials that are inherently radioactive or regulated for their chemical properties?[3] Are there multiple component materials in the M&E?
Accessibility	Identification of impacted, difficult-to-measure areas for performing conventional handheld measurements. Known or potential relationships among radionuclide concentrations or radioactivity in accessible and difficult-to-measure areas.	Are there issues with size or shape that limit accessibility (e.g., bottom of a large, bulky object)? Are there porous surfaces that could allow permeation of radioactivity? Are there seams, ruptures, or corroded areas where radioactivity could penetrate to difficult-to-measure areas?
Inherent Value	The inherent value of the M&E being investigated.	Can the M&E be reused or recycled? Can the M&E be repaired or remediated? What are the replacement and disposal costs?

2.4.1.1   Describe the Physical Dimensions of the M&E

It is important to understand the dimensions of the M&E being investigated in order to define the scale of decision making (Section 3.6 on identifying survey unit boundaries), support evaluation of measurement techniques (Sections 3.8 and 5.9), and identify any handling issues that may need to be addressed (Section 5.3). The dimensions generally are defined as the size and shape of the M&E being investigated. The size is primarily related to the scale of decision-making and may be defined as the length, width, and depth of an item, or as the quantity of M&E. Quantity may be expressed in terms of a number (e.g., 25 pumps) or a volume (e.g., 200 cubic yards of

---

[3] For example, materials regulated under the Resource Conservation and Recovery Act (40 CFR 261) or the Toxic Substances Control Act (40 CFR 700-766).

concrete rubble), and may be related to the mass of the M&E. An estimate of the total mass of the M&E should be provided. The shape of the M&E is primarily related to the evaluation of measurement techniques. The description of shape should consider surface conditions (e.g., clean or dirty, rough or smooth, curved or flat) that affect the surface efficiency for radiation instruments. An estimate of the total surface area of the M&E should be provided when the radionuclides of concern are, or could be, surficial.

2.4.1.2 Describe the Complexity of the M&E

The complexity of the M&E also affects the disposition survey design. Complexity refers to the number and types of components that make up the M&E, as well as the ability to segregate or combine the M&E into similar groups. M&E consisting of a single component is a simple case. Consider the situation where several hundred feet of pipe are being investigated and the entire pipe is made from steel.

A complex situation occurs when the M&E consist of a variety of component materials. Consider the same amount of pipe, but some pipe is steel, some is copper, and some is lined with rubber, lead, or PVC. Some types of process equipment (e.g., pipe originating from mineral processing industries) are internally lined with rubber, lead, or PVC. The presence of such liners can complicate the initial categorization, as well as subsequent characterization and survey of such equipment. The presence of lead can complicate the final disposition of process equipment (e.g., recycling as ferrous steel or disposal in landfills).

Equipment once used in process plants or systems should be checked for the presence of internally deposited sediment, sludge, oil, grease, water, and presence of process chemicals and reagents. The presence of such residues may require the implementation of special worker health and safety measures, procedures to collect and properly dispose of such hazardous material, and may restrict possible disposition options.

Complexity also comes from the ability to break down or combine the M&E into similar groups. A steel I-beam represents a simple case, where there is one material that can be cut into the desired lengths. Dispersible bulk materials represent a situation that is slightly more complex, especially when different types of materials have been combined. One example is a pile of scrap metal, where the metal can be segregated by material (e.g., aluminum versus steel) or type (e.g., sheet metal versus pipe versus I-beams).

Equipment tends to be more complex, because it often contains a variety of components that can generally be broken down by disassembling the equipment. Consider the case of a power tool consisting of a casing, an electric motor, and controls. There are different types of metal, plastic, and possibly glass or ceramics that make up the item, but disassembly into the individual components may render the tool unusable and may expose component materials that are inherently radioactive or hazardous. Disassembly of certain items could also result in the release of radioactivity or hazardous chemicals to non-impacted areas, and may require engineering controls to prevent such releases. The disposition survey design often increases in complexity as the equipment increases in size and complexity. However, complex M&E may also allow the user to segregate impacted from non-impacted items or components. This segregation may reduce the amount of M&E requiring additional investigation. One example is a front loader used to move piles of potentially radioactive material at a decommissioning or cleanup site. The bucket and tires of the front loader may be identified as impacted while the engine and cab are identified as non-impacted, depending on the controls in place while the equipment was being used. However, there may be cases where an adequate survey design cannot be developed based on decisions made earlier in the planning process. In these cases, it may be necessary to revisit some of the decisions made earlier, for example, re-evaluating the cost to benefit analysis.

2.4.1.3 Describe the Accessibility of the M&E

Accessibility is the next attribute to consider when describing the M&E being investigated. Accessibility has a direct impact on measurability, so it is a critical issue for making technically defensible disposition decisions. Areas (including surfaces and individual items) are either accessible or difficult-to-measure. Accessible areas are areas where radioactivity can be measured, and the results of the measurement meet the DQOs and measurement quality objectives (MQOs) defined for the survey. During the IA it is necessary to distinguish areas that are accessible from areas that may be difficult to measure.

The determination of whether an area is physically accessible, for purposes of the IA, should be based on whether a measurement could be performed using a conventional hand-held radiation instrument such as a sodium iodide (NaI[Tl]) detector, or Geiger-Mueller (GM) pancake probe. If difficult-to-measure areas are identified and these areas are categorized as impacted, the IA should attempt to identify if there are any known or potential relationships among radionuclide concentrations or radioactivity in accessible areas and radionuclide concentrations or radioactivity in difficult-to-measure areas. This information will be evaluated in Section 3.3.3 for the potential to use surrogate measurements as a method of estimating radionuclide concentrations or radioactivity in difficult-to-measure areas.

The potential for permeation and penetration of radioactivity should also be discussed as part of accessibility. Permeation describes the spread of radioactivity throughout a material and is usually associated with porous materials or surfaces (e.g., wood, concrete, unglazed ceramic). Certain chemical and physical forms can increase the permeation rate (e.g., liquids permeate faster than solids; small particles permeate faster than large particles). Penetration describes infiltrating into difficult-to-measure areas, and is generally associated with radioactivity entering through access points, seams, or ruptures. Corrosion of surfaces may also result in penetration of radioactivity into difficult-to-measure areas.

2.4.1.4 Describe the Inherent Value of the M&E

A part of describing M&E that is often overlooked during the IA is determining the inherent value of the materials or equipment being considered for release. Estimates of the value of materials and equipment should include the replacement cost, condition (i.e., can the materials or equipment be reused or recycled), and disposal cost. Replacement costs may consider increased productivity due to upgrades to existing facilities and equipment, decontamination costs for existing and new items, and the ultimate disposal of the replacements. Condition of the materials and equipment may include maintenance and repair costs to start or keep the items operational, as well as costs to decontaminate and release the items from radiological controls. Disposal costs may include shipping and handling of potentially hazardous material. The limited capacity of existing radiological waste disposal facilities may need to be considered along with the monetary cost of disposal.

2.4.2   Describe the Radiological Attributes of the M&E

A description of the radioactivity potentially associated with M&E being investigated is required to design a disposition survey. The review of historical documents (Section 2.2.2) and process knowledge (Section 2.2.3) are the primary sources of information on radioactivity associated with M&E. Sentinel measurements (Section 2.2.4) and preliminary surveys (Section 2.3) may also provide information, such as types of radiations and identity of radionuclides. The information describing the radioactivity is used to support a decision of whether the M&E are impacted and supports the development of a disposition survey for impacted M&E. The description of the radioactivity is divided into four attributes: radionuclides, activity, distribution, and location.

Table 2.2 lists the four attributes to be addressed when describing radioactivity potentially associated with the M&E being investigated. Questions related to the evaluation of the attributes are provided, along with a list of minimum information expected to be provided by the IA. The planning team should consider designing and implementing preliminary surveys (Section 2.3) to obtain information that is not provided by the IA.

Table 2.2 Radiological Attributes Used to Describe M&E

Attribute	Minimum Information	Questions for Consideration
Radionuclides	List of radionuclides of potential concern, including major radiations and energies.	What were the potential sources and mechanisms for the radioactivity to come into contact with the M&E?
Activity	List of expected radionuclide concentrations or radioactivity (e.g., average, range, variance) associated with the M&E List of known and potential relationships among radionuclide activities (e.g., activation and corrosion products, fission products, natural decay series).	What is the basis for the expected radionuclide concentrations or radioactivity? What is the basis for the known and potential relationships (e.g., process knowledge of similar sources, measurements of equilibrium conditions)?
Distribution	List of areas where the radioactivity is uniformly distributed. List of areas where the distribution of radioactivity is clustered. List of areas where the distribution is unknown.	Can the M&E be divided into sections where the distribution of radioactivity is uniform? Are there areas where small areas of elevated activity are a concern?
Location	State whether the radioactivity is surficial, volumetric, or a combination of both. State whether surficial radioactivity is fixed or removable.	Is the volumetric activity uniformly distributed, is there a gradient, or is the activity random or clustered?

2.4.2.1   Identify the Radionuclides of Potential Concern

Identification of the radionuclides of potential concern is a critical step in making disposition decisions. At a minimum, the planning team should review the information available from Section 2.2 to identify the radionuclides of potential concern. The quality and completeness of the existing information should be evaluated. Information on known or expected relationships among radionuclides of potential concern should be identified and evaluated for applicability to current conditions. If necessary, a study to identify a complete list of radionuclides of potential concern and determine relationships among radionuclides may be initiated before designing the disposition survey.

A list of radionuclides of potential concern should be developed based on existing data. The list should consider all potential sources of radioactivity, but only include radionuclides that are actually of concern for the M&E being investigated.

The list is designed to help focus the disposition decision. The list of radionuclides of potential concern should include the major types of radiation (e.g., alpha, beta, photon) and their corresponding energies. A discussion of the sources of radionuclides of potential concern, and their chemical and physical form should also be included, if possible.

Include a description of how the M&E became impacted if it is known. For example, it is important to document whether the potential radioactivity resulted from deposition of airborne particulate material, or from placing the M&E in an area of neutron flux that resulted in activation. All potential mechanisms for radioactivity that is associated with the M&E should be described.

The description of potential radioactivity from the IA may also identify known or suspected relationships among radionuclides (e.g., equilibrium conditions for natural decay series, relative activities of fission products or activation products based on process knowledge). Additional investigations (e.g., preliminary surveys) may be performed to verify the presence of radionuclides of potential concern and provide estimates of the activity relationships among radionuclides. These investigations may include field measurements and sample collection with laboratory analysis.

The identification of radionuclides of potential concern may impact other decisions made during development of a disposition survey design. Since the sources of action levels are radionuclide or radiation-specific, the identification of radionuclides of potential concern directly affects the selection of an appropriate action level. The planning team should consider the impact of the list of radionuclides of potential concern on other decisions (e.g., selection of measurement techniques or instruments) as well as the impact of other decisions on the action levels when considering potential sources of action levels. For example, the identification of available measurement techniques (Section 3.8) is also directly related to the radionuclides of potential concern. The determination of surficial or volumetric radioactivity (Section 2.4.2.4) may be based on the energy and penetrating power of the radiation emissions, which would be indirectly related to the radionuclides of potential concern. Caution must be used in evaluating radionuclide concentrations or radioactivity for M&E with high levels of inherent background radioactivity.

2.4.2.2   Describe the Radionuclide Concentrations or Radioactivity Associated with the M&E

A description of expected radionuclide concentrations or radioactivity is also important for supporting disposition decisions for M&E. Radionuclide concentrations or radioactivity in excess of background (see Section 3.9 and Appendix B) support a finding that the M&E are impacted. Historical records (Section 2.2.2) and process knowledge (Section 2.2.3) are sources of information on radionuclide activities associated with M&E. In addition, sentinel measurements (Section 2.2.4) can provide information on radionuclide concentrations or radioactivity. A description of the expected radionuclide concentrations or radioactivity should be developed for each of the radionuclides of potential concern. At a minimum, the average expected activity should be provided. Some assumption regarding the expected activity will be required in order to design a disposition survey using the guidance in Chapter 4. If no assumption can be made, a preliminary survey should be performed. If possible, information on the expected range and uncertainty (σ, as described in Sections 3.8.1 and 5.6) of the activity should be provided. The description of the expected activity should include the units, an estimate of uncertainty in the values, and a summary of how the data were obtained (e.g., purpose of data collection efforts, actual measurements, instrument used, count time, or process knowledge). Any known or suspected relationships among concentrations for individual radionuclides should be included in the description. For example, there is an expected relationship among fission products from a nuclear reactor because of the common source of the radionuclides (i.e., nuclear fission). Similarly, there is an expected relationship for activation and corrosion products. Members of the natural decay series (i.e., thorium series, uranium series, actinium series; see Appendix B) are also expected to have a relationship for activities based on equilibrium conditions.

2.4.2.3   Describe the Distribution of Radioactivity

The distribution of radioactivity is primarily concerned with whether the activity is clustered or more uniformly distributed throughout the item. A uniform distribution of activity has little spatial variability, so the radionuclide concentrations or levels of radioactivity are fairly constant. A clustered distribution of activity has high spatial variability, and small areas of elevated activity are present as well as areas with little or no activity above background. The expected distribution of radioactivity could include areas with uniform radionuclide concentrations or levels of radioactivity and areas where the radionuclide concentrations or radioactivity is non-uniform. For example, airborne deposition could have produced a uniform distribution of radioactivity on horizontal exterior surfaces, while penetration through seams and access points could result in clustered radioactivity on interior surfaces. In addition, the interior surfaces could have a uniform distribution of radioactivity over localized areas (e.g., areas around a vent or cooling fan). Concentrations of radionuclides on M&E can change over time due to in-growth, decay, or diffusion.

2.4.2.4   Describe the Location of Radioactivity

The location of radioactivity is primarily concerned with whether the activity is located on the surface or distributed throughout the volume of the M&E. Surficial radioactivity is restricted to the surface of the M&E and is further described as removable, fixed, or some combination of these two. Removable (or non-fixed) radioactive material is radioactive material that can be readily removed from a surface by wiping with an absorbent material. Fixed radioactive material is not readily removed from a surface by wiping. Surficial radioactivity is generally associated with non-permeable solid M&E. Volumetric radioactivity is not restricted to the surface of the M&E and is usually associated with permeable materials, surfaces, or activation by neutrons or other particles.

The question of surficial versus volumetric radioactivity is a complicated issue that may or may not have a significant impact on the disposition survey design. The description of the location of radioactivity used to design the survey may be independent of where the radioactivity is physically located. For example, consider two different methods for surveying ^{60}Co activity concentrations distributed on the surface of several thousand small bolts. First, the bolts may be surveyed in a container using in situ gamma spectroscopy assuming the radioactivity is volumetrically distributed.[4] If the same bolts are surveyed individually using a conveyorized survey monitor the conceptual model may describe the ^{60}Co as surficial radioactivity.

In some cases, the location of the residual radioactivity may be well known. For example, surface deposition of radioactivity on a non-porous material (e.g., smooth stainless steel) will not penetrate into the material to a significant extent under most conditions, so the residual radioactivity could be identified as surficial. Activated materials and bulk quantities of materials usually have volumetric residual radioactivity, although surficial radioactivity may also be present. On the other hand, the actual location of the residual radioactivity may be less well known or unknown.

Process knowledge is the primary source of information on the location of residual radioactivity. The planning team should review the information from Section 2.2.3 to determine the expected location of residual radioactivity and the level of knowledge (i.e., well known, less well known, unknown) associated with the information.

When the location of the residual radioactivity is well known, the planning team should proceed with a survey design based on the appropriate assumption, surficial or volumetric. When the location is less well known or unknown, the planning team may choose to proceed with multiple survey designs to determine the possible effect the location of the residual radioactivity may have on the design of the disposition survey.

2.4.3   Finalize the Description of the M&E

A final description of the M&E should be prepared following implementation of any preliminary surveys. The description of the M&E should consider the information in Table 2.1 and Table 2.2 and provide sufficient information to design the disposition survey.

---

[4] This example does not imply that any measurement technique should be applied to every situation. The information in Section 3.8 should be used to develop the measurement quality objectives (MQOs) for a project. The MQOs can be used to evaluate measurement techniques against the action levels and select the techniques best suited for a specific application.

2.5   Select a Disposition Option

The disposition of the materials and equipment will be a key factor in designing the disposition survey. MARSAME broadly considers two types of disposition decisions: release and interdiction. Release surveys are used to determine whether radiological controls can be reduced, removed, maintained at the current level, or transferred to another qualified user. Interdiction surveys are used to initiate radiological control, or to decide current radiological controls are adequate.

Examples of potential disposition options for release of impacted M&E include—

1. Reuse in a controlled environment.
2. Reuse without radiological controls (i.e., clearance).
3. Recycle for use in a controlled environment (i.e., authorized disposition).
4. Recycle without radiological controls.
5. Disposal as industrial or municipal waste.
6. Disposal as low-level radioactive waste.
7. Disposal as high-level radioactive waste.
8. Disposal as transuranic (TRU) waste.
9. Maintain current radiological controls.

Examples of potential disposition options for interdiction of impacted M&E include—

1. Remove M&E from general commerce and initiate radiological controls.
2. Decide to accept M&E for a specific application.
3. Decide not to accept M&E for a specific application.
4. Continue unrestricted use of M&E (no action).

The selection of a disposition option should be based on the information available at the end of the IA. The disposition option (e.g., reuse, recycle, disposal, initiation of control, or refusal) defines the action level (Section 3.3). The expected radionuclide concentrations or levels of radioactivity associated with the M&E (Section 2.4.2) are compared to the action level to determine whether the M&E will be controlled or uncontrolled following the disposition survey. The disposition option also defines the alternative actions for the decision rule to be developed in Section 3.6. Different disposition options may be applied to separate parts of equipment. If so, implementation of the different dispositions implies the necessity for total or partial disassembly. For example, it may be possible to remove a bucket from a backhoe for disposal and allow reuse of the rest of the equipment.

2.6   Document the Results of the Initial Assessment

The results of the IA should be documented to the extent necessary to support the decisions made. The level of documentation required will depend on the amount of information collected, the quantity of M&E covered by the IA, the type of assessment (e.g., standardized or project-specific), and, as applicable, administrative and regulatory requirements. Two options for documenting the assessment results are the Standardized IA and the conceptual model as described in the following sections. Figure 2.3 illustrates the documentation of the IA.

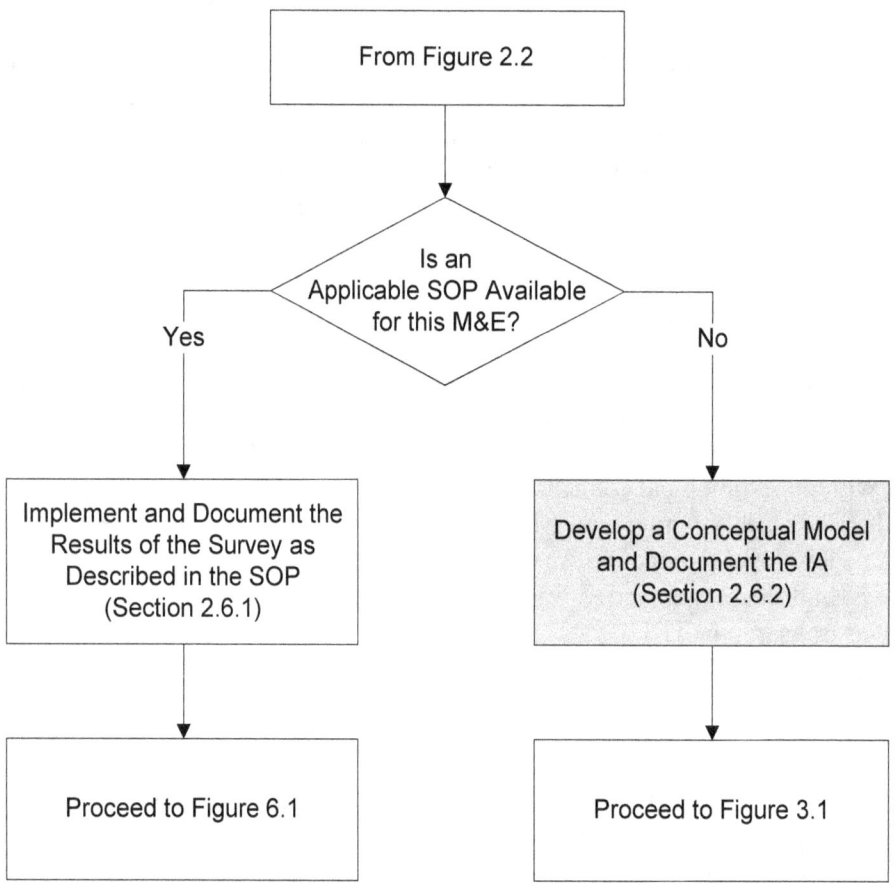

NOTE: Shaded box represents important milestone.

Figure 2.3 Documentation of the Initial Assessment

2.6.1   Document a Standardized Initial Assessment

A standardized IA is a set of instructions or questions that are used to perform the IA. These instructions are usually documented in an SOP. The SOP should be developed, reviewed, and documented in accordance with an approved quality system. Information on developing and documenting a functional quality system can be found in EPA QA/G-1 (EPA 2002c). Guidance on developing SOPs as part of a quality system can be found in EPA QA/G-6 (EPA 2001).

A standardized IA is generally associated with facilities or processes that regularly evaluate similar types of M&E. The release of small tools and personal items from an operating nuclear plant is one example of such a process. Another example, this time describing an interdiction process, would be evaluating truckloads of scrap metal entering a recycle facility. SOPs may be developed to describe repeated routine disposition surveys of similar M&E for both situations.

The documentation of the IA results is described in the SOP. The documentation should be sufficient to demonstrate that trained personnel using an approved SOP evaluated all potentially impacted M&E. For a standardized IA, all these records are maintained but may not be directly associated with the IA. Individual records for each item evaluated by an IA are not required.

The SOP should clearly describe its scope and the applicable types of M&E. This information may be useful for determining whether the M&E are impacted as well as whether the SOP can be used to evaluate the M&E. For example, if the SOP is applicable to all M&E used for a certain process or within a certain part of a facility, this defines what M&E can be considered impacted by that process.

The SOP should also describe the M&E that were used to develop the instructions. The description of the M&E being investigated (Sections 2.2 and 2.3) should be compared to the assumptions used to develop the instructions to determine if the SOP is appropriate. For example, it may be appropriate to apply an SOP developed for scrap metal to evaluate hand tools, since both are made from metal and may have similar surface radioactivity. Alternatively, it may not be appropriate to use an SOP developed for scrap metal to evaluate dry active waste or concrete rubble, since they may have volumetric activity and different surface efficiencies. At a minimum, the rationale for applying the SOP to M&E other than specified in the SOP should be documented.

The SOP should include the training requirements for personnel implementing the SOP. Personnel performing the IA should be familiar with the SOP being implemented, as well as the potential disposition options implied or explicitly stated in the SOP.

Additional documentation may be needed when the SOP is applied to situations other than those considered during development of the SOP. The purpose of the additional documentation is to determine whether the SOP may be applicable to a wider range of M&E. This documentation will help provide technical support for modifying the SOP. If incorrect decisions are made concerning the determination of whether M&E are impacted, or inappropriate recommendations are made for disposition options, it may be necessary to modify the SOP to reduce the number of decision errors. The additional documentation will help identify the source of the decision errors and help provide technical support for modifying or revising the SOP.

2.6.2 Document a Conceptual Model

If a standardized IA approach is not available for the M&E being investigated, the results of the IA should be documented in a conceptual model. If the information in MARSAME is being used to develop a standardized survey design (e.g., a new SOP), the information on developing a conceptual model applies.

The conceptual model is applied in case-by-case situations and decisions. The conceptual model describes the M&E and radioactivity expected to be present for the project. The definition of impacted and non-impacted as it applies specifically to the project should be included in the conceptual model. The conceptual model describes the processes involving radioactive materials, as well as how the radioactivity could become associated with the M&E.

The description of the M&E documents the results of the IA investigation. At a minimum the conceptual model should include a description of the physical attributes of the M&E (see Section 2.4.1 and Table 2.1), the radiological attributes of the M&E (see Section 2.4.2 and Table 2.2), and a list of the applicable disposition options (Section 2.5). In addition, the conceptual model helps identify data gaps and develop potential collection strategies for filling data gaps.

The conceptual model will serve as the basis for the information and assumptions used to develop the disposition survey design in Chapter 4. In many cases the information in the conceptual model will be included in either the survey design documentation or in the documentation of the results of the disposition survey. The structure and content of the conceptual model should be based primarily on the future uses of the data.

The planning team should review the information on radionuclides of potential concern provided by the IA for consistency with the conceptual model. If the data appear incomplete or the quality of the data is not adequate for the disposition survey being designed, the planning team may decide that additional information needs to be collected using preliminary surveys before proceeding with the survey design.

3   IDENTIFY INPUTS TO THE DECISION

3.1   Introduction

This chapter identifies sources of information needed to evaluate the disposition option, or options, selected during the initial assessment (IA). During implementation of an existing standard operating procedure (SOP), this information would have been considered during development of the SOP. This chapter discusses factors affecting the selection of survey units, provides guidance on defining spatial and temporal boundaries, and examines practical constraints on collecting data. Figure 3.1 depicts the process of identifying the inputs to the decision. The expected output from this chapter is a decision rule, or multiple decision rules. A decision rule is a theoretical "if...then..." statement that defines how the decision maker would choose among alternative actions if the true state of nature could be known with certainty (EPA 2006a). There are three parts to a decision rule (Section 3.7):

- An action level that causes a decision-maker to choose between the alternative actions (Section 3.3),
- A parameter of interest that is important for making decisions about the target population (Section 3.4), and
- Alternative actions that could result from the decision (Section 3.5).

Other inputs to the decision discussed in this chapter include selecting radionuclides or radiations of concern (Section 3.2), developing survey unit boundaries (Section 3.6), inputs for selecting provisional measurement methods (Section 3.8), and identifying reference material (Section 3.9). Also discussed in this chapter is the evaluation of an existing survey design to determine if it will meet the data quality objectives (DQOs; Section 3.10).

This chapter provides guidance on performing Step 3, Step 4, and Step 5 of the DQO process (EPA 2006a) for designing a disposition survey. These steps build on the IA where members of the planning team were identified and M&E under investigation were identified as impacted (non-impacted M&E do not require additional investigation). A disposition option was selected (Section 2.5) and documented (Section 2.6).

It is important to remember the DQO process is an iterative process. This means new information can be incorporated into the planning process and outputs from previous steps can be modified to incorporate the new information. For example, if no measurement methods are identified in Section 3.8 that meet the data requirements for a specific disposition option, the planning team may return to Section 2.5 to select a different disposition option. Alternatively, the selection of an action level or survey unit boundary may be affected by the available measurement techniques. The issues associated with surficial vs. volumetric radioactivity (see Section 2.4.2) affect the kinds of information (i.e., action level, survey unit identification, and measurement techniques) as well as the definition of study boundaries (i.e., target population, spatial boundaries, practical constraints on collecting data, subpopulation for which separate decisions will be made).

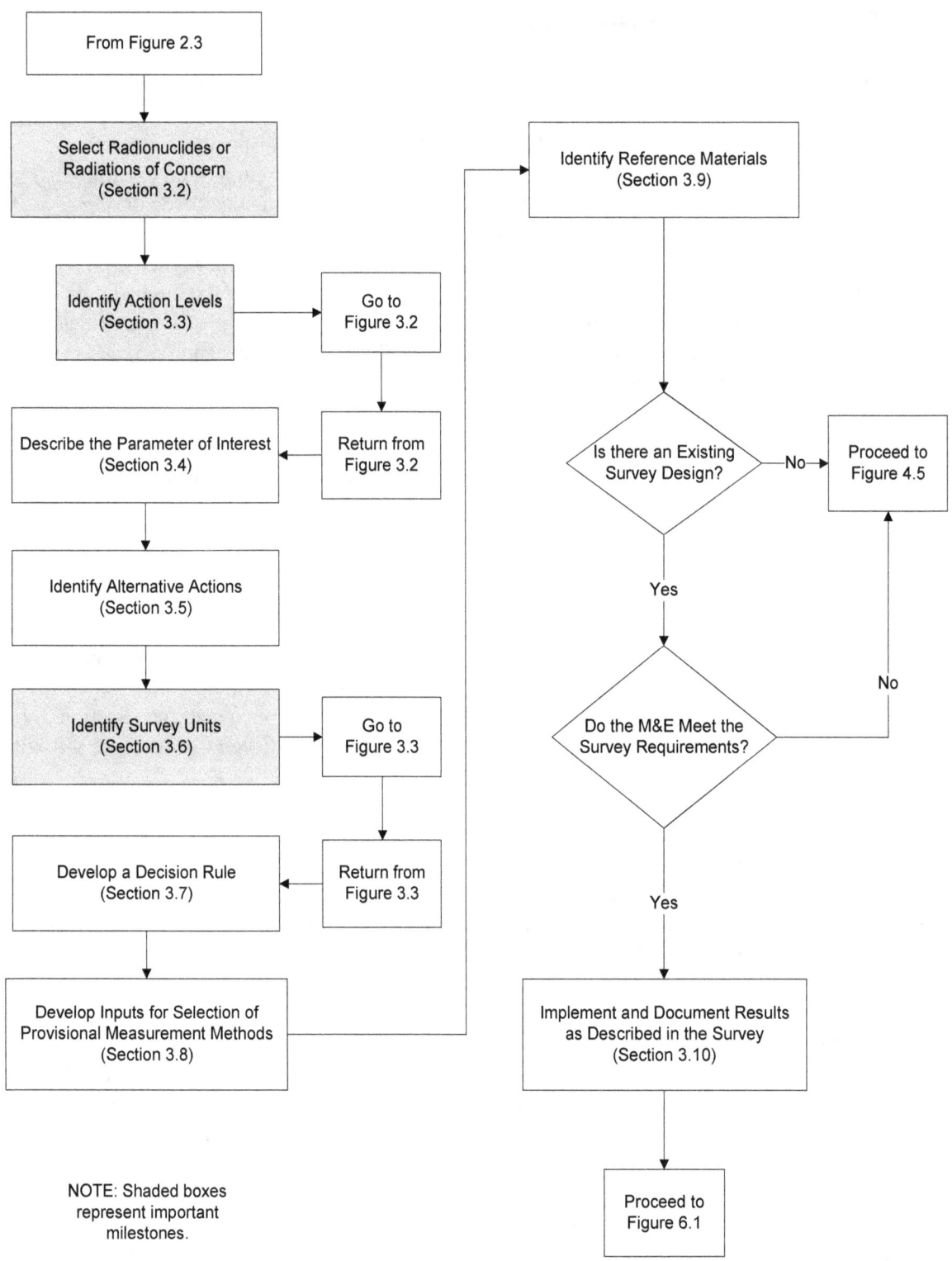

Figure 3.1 Identifying Inputs to the Decision

At the end of this chapter, the planning team should have the information required to design the disposition survey and know whether appropriate measurement techniques are available. Spatial and temporal boundaries will be identified, along with any practical constraints on data collection activities. Examples of practical constraints on data collection include time, budget, personnel, or equipment. For example, a box counter is selected to perform measurements for clearance of items from a radiologically controlled area. Assume a five-minute count time is required to achieve the survey objectives, and another minute is required to swap items in the detector. This means that ten measurements can be performed each hour. If more than 240 items require clearance each day, this measurement method would be impractical since a single box counter cannot clear all of the M&E. The decision rule(s) developed at the end of this chapter will be used to develop survey designs in Chapter 4.

3.2   Select Radionuclides or Radiations of Concern

A list of radionuclides of potential concern was developed in Section 2.4.2.1 as part of the description of radiological attributes associated with the M&E. Before a decision rule can be developed or a disposition survey designed, a final list of radionuclides or radiations to be measured must be prepared.

The selection of radionuclides or radiations of concern is linked to several inputs to the decision. For example, the identification of an action level (Section 3.3) may determine if the survey results need to be radionuclide-specific, forcing the planning team to identify individual radionuclides of concern. On the other hand, the selection of a non-radionuclide specific measurement method may allow the selection of a radiation of concern (i.e., alpha [α], beta [β], gamma [γ], x-ray, or neutron radiation) without ever finalizing a list of radionuclides of concern.

Finalizing the list of radionuclides or radiations of concern is an example of the iterative nature of the survey design process. The planning team is expected to evaluate different survey techniques and measurement methods. Evaluating these different survey techniques and measurement methods will require the planning team to return to the list of radionuclides of potential concern and go through the selection of radionuclides or radiations of concern. Actually, the final selection of radionuclides or radiations of concern may not occur until the disposition survey is optimized; the last step (Step 7) of the DQO process (Section 4.4.5).

3.3   Identify Action Levels

The action level is the numerical value or values that cause a decision maker to choose one of the alternative actions. The radionuclides of concern and disposition options selected at the completion of the IA define the alternative actions for the disposition survey.

Figure 3.2 shows the process for selecting action levels. As shown , the iterative nature of the DQO process may result in changes to the action levels or disposition options based on other factors (e.g., availability of appropriate measurement techniques, measurability, surficial vs. volumetric activity). The planning team should consider the effect of action levels on other steps in the survey design process, as well as any effects these other steps might have on the action levels.

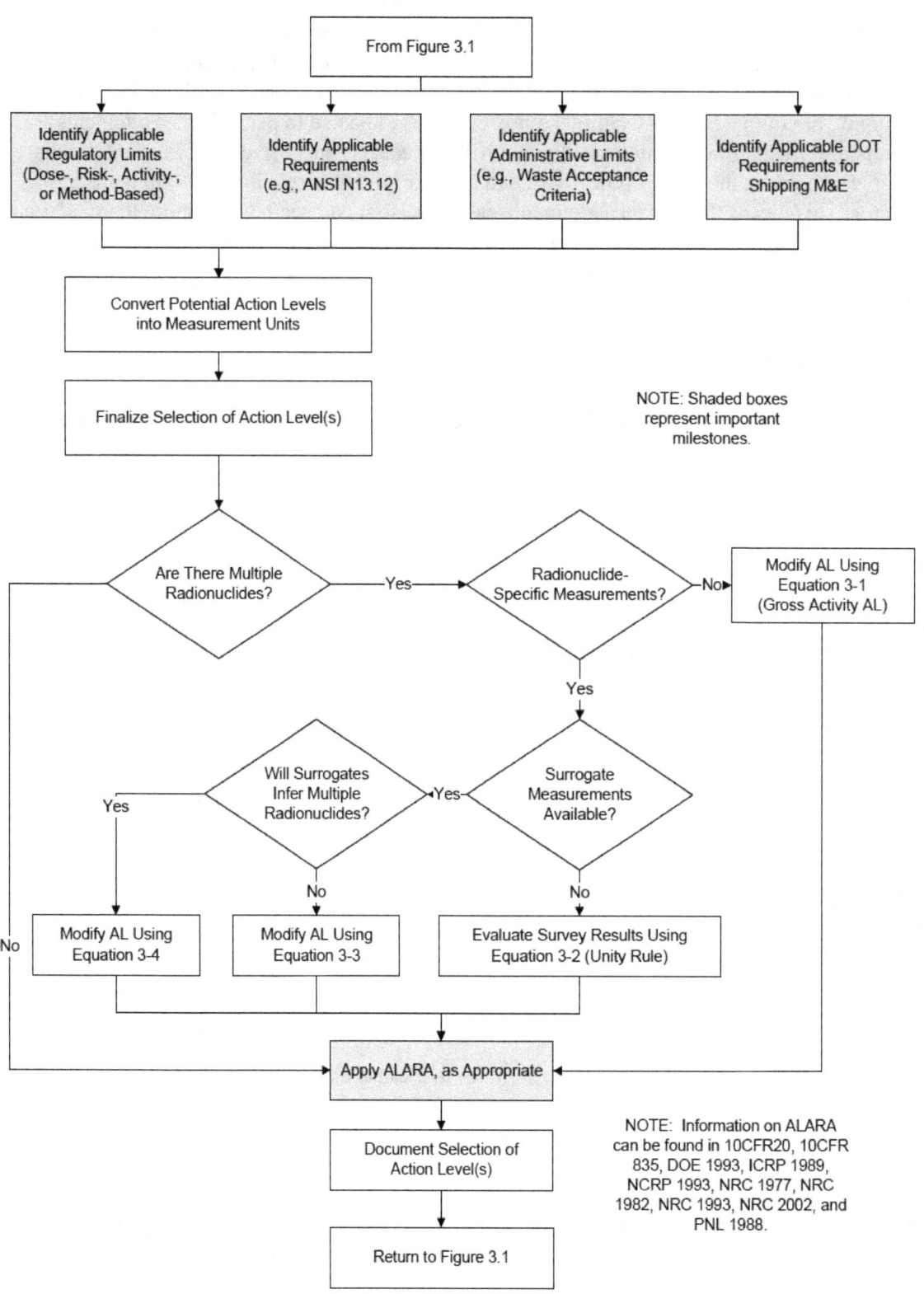

Figure 3.2 Identifying Action Levels
(Apply to Each Disposition Option Selected in Section 2.5)

Action levels are radionuclide- or radiation-specific and in units of concentration or activity (e.g., Bq/kg of ^{137}Cs, Bq/m^2 of alpha radiation, Bq of ^{60}Co). Action levels may be provided, derived from dose- or risk-based standards, or converted into more convenient units for a specific measurement technique.[1]

More than one action level may be required to demonstrate compliance with a specific standard. For example, DOE Order 5400.5 Figure IV-1 (DOE 1993) provides limits for average total surface activity, maximum total surface activity, and maximum removable surface activity (see Appendix E). All three limits must be achieved to demonstrate compliance for disposition of the M&E. Sometimes multiple regulatory requirements may apply, for example transportation regulations combined with waste acceptance criteria and health protection standards.

Action levels may be established based on total activity or incremental activity levels relative to background. Examples of incremental action levels include activity levels based on dose or risk above background, or interdiction at some quantity above background. For these types of action levels it is important to establish a representative reference material (Section 3.9) for comparison.

Action levels may explicitly or implicitly require the use of a specific measurement technique (Section 5.9.1) or instrument (Section 5.9.2). These "method-based" requirements should be considered not only during survey design and implementation, but also during selection of disposition options and action levels. For example, U.S. Department of Transportation (DOT) regulations include package dose rate limits (49 CFR 173.441) as well as removable external radioactive contamination limits (49 CFR 173.443 and 177.843). Dose rate limits on external surfaces at one meter from package surfaces imply the use of in situ or direct measurements, although the selection of specific instrumentation is not specified. Measurements of removable contamination explicitly require the use of smears, but the procedure for collecting and analyzing the smear is not specified. The NRC regulations for transportation of radioactive packages (10 CFR 71) replicate DOT regulations and define limits for "surface contaminated objects," including fixed and non-fixed surface radionuclides (Appendix E.3.7).

At this point, it is important to identify action levels appropriate for the disposition survey. If multiple action levels are identified, the planning team may decide to continue with the development of multiple survey designs that will be evaluated in Section 4.4. The decision maker and the planning team will need to evaluate the action levels and select the action level that best meets the DQOs developed for the survey. The selected action levels are used to develop decision rules in Section 3.7. Alternatively, the planning team may decide to revisit the selection of disposition options from the IA to further limit the scope of the disposition survey and eliminate some of the action levels. In either case, the selection of action levels will be finalized in Section 4.4 with the development of a disposition survey design. Information supporting the selection of an action level(s) is discussed in Sections 3.3.1 through 3.3.4.

---

[1] Correctly converting action levels to counts or counts per minute (cpm) using the appropriate calibration may provide a useful comparison for real-time evaluation of field measurement results as long as field results (e.g., cpm) are converted to and recorded in the same radiological units as the action levels.

3.3.1   Identify Sources of Action Levels

There are many potential sources of action levels available for use in developing disposition surveys. An action level may be based on—

- Dose- or risk-based regulatory standard (i.e., disposition criterion),
- Waste acceptance criteria at a disposal site,
- Regulatory threshold standard (e.g., indistinguishable from background or no detectable radioactivity),
- DOT regulations for shipping radioactive M&E,
- Activity-based standard,
- As low as reasonably achievable (ALARA) considerations,
- Administrative limits, or
- Limitations on technology (performance criteria for an analytical method).

Appendix E provides information on some of the federal sources of action levels that can be applied to M&E. The list of sources for action levels is not exhaustive, but is intended to provide examples of different types of action levels that are referred to throughout this supplement. National and International organizations have published recommendations for action levels (e.g., NCRP 2002, ANSI 1999). These recommendations may be a useful source of action levels if approved by the appropriate authorities.

As previously stated, in many cases the action levels will be dictated by the disposition option selected during the IA. For example, the action levels for M&E being considered for clearance may be a regulatory standard, whereas the action levels for M&E being considered for disposal as radioactive waste will often use the waste acceptance criteria for a disposal site.

Multiple sources of action levels may be identified for a single disposition option. Waste acceptance criteria can be evaluated for several potential burial sites.

In addition, a single source of action levels could be acceptable for more than one disposition option. Dose- and risk-based regulatory standards can be applied to both release and recycle scenarios, as well as for surficial or volumetric radioactivity. On the other hand, activity-based standards may have limited applicability, such as DOE Order 5400.5 (DOE 1993) Figure IV-1 that only applies to release of M&E with surficial radioactivity.

The identification of sources for action levels may affect other decisions made during development of a disposition survey design. Identification of survey units and spatial boundaries for a survey are often directly linked to the action levels. In addition, the expected levels of residual radioactivity identified during the IA (Section 2.6) will often suggest which disposition options are feasible.

At a minimum the planning team should identify at least one source of action levels applicable to the disposition option(s) selected during the IA. Any information related to the action levels that may affect other decisions should also be listed. A partial list of information that may be available from sources of action levels includes—

- Radionuclides of concern or types of radiation,
- Assumptions regarding surficial (fixed and removable) or volumetric residual radioactivity,
- Area or volume over which the residual radioactivity can be averaged,
- Assumptions about potential disposition of the M&E (e.g., exposure scenarios, reuse vs. recycle), and
- Conversions from dose or risk to activity or concentration (e.g., modeling and modeling assumptions).

3.3.2 Finalize Selection of Action Levels

In cases where more than one source of action levels is identified, it is necessary to select an action level as the basis for the disposition survey design. Generally, the source that provides the most restrictive action levels (i.e., the most protective of human health and the environment) will be appropriate for designing the disposition survey. If the planning team cannot determine which action levels are most restrictive, multiple survey designs should be developed and the selection of action levels will be determined by the selection of the most effective survey design (Section 4.4).

The expected location of residual radioactivity is an important factor in the selection of appropriate action levels. Some sources of action levels are only applicable for surficial radioactivity (e.g., Figure IV-1 of DOE 1993, DOT regulation 49 CFR 173.433). Other sources of action levels (e.g., ANSI 1999) or dose assessments for deriving action levels (e.g., NRC 2003a) make assumptions about whether the residual radioactivity is surficial or volumetric, or a combination of both. Section 2.4.2.4 discusses the location of radioactivity associated with the M&E.

While the location of residual radioactivity is important in determining the most restrictive action levels, other physical and radiological characteristics should also be considered. The final selection of action levels should be supported by the description of the M&E provided by the IA (Section 2.6).

3.3.3 Modify Action Levels When Multiple Radionuclides are Present

The implementation of action levels should be considered when evaluating whether they will be applied to a specific survey unit or project. Section 3.3.1 discusses potential sources for action levels, and Section 3.2 discusses the approach for selecting the radionuclides of concern. Calculating the relative ratios among multiple radionuclides and determining the state of equilibrium for decay series radionuclides is discussed in MARSSIM Section 4.3. This section describes how individual action levels can be combined and applied when more than one radionuclide is present.

Action levels are often provided for types of radioactivity or groups of radionuclides. For example, DOE Order 5400.5 Figure IV-1 (DOE 1993) provides surface activity action levels for four groups of radionuclides (Appendix E). For the simple case in which the activity is entirely attributable to one radionuclide, the action levels for that radionuclide are used for comparison to survey data. In these examples, the disposition survey data may be obtained from direct measurements of activity, scanning with data logging, conveyorized survey monitor surveys, or other appropriate methods.

Dose- or risk-based action levels may be radionuclide-specific. Each radionuclide-specific action level corresponds to the chosen disposition criterion (e.g., regulatory limit in terms of dose or risk). For example, ANSI 1999 provides surface and volumetric activity action levels for individual radionuclides. When multiple radionuclides are present at concentrations equal to the action levels, the total dose or risk for all radionuclides would exceed the disposition criterion. In these cases it is possible to modify the action levels based on relationships between the radionuclides of concern and still demonstrate compliance with the disposition criterion.

The method used to modify the action levels depends on the radionuclides of concern and the selected measurement method. If the measurement method reports total activity for a type of radiation (e.g., gross α, β, or γ assays) the method is non-radionuclide specific and the guidance in Section 3.3.3.1 should be applied. If the measurement reports activity for individual radionuclides (e.g., gamma spectroscopy, alpha spectrometry) the method is radionuclide specific and the guidance in Section 3.3.3.3 should be applied.

3.3.3.1 Modify Action Levels for Non-Radionuclide-Specific Measurement Methods

For situations in which there are radionuclide-specific action levels and multiple radionuclides are present, a gross activity action level can be developed. Gross activity action levels are also discussed in Section 4.3.4 of MARSSIM. This approach enables field measurement of gross activity (using static direct measurements or scans), rather than determination of individual radionuclide activity, for comparison to the action levels. The gross activity action level for M&E with multiple radionuclides is calculated as follows:

1. Determine the relative fraction (f) of the total activity contributed by the radionuclide.[2]
2. Obtain the action level for each radionuclide present.
3. Substitute the values of f and action levels in the following equation.

$$\text{Gross Activity AL} = \frac{1}{\left(\dfrac{f_1}{AL_1} + \dfrac{f_2}{AL_2} + \ldots \dfrac{f_n}{AL_n} \right)} \tag{3-1}$$

Where:

$\quad f_i \quad = \quad$ relative fraction of total activity contributed by radionuclide i ($i = 1, 2, \ldots, n$)
$\quad AL_i \quad = \quad$ action level for radionuclide i

---

[2] The determination of relative fractions may be based on process knowledge, empirical data, or a combination of both. It may be difficult or impractical to determine the relative fractions contributed by all radionuclides of concern. The alternatives are to analyze each radionuclide independently, or use conservative assumptions to determine the relative fractions. Additional guidance is provided in MARSSIM Section 4.3.

Example 1: Assume that 40% of the total radioactivity was contributed by a radionuclide with an action level of 1.4 Bq/cm^2 (8,400 dpm/100 cm^2). An additional 40% of the total radioactivity was contributed by a radionuclide with an action level of 0.28 Bq/cm^2 (1,700 dpm/100 cm^2), and the final 20% of the radioactivity was contributed by a radionuclide with an action level of 0.14 Bq/cm^2 (840 dpm/100 cm^2). Using Equation 3-1:

$$\text{Gross Activity AL} = \frac{1}{\left(\dfrac{0.40}{1.4} + \dfrac{0.40}{0.28} + \dfrac{0.20}{0.14}\right)} = 0.32 \text{ Bq/cm}^2 \ (1{,}900 \text{ dpm/100 cm}^2)$$

Equation 3-1 may not be appropriate for survey units with radioactivity from multiple radionuclides having unknown or highly variable concentrations of radionuclides. In these situations, the best approach may be to select the most restrictive surface activity action level from the mixture of radionuclides present.[3] If the mixture contains radionuclides that cannot be measured using field survey equipment, such as ^{3}H or ^{55}Fe, laboratory analyses of M&E samples may be necessary.

3.3.3.2 Modify Action Levels for Non-Radionuclide-Specific Measurements of Decay-Series Radionuclides

Demonstrating compliance with surface activity action levels for radionuclides of a decay series (e.g., radium, thorium, uranium) that emit both alpha and beta radiation may be demonstrated by assessing alpha, beta, or both radiations. However, relying on the use of alpha surface activity measurements often proves problematic because of the highly variable level of alpha attenuation by rough, porous, uneven, and dusty surfaces. Beta measurements typically provide a more accurate assessment of thorium and uranium (and their decay products) on most building surfaces because surface conditions cause significantly less attenuation of beta particles than alpha particles. Beta measurements, therefore, may provide a more accurate determination of surface activity than alpha measurements.

The relationship of beta and alpha emissions from decay chains or various enrichments of uranium should be considered when determining the surface activity for comparison with the action level value(s). When the initial member of a decay series has a long half-life, the radioactivity associated with the subsequent members of the series will increase at a rate determined by the individual half-lives until all members of the decay series are present at activity levels equal to the activity of the initial member. This condition is known as secular equilibrium. Pages 4-6 and 4-7 in MARSSIM also provide a discussion on secular equilibrium.

The difficulty with radionuclides that are part of a natural decay series is that time must pass for a sufficient number of half-lives of the longest-lived decay product that intervenes between a radionuclide and the initial member of a decay series in order to establish secular equilibrium. In the case of ^{232}Th, the time to establish secular equilibrium is almost 40 years. This is because ^{232}Th decays into ^{228}Ra, which has a half-life of 5.75 years. In the case of ^{238}U, the time to establish secular equilibrium is approximately 2 million years. This is because ^{234}U has a half-

---

[3] In Example 1, the most conservative action level is 0.14 Bq/cm^2.

life of approximately 250,000 years. ^{226}Ra, another member of the ^{238}U decay series, presents special problems. ^{226}Ra decays into ^{222}Rn, which is a noble gas that can escape the matrix and disrupt equilibrium. It is important to remember the reason for determining relationships between radionuclides. If the relationships are known or can be estimated,[4] the costs and amount of time required for performing measurements can be significantly reduced. The alternative to determining the relationships between radionuclides is performing radionuclide-specific measurements for each radionuclide of concern.

Example 2: the radionuclide of concern is ^{232}Th, and all of the decay products are in secular equilibrium. Assume that a gas proportional detector will be used for surface activity measurements. The detector's efficiency is dependent upon the radionuclide mixture measured and the calibration source area. Guidance from the International Organization for Standardization (ISO 1988) states:

> "The dimensions of the calibration source should be sufficient to cover the window of the instrument detector. Where, in extreme cases, sources of such dimensions are not available, sequential measurements with smaller distributed sources of at least 100 cm^2 active area shall be carried out. These measurements shall cover the whole window area or at least representative fractions of it and shall result in an average value for the instrument efficiency."

The concentration of ^{232}Th is inferred from a measurement that includes the initial member of the decay series and all of its decay products. The efficiency of such measurements, relative to each decay of ^{232}Th, can be greater than 100%. The efficiency, relative to each decay of ^{232}Th, is calculated by weighting the individual efficiencies from each of the radionuclides present (Table 3.1).

It is important to recognize that if the action level for ^{232}Th includes the entire ^{232}Th decay series, the total efficiency for ^{232}Th must account for all of the radiations in the decay series. The total weighted efficiency calculated in Table 3.1 may be used to modify action levels for non-radionuclide specific measurements using a gas proportional counter to measure thorium series radionuclides. The total weighted efficiency can be substituted into an equation (e.g., MARSSIM Equations 6-1, 6-2, 6-3, or 6-4) to convert the action level (e.g., activity units) into measurement units (e.g., counts or cpm). The modified action level can then be compared directly to the measurement results for a real time assessment of the data.

---

[4] There are risks and tradeoffs associated with using estimated values. The planning team should compare the consequences of potential decision errors with the resources required to improve the quality of existing data to determine the appropriate approach for a specific project.

Table 3.1 Example Detector Efficiency Calculation (^{232}Th in Complete Equilibrium with its Decay Products) Using a Gas Proportional Detector

Radionuclide	Energy* (keV)	Fraction	Instrument Efficiency	Surface Efficiency	Weighted Efficiency
^{232}Th	4.00 MeV alpha	1	0.40	0.25	0.1
^{228}Ra	7.2 keV beta	1	0	0	0
^{228}Ac	377 keV beta	1	0.54	0.50	0.27
^{228}Th	5.40 MeV alpha	1	0.40	0.25	0.1
^{224}Ra	5.67 MeV alpha	1	0.40	0.25	0.1
^{220}Rn	6.29 MeV alpha	1	0.40	0.25	0.1
^{216}Po	6.78 MeV alpha	1	0.40	0.25	0.1
^{212}Pb	102 keV beta	1	0.40	0.25	0.1
^{212}Bi	769 keV beta	0.64	0.66	0.50	0.211
^{212}Bi	6.05 MeV alpha	0.36	0.40	0.25	0.036
^{212}Po	8.78 MeV alpha	0.64	0.40	0.25	0.064
^{208}Tl	557 keV beta	0.36	0.58	0.50	0.104
				Total efficiency =	1.29

*Alpha energies are weighted averages based on relative abundance of major particle emissions totaling at least 90% of the total emissions. Beta energies are average energies. Source: Japanese Atomic Energy Research Institute data from NRC Radiological Toolbox Version 1.0.0 (NRC 2003b). Table adapted from NUREG-1761 Table 4.3 (NRC 2002a).

3.3.3.3   Modify Action Levels for Radionuclide-Specific Measurement Methods

In many cases action levels correspond to a disposition criterion (e.g., a regulatory limit) in terms of dose or risk. When multiple radionuclides are present at concentrations equal to the action levels, the total dose or risk for all radionuclides would exceed a dose- or risk-based disposition criterion. In this case, the individual action levels would need to be adjusted to account for the presence of multiple radionuclides contributing to the total dose or risk. The surrogate measurements discussed in this section describe adjusting action levels to account for multiple radionuclides when radionuclide-specific analyses of media samples or radionuclide–specific in situ measurements (e.g., in situ gamma spectroscopy) are performed. The use of surrogate measurements is also described in Section 4.3.2 of MARSSIM. Other methods used to account for the presence of multiple radionuclides include the use of the unity rule (MARSSIM Section 4.3.3) and development of a gross activity action level to adjust the individual radionuclide action levels (see Section 3.3.3.1 and MARSSIM Section 4.3.4).

The unity rule is satisfied when radionuclide mixtures yield a combined fractional concentration limit that is less than or equal to one. The unity rule can be described by Equation 3-2:

$$\frac{C_1}{AL_1} + \frac{C_2}{AL_2} + ... + \frac{C_n}{AL_n} \leq 1 \qquad (3\text{-}2)$$

Where:

C_i = concentration or activity value for each individual radionuclide (i = 1, 2, ..., n)[5]

AL_i = action level value for each individual radionuclide (i = 1, 2, ..., n)

For the disposition of M&E that contain multiple radionuclides, it may be possible to measure just one of the radionuclides and still demonstrate compliance for all of the radionuclides present in the M&E through the use of surrogate measurements. In the use of surrogates, it is often difficult to establish a "consistent" ratio between two or more radionuclides. Rather than follow prescriptive guidance on acceptable levels of variability for the surrogate ratio, the planning team should review the data collected to establish the ratio (e.g., from preliminary surveys or process knowledge) and account for the variability as a measurement quality objective (MQO) during selection of a measurement method (see Sections 3.8, 5.5, and 7.3). The action levels must then be modified to account for the fact that one radionuclide is being used to account for the presence of one or more other radionuclides.

Action levels for the measured radionuclide are modified ($AL_{meas,mod}$) to account for a single inferred radionuclide (e.g., inferring ^{55}Fe based on the presence of ^{60}Co) using Equation 3-3 (modified from Equation 6.2 in Abelquist 2001):

$$AL_{meas,mod} = (AL_{meas}) \left(\frac{AL_{infer}}{\left(\frac{C_{infer}}{C_{meas}} \right) AL_{meas} + AL_{infer}} \right) \tag{3-3}$$

Where:

$AL_{meas,mod}$ = modified action level for the radionuclide being measured

AL_{meas} = action level for the radionuclide being measured

AL_{infer} = action level for the inferred radionuclide (i.e., not measured)

C_{infer}/C_{meas} = surrogate ratio of the inferred to the measured radionuclide

When the measured radionuclide will be used as a surrogate for more than one radionuclide, $AL_{meas,mod}$ can be calculated using Equation 3-4 (MARSSIM Equation I-14):

$$AL_{meas,mod} = \frac{1}{\left(\frac{1}{AL_1} + \frac{R_2}{AL_2} + \frac{R_3}{AL_3} + ... \frac{R_n}{AL_n} \right)} \tag{3-4}$$

Where:

AL_1 = action level for the measured radionuclide by itself

AL_2 = action level for the second radionuclide (or first radionuclide being inferred) that is being inferred by the measured radionuclide

R_2 = ratio of concentration of the second radionuclide to that of the measured radionuclide

---

[5] C (radionuclide concentration) must be in the same units as the action level. If the action level is provided in activity units, C will also be in units of activity.

AL_3 = action level for the third radionuclide (or second radionuclide being inferred) that is being inferred by the measured radionuclide

R_3 = ratio of concentration of the third radionuclide to that of the measured radionuclide

AL_n = action level for subsequent radionuclides being inferred by the measured radionuclide

R_n = ratio of concentration of subsequent radionuclides to that of the measured radionuclide

Recall that the benefit of using surrogates is the avoidance of costly laboratory-based analytical methods to provide estimates of activity for individual radionuclides of concern. Surrogates often emit γ-rays, which enable the use of noninvasive and nondestructive methods. However, α- and β-emitting radionuclides can also be used as surrogates, depending on the objectives of the survey and project-specific information. The surrogates come in two forms: (1) surrogates by virtue of a decay series, and (2) surrogates by virtue of association. Surrogates that are part of a decay series are discussed in Section 3.3.3.2. Radionuclides that are not part of a decay series have the potential to be surrogates when they are produced by the same nuclear process (usually fission or activation) and have similar chemical properties and release mechanisms. However, this type of surrogate needs special attention because there must be a consistent ratio between the measured radionuclide and surrogate, which is not always easy to demonstrate. For example, in the case of nuclear power reactors, ^{60}Co can be used as a surrogate of ^{55}Fe and ^{63}Ni because both are activation-corrosion products with similar chemical properties. Similarly, ^{137}Cs can be used as a surrogate for the β-emitting ^{90}Sr because both are fission products and generally are found in soluble cationic forms. While ^{137}Cs has been suggested as a possible surrogate for ^{99}Tc, it must be noted that ^{99}Tc has different chemical properties and, in nuclear power reactors, it has different release mechanisms. Additional information is available on surrogates and establishing ratios (MARSSIM 2002, NRC 2000, and EPRI 2003).

3.3.4 Evaluate Interface With Exposure Pathway Models

Disposition criteria may be provided in units that cannot be measured directly, for example total effective dose equivalent (TEDE) or lifetime risk of cancer incidence. These criteria are usually converted into action levels with concentration or activity units. This conversion is typically accomplished using exposure pathway models, such as RESRAD-Recycle for metals (DOE 2005). While the selection and application of these models is outside the scope of MARSAME, the assumptions used to develop action levels should be considered during development of a disposition survey design.

Alternatively, disposition criteria may be provided in units more easily measured. In general, there are assumptions used in the development of these types of action levels. It is the responsibility of the authority issuing the action levels to ensure regulatory involvement in their development and to document and make assumptions available to users.

The assumptions used to design the disposition survey (Section 4.4) need to match the assumptions used to develop the action levels. Examples of parameters that could affect disposition survey designs and could be inputs to exposure pathway models include—

- Volume, mass, or surface area of M&E;
- Accessibility;
- Physical and chemical characteristics of radionuclides or radiations of concern (types of emissions, energies, half-lives, known or expected relationships);
- Distribution of radioactivity (uniform or variable);
- Location of radioactivity (surficial or volumetric); and
- Fixed or removable radioactivity, or some combination of both.

3.4   Describe the Parameter of Interest

The parameter of interest is the population parameter (e.g., mean, median, percentile, or total amount) that the planning team considers to be important for making decisions about the target population (EPA 2006a). The target population is the collection of all possible measurement results that could be used to support a disposition decision concerning the M&E being investigated. The target population is defined by the selection of survey unit boundaries (see Section 3.6), since a separate disposition decision will be made for each survey unit.

The parameter of interest may be specified as part of the action level. For example, DOE Order 5400.5 Figure IV-1 (DOE 1993) lists action levels (i.e., surface concentration limits in dpm per 100 cm^2), parameters of interest (i.e., mean and maximum values), and target populations (i.e., 1 m^2 for average concentration and 100 cm^2 for maximum and removable limits).

Alternatively, the planning team may need to select the parameter of interest based on project-specific needs and considerations. The most common parameter used in decision-making is the mean because the mean is frequently used to model random exposure to environmental contamination (EPA 2006a). The more complex the parameter of interest, the more complex will be the decision rule (Section 3.7) and accompanying survey design.

3.5   Identify Alternative Actions

Before decision rules can be developed, the planning team needs to identify the alternative actions based on the disposition options identified in Section 2.5. Alternative actions are the possible actions that may be taken for disposition of M&E, including an alternative that requires no action. Table 3.2 lists examples of alternative actions for disposition options provided in Section 2.5.

Table 3.2 Example Alternative Actions

Disposition Option	Alternative Actions
Release for reuse	Reuse without radiological controls
	Reuse with radiological controls
	Maintain current level of radiological control and do not reuse (no action)
Release for recycle	Recycle without radiological controls
	Recycle with radiological controls
	Maintain current level of radiological control and do not recycle (no action)
Release for disposal	Dispose of M&E as municipal or industrial waste
	Dispose of M&E as low-level radioactive waste
	Dispose of M&E as high-level radioactive waste
	Dispose of M&E as transuranic (TRU) waste
	Maintain current level of radiological control without disposal (no action)
Interdiction	Remove M&E from general commerce and initiate radiological controls
	Decide to use or accept M&E for a specific application
	Decide not to use or accept M&E for a specific application
	Continue unrestricted use of M&E (no action)

3.6   Identify Survey Units

To make a decision concerning the disposition of M&E it is necessary to describe the total collection of M&E being investigated and define what segment of the total will be considered for individual decisions. In other words, the planning team must specify the amount of M&E for which a separate disposition decision will be made. When the M&E consist of discrete items surveyed individually (e.g., hand tools) this task is simple. However, disposition decisions are often required for more complex situations (e.g., bulk dispersible materials, excavation equipment).

Survey unit boundaries should be clearly defined in order to know exactly what amount of M&E is covered by a single decision. This clear and unambiguous definition will make data interpretation more straightforward.

An M&E survey unit is the specific lot, amount, or piece of M&E on which measurements are made to support a disposition decision concerning that specific lot, amount, or piece of M&E. The purpose of this section is to identify the information that will be used to define the survey unit boundaries. The expected output from this section is the identification of survey unit boundaries that will be used to develop the decision rule in Section 3.7. Figure 3.3 shows the process used to develop survey unit boundaries.

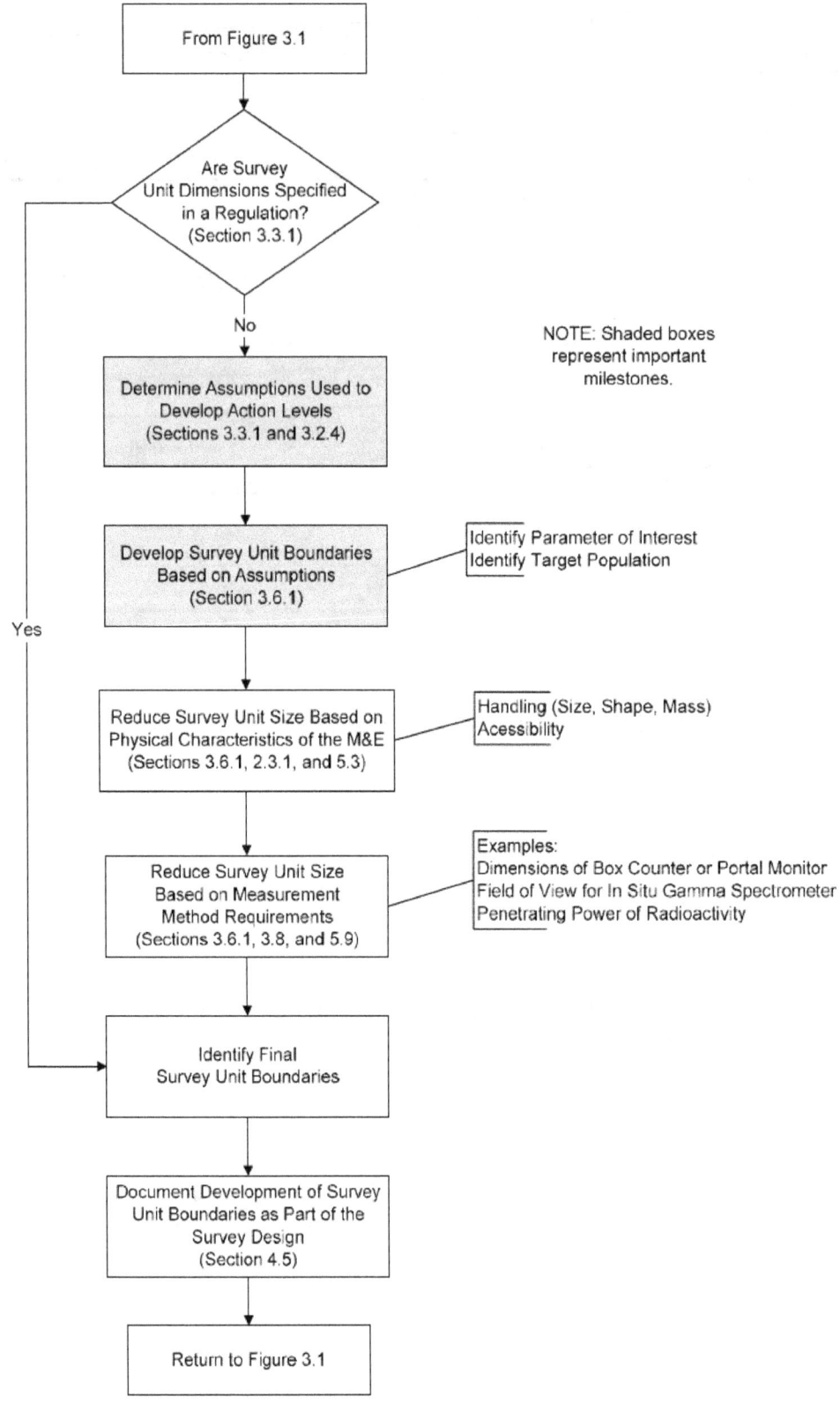

Figure 3.3 Developing Survey Unit Boundaries
(Apply to All Impacted M&E for Each Set of Action Levels Identified in Section 3.3)

Survey unit boundaries are affected by many variables associated with the action level, physical properties of the M&E, characteristics of the radionuclides of concern, and available measurement techniques. Variables affecting the definition of survey units include—

- Action Level (Section 3.3)
  - Assumptions used to develop the action level (e.g., surficial [fixed or removable] or volumetric, Section 3.3.1)
  - Modeling assumptions used to convert from dose or risk to concentration or activity (Section 3.3.4)
- Physical Properties of the M&E (Section 2.4.1)
  - Dimensions (i.e., size, shape, surface area)
  - Complexity (i.e., number and type of components)
  - Accessibility (i.e., measurability)
  - Inherent value
- Radiological Attributes of the M&E (Section 2.4.2)
  - Radionuclides of concern (e.g., major radiations and energies, half-life)
  - Expected activity levels (e.g., average, range, variance, known or potential relationships)
  - Distribution (i.e., uniform or non-uniform)
  - Location (i.e., surficial [fixed or removable] or volumetric)
- Available Measurement Methods (Section 3.8, Section 5.9)
  - Measurement quality objectives (Section 3.8, Section 5.5, Section 7.3)
  - Measurement performance characteristics (Section 5.5)

3.6.1 Define Initial Survey Unit Boundaries

Initial survey unit boundaries should be developed based on one primary factor and modified, as needed, using additional variables. MARSAME recommends using the assumptions used to develop the action levels as the primary factor used to develop survey unit boundaries. The modifying variables will usually be specific to a measurement technique, or determined by the M&E being investigated.[6]

In many cases the action levels will define the survey unit boundaries. For example, DOE Order 5400.5 Figure IV-1 (DOE 1993) provides action levels for surface activity. The survey unit boundaries are restricted to the surface of the M&E being investigated. Alternatively, NUREG-1640 (NRC 2003a) provides modeling assumptions used to develop the action levels for different materials. Radionuclide-specific action levels are provided for separate materials (e.g., ferrous metals, concrete) for both surficial and volumetric radioactivity. In addition, each action level lists the limiting exposure scenario. For example, exposure scenarios for concrete (NRC 2003a) include—

---

[6] This approach differs from guidance found in MARSSIM Section 4.6. While MARSSIM also uses the assumptions used to develop the action levels (i.e., derived concentration guideline levels [DCGLs] in MARSSIM) as the primary factor in developing survey unit boundaries, the modifications are different. MARSSIM guidance allows increasing and decreasing survey unit size based on classification. In MARSSIM, Class 1 survey units generally are smaller than the area assumed in the exposure pathway model, while MARSSIM allows Class 3 survey units to be larger in area. Additional modifications to survey unit boundaries in MARSSIM can be made based on site-specific variables (e.g., room size, topography).

- Worker processing concrete rubble at a satellite facility,
- Truck driver hauling concrete rubble,
- Worker building a road using recycled concrete,
- Driver on a road built using recycled concrete,
- Worker handling concrete rubble at an industrial landfill,
- Worker handling concrete rubble at a municipal landfill,
- Individual drinking groundwater contaminated with leachate from an industrial landfill, and
- Individual drinking groundwater contaminated with leachate from a municipal landfill.

Each exposure scenario assumes different conditions that help define survey unit boundaries. For example, a truck driver hauling concrete rubble would be exposed to one truckload of concrete rubble, so the survey unit boundaries would be defined by a truckload of concrete rubble (i.e., 2×10^4 kg [22 tons] or 8.3 m^3; NRC 2003a).

3.6.2 Modify Initial Survey Unit Boundaries

Modifications to survey unit boundaries are expected based on practical constraints for data collection activities. In most cases smaller survey units will be acceptable, since a reduction in size would not result in an increased dose or risk. Increasing the size of the survey unit may result in increased dose or risk, and therefore requires approval of the planning team and stakeholders.

Constraints on collecting data are often associated with specific measurement techniques, which could affect the survey unit boundaries. For example, using in situ gamma spectroscopy may restrict survey unit sizes based on the field of view of the detector, the penetrating power of the gamma energies being measured, or the assumptions used to develop the instrument efficiency. Alternatively, using a box counter or portal monitor may restrict survey unit sizes based on what will fit inside or through the detector. Information on measurement parameters affecting disposition survey design is provided in Section 3.8. Section 5.9 and Appendix D provide detailed information on specific measurement methods.

The M&E being investigated may also cause modifications to survey unit boundaries. These modifications are often associated with physical characteristics (e.g., size, shape). Identification of actual survey units as part of the final disposition survey design is discussed in Chapter 4.

3.7 Develop a Decision Rule

In order to design a disposition survey, the user should define a decision rule describing the conditions for selecting between alternative actions. The planning team should assume that ideal data are available and there is no uncertainty in the decision making process. The available data are integrated into an "if...then..." statement, which is the theoretical decision rule.[7]

---

[7] This is called a *theoretical decision rule* because it is stated in terms of the true value for the parameter of interest, even though in reality this value cannot be known. An operational decision rule that is based on an estimate of the target population parameter of interest will be incorporated as part of the final disposition survey design selected and documented in Chapter 4.

The theoretical decision rule is constructed by combining the action level (Section 3.3) and the parameter of interest (Section 3.4) with the alternative actions (Section 3.5) in an "if...then..." statement.

For example:

> Hypothetically, if the mean concentration of ^{226}Ra in 20,000 kg (8.3 m^3, one truckload) of concrete rubble is less than the clearance action level of 0.34 Bq/g for volumetric radioactivity, then the concrete rubble can be cleared, otherwise radiological control of the concrete will continue.

It may be necessary to develop more than one decision rule. For example, if more than one action level is selected in Section 3.3, a separate decision rule needs to be developed for each action level. In addition, selection of multiple disposition options in Section 2.5 (e.g., release and disposal as low-level radioactive waste) may result in multiple alternative actions requiring multiple decisions and multiple decision rules. For example,

> Hypothetically, if the mean concentration of ^{226}Ra in 20,000 kg (8.3 m^3, one truckload) of concrete rubble is less than the clearance action level of 0.34 Bq/g for volumetric radioactivity, then the concrete rubble can be cleared, otherwise the concrete will be considered for disposal as low-level radioactive waste. If the concrete rubble meets the waste acceptance criteria for the low-level radioactive waste disposal facility (e.g., mean and total activity levels, chemical and physical form, toxicity) the concrete will be packaged and transported for disposal, otherwise radiological control of the concrete will continue.

3.8   Develop Inputs for Selection of Provisional Measurement Methods

The identification and evaluation of provisional measurement methods is an important step in developing a disposition survey design. A measurement method is the combination of instrumentation (e.g., GM detector, NaI[Tl] scintillation detector, gamma spectrometer) with a measurement technique (i.e., scan, in situ, sample collection). The selection of a measurement method is discussed in more detail in Section 5.9. The availability of measurement methods and the amount of resources required to implement specific measurement methods is an important factor in selecting between different survey designs, or in reducing the number of options to be considered when developing potential disposition survey designs.

There are two potential results of this evaluation of provisional measurement methods. First, the evaluation may identify specific measurement methods that will be included in the final documentation of the selected disposition survey design (see Section 4.5). For example, scanning 100% of a piece of equipment using a 2-inch by 2-inch NaI(Tl) detector at a specified height above the surface using a specified scan speed may be identified as the measurement method. Second, the evaluation may identify characteristics of a measurement method required to meet the objectives of a survey. These characteristics are called measurement quality objectives (MQOs). Section 5.5 and Section 7.3 provide additional information on MQOs applied to disposition surveys.

Examples of MQOs are described in the following sections. A list of minimum MQOs required for a survey can be developed and documented in the final disposition survey design (see Section 4.5). The selection of a measurement technique that meets the MQOs is accomplished during implementation of the survey design.

This section focuses on measurability. Most of the variables that need to be considered for the identification of measurement techniques have been discussed earlier in this chapter. The identification of measurement methods is directly or indirectly related to—

- Identification of radionuclides of concern,
- Location of residual radioactivity,
- Application of action levels,
- Physical properties of the M&E,
- Distribution of residual radioactivity,
- Expected levels of residual radioactivity,
- Relationships between radionuclide activities,
- Equilibrium status of natural decay series, and
- Background radioactivity.

Measurable radioactivity is radioactivity that can be quantified and meets the DQOs and MQOs established for the survey. Radioactivity that is quantified using known or predicted relationships developed from process knowledge or preliminary measurements is considered measurable as long as the relationships are developed and verified as specified in the DQOs and MQOs. The *Multi-Agency Radiological Laboratory Analytical Protocols* manual (MARLAP 2004)[8] lists method performance characteristics that should be considered when establishing MQOs for a project. This list is not intended to be exhaustive:

- The method uncertainty at a specified concentration (expressed as a standard deviation);
- The method's detection capability (expressed as the minimum detectable concentration, or MDC);
- The method's quantification capability (expressed as the minimum quantifiable concentration, or MQC);
- The method's range, which defines the method's ability to measure the radionuclide of concern over some specified range of concentration;
- The method's specificity, which refers to the ability of the method to measure the radionuclide of concern in the presence of interferences; and
- The method's ruggedness, which refers to the relative stability of method performance for small variations in method parameter values.

Project-specific method performance characteristics should be developed as necessary and may or may not include the characteristics listed here. Once lists of performance characteristics that affect measurability have been identified, the planning team should develop MQOs describing

---

[8] MARLAP was developed for selecting laboratory protocols. Applying the framework and performance-based approach for planning and conducting radiological work from MARLAP to the selection of field measurement techniques is an expansion of the original scope and purpose of MARLAP.

the project-specific objectives for potential measurement techniques. Potential measurement techniques should be evaluated against the MQOs to determine if they are capable of meeting the objectives for measurability.

3.8.1 Measurement Method Uncertainty

The required measurement method uncertainty is perhaps the most important MQO to be established during the planning process. Section 4.2 discusses the rationale involved in setting the required measurement method uncertainty and developing statistical hypothesis tests for the implementation of disposition decision rules using measurement data. Section 5.5 discusses the application of MQOs, including the measurement method uncertainty, to disposition surveys for M&E. Section 7.3 discusses procedures for determining the required measurement method uncertainty and whether or not it has been achieved.

MARLAP uses the term method uncertainty to refer to the predicted uncertainty of a measured value that would likely result from the performance of a measurement at a specified concentration, typically the action level. Reasonable values for method uncertainty can be predicted for a particular measurement technique based on typical values for specific parameters (e.g., count time, efficiency) and process knowledge for the M&E being investigated (see Sections 5.5 and 7.3). The MQO for measurement method uncertainty is related to the width of the gray region (Section 4.2.2). The required measurement method uncertainty is directly related to the MDC and the MQC discussed below.

The distinction between imprecision and bias as a data quality indicator depends on context. Additional information on data quality indicators can be found in MARSSIM Appendix N and EPA QA/G-5 (EPA 2002a). A reliable estimate of bias requires a data set that includes many measurements, so MARSAME and MARLAP focus on developing an MQO for measurement method uncertainty. Measurement method uncertainty effectively combines imprecision and bias into a single parameter whose interpretation does not depend on context. This approach assumes that all potential sources of bias present in the measurement process have been considered in the estimation of the measurement uncertainty and, if not, that any appreciable bias would only be detected after a number of measurements of quality control (QC) and performance evaluation samples have been performed (see the QC discussion in Section 5.10). MARLAP Appendix C provides examples on developing MQOs for measurement method uncertainty of laboratory measurement techniques.

3.8.2 Detection Capability

The MDC (see Sections 5.7 and 7.5) is recommended as the MQO for defining the detection capability, and is an appropriate MQO when decisions are to be made based on a single measurement as to whether excess radioactivity is present or not. The MDC must not exceed the action level if the MDC is to be used as a decision parameter. Chapter 5 provides guidance on implementation of the selected measurement technique, including calculation of the MDC. Additional information on calculating the MDC can be found in MARSSIM (Section 6.7, examples in Appendix H) and MARLAP (Chapter 19, Appendix C).

3.8.3 Quantification Capability

When the average of several measurements will be compared to a disposition criterion, an MQO more stringent than the MDC is required. The MQC (see Sections 5.8 and 7.6) is recommended as the parameter for defining the measurement capability for making quantitative comparisons of averages to a limit. An MQO for the required measurement method uncertainty (Section 5.6) is related to an MQO for the quantification capability because an MQC is defined as the concentration at which a specified relative standard uncertainty is achieved. MARLAP presents three reasons why it is important to consider this measurement method performance characteristic:

1. To emphasize the importance of the quantification capability of a measurement technique for instances when the issue is not whether a radionuclide is present or not (e.g., measuring ^{238}U in soil where the activity is inherent) but rather how precisely the radionuclide can be measured,
2. To promote the MQC as an important measurement method performance characteristic for comparison of measurement techniques, and
3. To provide an alternative to the overemphasis on establishing required MDCs in instances where detection (i.e., reliably distinguishing a radionuclide concentration from zero) is not the key analytical issue.

The MQC must not exceed the action level if the MQC is to be used as a decision parameter. Chapter 5 provides guidance on implementation of the selected measurement technique, including calculation of the MQC. Section 5.8 discusses issues related to measurement quantifiability. Section 7.6 provides information on the statistical basis of the MQC calculation including example calculations. Additional information on calculating the MQC can be found in MARLAP Chapter 19, with examples in MARLAP Appendix C.

3.8.4 Range

The expected concentration range for a radionuclide of concern (see Section 2.4.2) may be an important measurement method performance characteristic. Most radiation measurement techniques are capable of measuring over a wide range of radionuclide concentrations. However, if the expected concentration range is large, the range should be identified as an important measurement method performance characteristic and an MQO should be developed. The MQO for the acceptable range should be a conservative estimate. This will help prevent the selection of measurement techniques that cannot accommodate the actual concentration range.

3.8.5 Specificity

Specificity is the ability of the measurement method to measure the radionuclide of concern in the presence of interferences. To determine if specificity is an important measurement method performance characteristic, the planning team will need information on expected concentration ranges for the radionuclides of concern and other chemical and radionuclide constituents, along with chemical and physical attributes of the M&E being investigated (see Section 2.4). The importance of specificity depends on—

- The chemical and physical characteristics of the M&E being investigated,
- The chemical and physical characteristics of the residual radioactivity, and
- The expected concentration range for the radionuclides of concern.

If potential interferences are identified (e.g., inherent radioactivity, similar radiations), an MQO should be established for specificity.

If inherent radioactivity is associated with the M&E being investigated, a method that measures total activity may not be acceptable. Consider concrete, which contains measurable levels of naturally occurring radioactivity and emits radiation in the form of alpha particles, beta particles, and photons. If the action level for the radionuclide of concern is close to background (e.g., within a factor of 3) gross measurement methods may not meet the survey objectives. Performing gross alpha measurements using a gas proportional detector may not provide an acceptable MDC or MQC for plutonium isotopes, where a more specific measurement method such as alpha spectrometry following radiochemical separation would be acceptable.

Radionuclides have similar radiations if they emit radiations of the same type (i.e., alpha, beta, photon) with similar energies. For example, both ^{226}Ra and ^{235}U emit a gamma ray with energy of approximately 186 keV. Gamma spectroscopy may not be able to resolve mixtures of these two radionuclides, which are both associated with naturally occurring radioactivity. More specific methods involving ingrowth of ^{226}Ra decay products or chemical separation prior to measurement can be used to accurately quantify the radionuclides.

Documented measurement methods should include information on specificity. MARSSIM Table 7.2 lists examples of references providing laboratory measurement methods. NUREG-1506 (NRC 1995) provides generic information on field measurement techniques, but most field measurement methods are documented in proprietary SOPs. If specificity is identified as an important issue for a project, consultation with an expert in radiometrics or radiochemistry is recommended.

3.8.6   Ruggedness

For a project that involves field measurements that are performed in hostile, hazardous, or variable environments, or laboratory measurements that are complex in terms of chemical and physical characteristics, the measurement method's ruggedness may be an important method performance characteristic. Ruggedness refers to the relative stability of the measurement technique's performance when small variations in method parameter values are made. For field measurements the changes may include temperature, humidity, or atmospheric pressure. For laboratory measurements, a change in pH or the quantity of a reagent may be important. In order to determine if ruggedness is an important measurement method performance characteristic, the planning team needs detailed information on the chemical and physical characteristics of the M&E being investigated and operating parameters for the radiation instruments used by the measurement technique. Information on the chemical and physical characteristics of the M&E is available as outputs from the IA. Information on the operating parameters for specific instruments should be available from the instrument manufacturer. Generic information for radiation detector operating parameters may be found in consensus standards. A limited list of examples of consensus standards is provided in Table 3.3.

Table 3.3 Examples of Consensus Standards for Evaluating Ruggedness

Standard Number	Title
ANSI N42.12-1994	American National Standard Calibration and Usage of Thallium-Activated Sodium Iodide Detector Systems for Assay of Radionuclides
ANSI N42.17A-2003	American National Standard Performance Specifications for Health Physics Instrumentation – Portable Instrumentation for Use in Normal Environmental Conditions
ANSI N42.17C-1989	American National Standard Performance Specifications for Health Physics Instrumentation – Portable Instrumentation for Use in Extreme Environmental Conditions
ANSI N42.34-2003	American National Standard Performance Criteria for Hand-held Instruments for the Detection and Identification of Radionuclides
IEEE 309-1999/ ANSI N42.3-1999	Institute of Electrical and Electronics Engineers, Inc. Standard Test Procedures and Bases for Geiger Mueller Counters
ASTM E1169-2002	Standard Guide for Conducting Ruggedness Tests

If it is determined that measurement method ruggedness is an important performance characteristic, an MQO should be developed. The MQO may require performance data that demonstrate the measurement technique's ruggedness for specified changes in select measurement method parameters. Alternatively, the MQO could list the acceptable ranges for select measurement method parameters and monitor the parameters as part of the QC program for the project (Section 5.10). For example, sodium iodide detectors are required to perform within 15% of the calibrated response between 0 and 40 °C (32 and 104 °F, respectively) (ANSI 1994). The disposition survey design may call for a work stoppage at temperatures outside this range, or an increase in the frequency of QC measurements at temperatures outside this range.

3.9   Identify Reference Materials

Action levels may be developed that are related to background radioactivity, either based on an incremental dose or risk above background, as an administrative limit based on background, or as a limit on technology (e.g., minimum detectable concentration). For situations where the action levels are incremental above background, reference materials should be identified to provide an estimate of background. MARSSIM Section 4.5 provides guidance on determining when a reference material is required.

Reference materials are used to develop an estimate of the distribution of background radioactivity that can be compared to the measurements performed in a comparable survey unit. The reference material is selected to provide information on the level of radioactivity that would be present if the M&E being investigated had not been radiologically impacted.

Whenever possible, reference data should be obtained by performing a survey of the M&E before it comes in contact with radiological materials. The M&E can then be surveyed prior to leaving the area to determine the level of residual radioactivity. This works especially well for decommissioning or cleanup applications where M&E are brought into a radiologically controlled area for a limited time and a specific application. Unfortunately, there are numerous situations where pre-contact surveys are not possible.

If the M&E cannot be used as its own reference material, it is necessary to identify reference material that is representative of the M&E being investigated. Non-impacted M&E that closely resembles the impacted M&E being investigated (i.e., similar chemical, physical, and radiological characteristics) will generally be acceptable as reference material. For example, if the conceptual model shows that only surficial activity is expected, the impacted surface may be removed and the non-impacted volume used as the reference material. When similar materials are not available, the best match available should be used as reference material. It may be necessary to evaluate more than one source of reference material before an acceptable match is identified. It may be important to perform reference material surveys in areas of low ambient background. Consider M&E consisting of individual objects that are small relative to the size of the detector used to perform the measurements. When each object receives a separate measurement, the ambient background may have a larger impact on the measurement than the background contributed by the M&E itself.

As shown in Table B.1 in Appendix B, background radionuclide concentrations for materials can vary significantly. For example, concentrations for thorium series radionuclides in concrete can range from 15 to 120 Bq/kg (Eicholz 1980), so it is important to identify an appropriate reference material.

The planning team should understand that background is variable. Ambient background can change with location and over time. It may be possible to simply move the M&E being investigated to an area with a lower ambient background to improve the detection capability of a measurement method. Local conditions (e.g., temperature, barometric pressure, precipitation) can cause variations in ambient background as discussed in NUREG-1501 (NRC 1994). NUREG-1505 (NRC 1998a) Chapter 13 provides information on accounting for variability in background.

The planning team should evaluate the process knowledge from the IA and use professional judgment to identify M&E that require reference materials, and identify potential reference materials to support the disposition survey.

3.10 Evaluate an Existing Survey Design

It is not necessary to develop a new survey design for all M&E being investigated. Existing survey designs are often available for routine or repetitive applications. If an existing survey design is identified, the planning team or decision maker should evaluate the applicability of the existing design to the current investigation.

Standardized survey designs for operating facilities are often documented in the form of standard operating procedures (SOPs, see Section 4.5.1). In other cases, existing survey designs may have been developed for similar projects. A description of the M&E that can be measured should be included in each existing SOP or survey design. If the description matches the M&E being investigated, the existing SOP or survey design can be used to perform the disposition survey. If the description of the M&E is incomplete or vague, or the M&E do not match the description, a more detailed evaluation may be performed to determine the acceptability of the existing survey design.

Personnel familiar with the existing survey design and the proposed application should perform the detailed evaluation of an existing survey design. All supporting documentation used to develop the existing survey design should be available for the evaluator(s), not just the SOP or survey design being reviewed.

The detailed evaluation should determine whether the M&E are measurable using the existing survey design. If the M&E are measurable, the existing survey design can be used. Detailed evaluations should include a review of each step in the survey development process, including–

- Selection of a disposition option (Section 2.5),
- Identification of action levels (Section 3.3),
- Specification of the population parameter of interest (Section 3.4),
- Development of survey unit boundaries (Section 3.6),
- Selection of measurement methods (Section 3.8 and Section 5.9),
- Identification of alternative actions (Section 3.5), and
- Development of a decision rule (Section 3.7 and Section 4.2.6).

The results of the evaluation should be documented. The documentation may require a modification to the existing survey design. For example, the description of M&E that can (or cannot) be measured using a specific SOP may be expanded for M&E that are routinely or repeatedly surveyed. Alternatively, the documentation may consist of a notation in a survey log (including a name, title, and date) for unique items.

4   DEVELOP A SURVEY DESIGN

4.1   Introduction

Once a decision rule has been developed, a disposition survey can be designed for the impacted materials and equipment (M&E) being investigated. The disposition survey incorporates all of the available information to determine the quantity and quality of data required to support a disposition decision. This chapter provides information on selecting the type, number, and location of measurements required to support a decision regarding the disposition of the M&E. Facilities or installations can use the process in this chapter and following chapters to develop a standard operating procedure (SOP) so multiple surveys can be performed for similar M&E to avoid costly and time-consuming development of redundant survey designs. The evaluation of existing SOPs for usability is discussed in Section 3.10. The output from this chapter is a documented disposition survey design that integrates measurement, data collection, and data analysis techniques.

The information in this chapter builds on the information collected and decisions made in Chapters 2 and 3. The disposition option selected in Section 2.5 and the action levels (ALs) identified in Section 3.3 are incorporated into the decision rules developed in Section 3.7. A decision rule is the basis for the disposition survey design. If multiple survey designs address the same decision rule and meet the data quality objectives (DQOs), the decision-maker needs to determine the most effective design for that decision rule. If none of the survey designs meet the DQOs for a specific decision rule, it may be necessary to reconsider decisions made earlier in the survey design process and adjust the DQOs.[1] If there are multiple decision rules (e.g., one for total radioactivity and one for removable radioactivity) more than one survey design may need to be developed to meet all of the DQOs for the project or a single survey design may be developed to incorporate all of the decision rules.

The complexity of a survey design generally reflects the complexity of the statistics used to interpret the results (Chapter 6). Survey designs range from simple (e.g., scan 100% of the M&E for surface radioactivity at a specified AL) to complex (e.g., develop a MARSSIM-type survey design). Simple survey designs typically require few resources for planning, but may require significant resources to implement. Complex survey designs typically require more resources during planning, with fewer resources required during implementation. If the planning and implementation portions of the data life cycle are performed correctly, the assessment and decision making stages should require few resources. This chapter provides information on statistical decision-making and how it is used during development of survey designs.

4.2   Making Decisions Using Statistics

In Section 3.6, the planning team assumed the levels and distribution of radioactivity associated with the M&E were known with no uncertainty. A theoretical decision rule was developed using this assumption to help focus the attention of the planning team on *how* they would make

---

[1] Refer to Section 2.3 for information on performing preliminary surveys to help ensure at least one survey design will meet the DQOs.

decisions. In this chapter the planning team accounts for uncertainty in decisions when ideal data are not available by establishing a statistical test to implement the decision rule. Decisions regarding the disposition of M&E are based on data with uncertainties. Through the use of statistics, the disposition survey design attempts to control the probability of making a decision error because of these uncertainties. MARSSIM Section 2.3 provides additional discussions on the use of statistics for making decisions based on environmental data. These steps are discussed briefly below and in further detail in Section 7.1. MARSAME recommends the planning team complete the following steps:

- Select a null hypothesis (Section 4.2.1),
- Choose a discrimination limit (Section 4.2.2),
- Define Type I and Type II decision errors (Section 4.2.5),
- Set a tolerable Type I decision error rate at the action level (Section 4.2.5), and
- Set a tolerable Type II decision error rate at the discrimination limit (Section 4.2.5).

4.2.1   Null Hypothesis

In hypothesis testing, two assertions about the actual level of radioactivity associated with the M&E are formulated. The two assertions are called the null hypothesis (H_0) and the alternative hypothesis (H_1). H_0 and H_1 together describe all possible radionuclide concentrations or levels of radioactivity under consideration. The survey data are evaluated to choose which hypothesis to reject or not reject, and by implication which to accept.[2] In any given situation, one and only one of the hypotheses must be true. The null hypothesis is assumed to be true within the established tolerance for making decision errors (Section 4.2.5). Thus, the choice of the null hypothesis also determines the burden of proof for the test.

4.2.2   Discrimination Limit

Action levels were defined in Section 3.3 based on the selected disposition option and applicable regulatory requirements. The planning team also chooses another radionuclide concentration or level of radioactivity that can be reliably distinguished from the action level by performing measurements (i.e., direct measurements, scans, in situ measurements, samples and laboratory analyses). This radionuclide concentration or level of radioactivity is called the discrimination limit (DL). An example where the discrimination limit is defined is provided in Section 8.4.5. The gray region is defined as the interval between the action level and the discrimination limit (Figures 4.1, 4.2, and 4.3 provide visual descriptions of the gray region). The width of the gray region is called the shift and denoted as Δ. The objective of the disposition survey is to decide whether the concentration of radioactivity is more characteristic of the DL or of the AL, i.e., whether action should be taken, or if action is not necessary. The width of the gray region expressed as a multiple of the measurement standard deviation, σ, is called the relative shift, $\bar{\Delta}\sigma$. Survey effort will increase as the relative shift decreases.

---

[2] In hypothesis testing, to "accept" the null hypothesis only means not to reject it. For this reason many statisticians avoid the word "accept." A decision not to reject the null hypothesis does not imply the null hypothesis has been shown to be true.

In Figure 4.1, it may be seen that a large σ can be tolerated because Δ is large enough that the resulting relative shift Δ/σ is large. In Figure 4.2, even though σ is small, either more accuracy or more samples are needed because Δ/σ is also small.

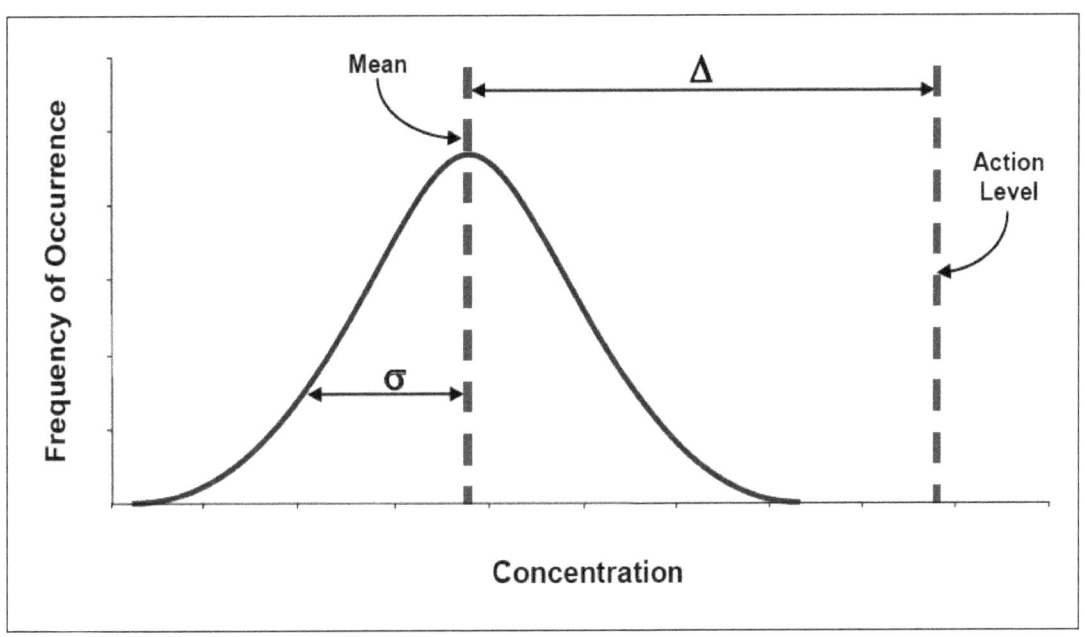

Figure 4.1 Relative Shift, Δ/σ, Comparison for Scenario A:
σ is Large, but the Large Δ Results in a Large Δ/σ and Fewer Samples

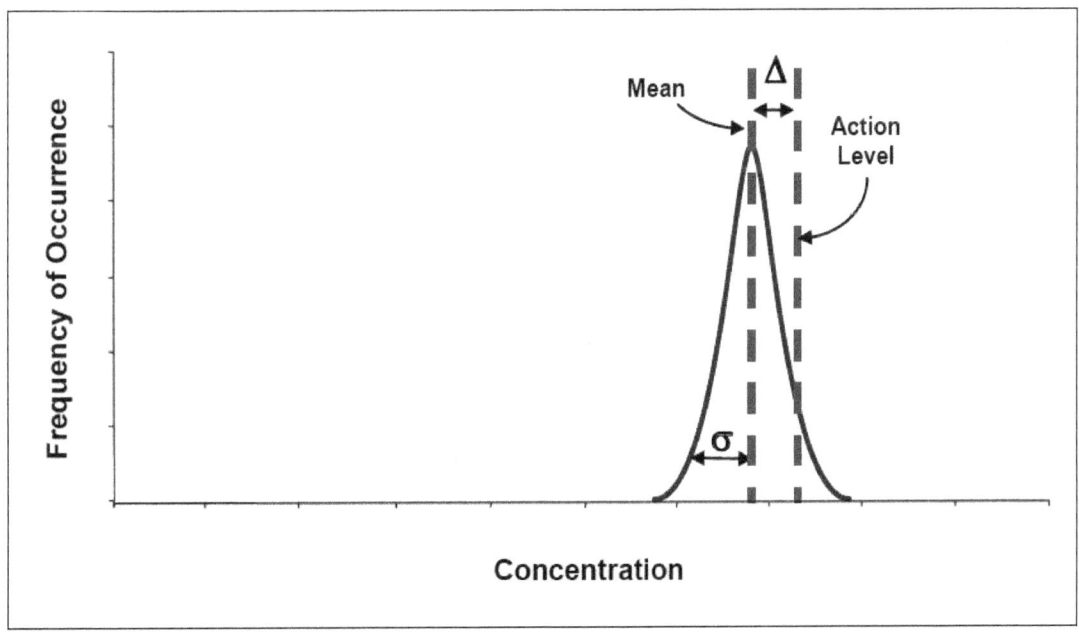

Figure 4.2 Relative Shift, Δ/σ, Comparison for Scenario A:
σ is Small, but the Small Δ Results in a Small Δ/σ and More Samples

4.2.3   Scenario A

The null hypothesis for Scenario A specifies that the radionuclide concentration or level of radioactivity associated with the M&E is equal to or exceeds the action level. For Scenario A (H_0: X > AL), the upper bound of the gray region (UBGR) is equal to the AL and the lower bound of the gray region (LBGR) is equal to the DL. As a general rule for applying Scenario A, the DL should be set no higher than the expected radionuclide concentration associated with the M&E. The DL and the AL should be reported in the same units. Figure 4.3 illustrates Scenario A.

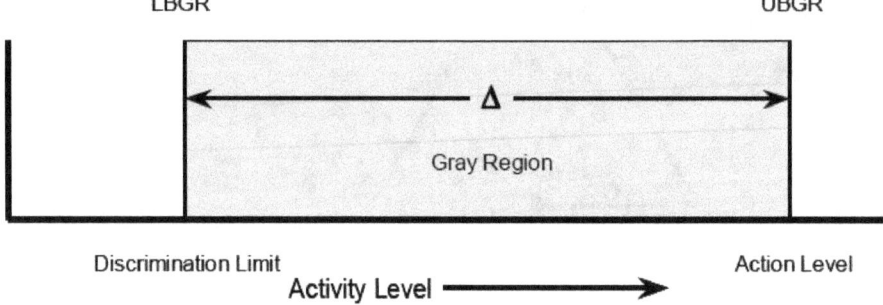

Figure 4.3 Illustration of Scenario A

4.2.4   Scenario B

The null hypothesis for Scenario B specifies the radionuclide concentration or level of radioactivity associated with the M&E is less than or equal to the action level. For Scenario B (H_0: X ≤ AL), the UBGR is equal to the DL and the LBGR is equal to the AL. The DL defines how hard the surveyor needs to look, and is determined through negotiations with the regulator.[3] In some cases the DL will be set equal to a regulatory limit (e.g., 10 CFR 36.57 and DOE 1993).

The DL and the AL should be reported in the same units. Figure 4.4 illustrates Scenario B. This description of Scenario B is based on information in MARLAP and is fundamentally different from the description of Scenario B in NUREG-1505 (NRC 1998a).

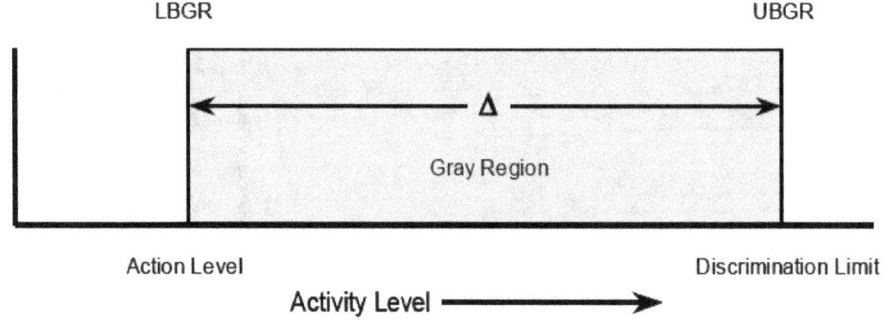

Figure 4.4 Illustration of Scenario B

---

[3] In some cases, setting the discrimination limit may include negotiations with stakeholders.

In NUREG-1505 (NRC 1998a) the gray region is defined to be below the AL in both Scenario A and Scenario B. In MARSAME and MARLAP the gray region is defined to be above the AL in Scenario B. The difference lies in how the action level is defined.

4.2.5   Specify Limits on Decision Errors

There are two possible types of decision errors:

- Type I error: rejecting the null hypothesis when it is true, and
- Type II error: failing to reject the null hypothesis when it is false.

Because there is always uncertainty associated with the survey results, the possibility of decision errors cannot be eliminated. So instead, the planning team specifies the maximum Type I decision error rate (α) that is allowable when the radionuclide concentration or level of radioactivity is at or above the action level. This maximum usually occurs when the true radionuclide concentration or level of radioactivity is exactly equal to the action level. The planning team also specifies the maximum Type II decision error rate (β) that is allowable when the radionuclide concentration or level of radioactivity equals the discrimination limit. Equivalently, the planning team can set the "power" ($1-\beta$) when the radionuclide concentration or level of radioactivity equals the discrimination limit. See MARSSIM Appendix D, Section D.6 for a more detailed description of error rates and statistical power.

It is important to clearly define the scenario (i.e., A or B) and the decision errors for the survey being designed. Once the decision errors have been defined, the planning team should determine the consequences of making each type of decision error. For example, incorrectly deciding the activity is less than the action level may result in increased health and ecological risks. Incorrectly deciding the activity is above the action level when it is actually below may result in increased economic and social risks. The consequences of making decision errors are project specific.

4.2.6   Develop an Operational Decision Rule

The theoretical decision rule developed in Section 3.6 was based on the assumption that the true radionuclide concentrations in the M&E were known. Because the disposition decision will be made based on measurement results and not the true but unknown concentration, an operational decision rule needs to be developed to replace this theoretical decision rule. The operational decision rule is a statement of the statistical hypothesis test, which is based on comparing some function of the measurement results to some critical value. The theoretical decision rule is developed during Step 5 of the DQO process (Chapter 3), while the operational decision rule is developed as part of Step 6 and Step 7 of the DQO process. For example, a theoretical decision rule might be "if the results of any measurement identify surface radioactivity in excess of background, the front loader will be refused access to the site; if no surface radioactivity in excess of background is detected, the front loader will be granted access to the site." The related operational decision rule might be "any result that exceeds the critical value associated with the MDC set at the discrimination limit will result in rejection of the null hypothesis, and the front loader will not be allowed on the site" (see more examples in Chapter 7).

4.3   Classify the Materials and Equipment

Classification is used to determine the level of survey effort for the disposition survey. The level of survey effort is linked to the potential to exceed the action level(s) (i.e., classification), and is a graded approach to survey design. Impacted M&E with the highest potential to exceed the action level(s) (i.e., Class 1) receive the greatest effort for the disposition survey, while M&E with a lower potential to exceed the action level(s) (i.e., Class 2 or Class 3) require less survey effort. Classification in MARSAME is analogous to classification in MARSSIM. The planning team needs to remember that classification is based on estimated radionuclide concentrations or radioactivity relative to the AL.

There are tradeoffs (costs and benefits) associated with classification based on estimated[4] or known radionuclide concentrations or levels of radioactivity relative to the action levels. This means that some knowledge of radionuclide concentrations is required before M&E can be classified. Known radionuclide concentrations or levels of radioactivity may be available from historical data identified during the initial assessment (IA; see Section 2.2), or performance of preliminary surveys (Section 2.3). Estimates of radionuclide concentrations can be developed based on historical data or process knowledge (Section 2.2). In the absence of information on the radionuclide concentrations, the default assumption is that all impacted M&E are Class 1.

Because classification of impacted M&E is based in part on an action level, classification cannot be performed until potential action levels have been identified (Section 3.3). For projects where multiple potential action levels have been identified, classification and selection of an appropriate action level may be an iterative process used to reduce the number of survey options. Alternatively, multiple survey designs can be developed to address all potential action levels. In the final step of the DQO process the most resource efficient survey design that meets the survey objectives is selected (Section 4.4.4).

4.3.1   Class 1

Class 1 M&E are impacted M&E that have, or had, the following: (1) highest potential for, or known, radionuclide concentration(s) or radioactivity about the action level(s); (2) highest potential for small areas of elevated radionuclide concentration(s) or radioactivity; and (3) insufficient evidence to support reclassification as Class 2 M&E or Class 3 M&E. Such potential may be based on historical information and process knowledge, while known radionuclide concentration(s) or radioactivity may be based on preliminary surveys. This class of M&E might consist of processing equipment, components, or bulk materials that may have been affected by a liquid or airborne release, including, for example, inadvertent effects from spills.

Class 1 M&E are those that may have been in direct contact with radioactive materials during operations or may have become activated and are likely to exceed the action level. Additionally, M&E that have been cleaned to remove residual radioactivity above the action level generally are considered to be Class 1. An exception to Class 1 classification may be considered if there

---

[4] There are risks and tradeoffs associated with using estimated values. The planning team should compare the consequences of potential decision errors with the resources required to improve the quality of existing data to determine the appropriate approach for a specific project.

are no difficult-to-measure areas and any residual radioactivity is readily removable using cleaning techniques. Examples of such methods may include vacuuming, wipe downs, or chemical etching that quantitatively remove sufficient amounts of radionuclides such that surficial activity levels would be less than the release criteria. Documented process knowledge of cleaning methods directly applicable to the particular M&E should be provided to justify this exception.

4.3.2   Class 2

Class 2 M&E are impacted M&E that have, or had, (1) low potential for radionuclide concentration(s) or radioactivity above the action level(s); and (2) little or no potential for small areas of elevated radionuclide concentration(s) or radioactivity. Such potential may be based on historical information, process knowledge, or preliminary surveys. This class of materials might consist of electrical panels, water pipe, conduit, ventilation ductwork, structural steel, and other materials that might have come in contact with radioactive materials. Radionuclide concentration(s) and radioactivity above the action level, including small areas of elevated radionuclide concentration(s) or radioactivity, are not expected in Class 2 M&E.

4.3.3   Class 3

Class 3 M&E are impacted M&E that have, or had, (1) little, or no, potential for radionuclide concentration(s) or radioactivity above background; and (2) insufficient evidence to support categorization as non-impacted. Radionuclide concentration(s) and radioactivity above a specified small fraction of the UBGR are not expected in Class 3 M&E. The specified fraction should be developed by the planning team using a graded approach and approved by the regulatory authority.

4.3.4   Other Classification Considerations

The planning team should review any historical data used to provide information on radionuclide concentrations or radioactivity and evaluate whether or not the data meet the objectives of the disposition survey, as illustrated in the following examples. Representativeness (see MARSSIM Appendix N) is a key data quality indicator when evaluating historical data. Ideally, the IA should provide information on the radionuclides of potential concern, expected radionuclide concentrations or radioactivity, distribution of radioactivity, and locations where radioactivity is expected (e.g., surficial or volumetric, see Section 2.4.3). In addition, the data should meet the criteria for measurability (e.g., MQC) or detectability (e.g., MDC) established for the project (see Sections 3.8 and 5.5). Historical data that do not meet the objectives of the disposition survey may still be used to provide estimates for radionuclide concentrations or levels of radioactivity.

The results of the IA may provide estimated radionuclide concentrations or levels of radioactivity based on process knowledge, historical data, sentinel measurements, or preliminary surveys. In some cases, a survey is performed to develop adequate estimates for levels and variability of radionuclide concentrations or radioactivity. Again, the planning team should evaluate the data used to develop the estimated radionuclide concentrations or levels of radioactivity. In general, estimated data will have a higher associated uncertainty than known data that meet the objectives of the project. The planning team should keep this in mind when developing estimates for radionuclide concentrations or radioactivity to be used in classifying M&E.

If the action level is defined in terms of average activity, the average radionuclide concentration or radioactivity should be compared to the action level to determine the appropriate classification. Similar comparisons should be developed for action levels provided in terms of maximum activity or total activity. For example, DOE Order 5400.5 (DOE 1993) provides three surface activity action levels for each group of radionuclides: average total surface activity, maximum total surface activity, and maximum removable surface activity. These action levels must be evaluated prior to disposition of the M&E. Classification would be determined by comparing the average total surface activity, maximum total surface activity, and maximum removable surface activity (or appropriate conservative estimates) to the corresponding action level. The overall classification would be determined by the most restrictive case. If the maximum total surface activity indicates the M&E is Class 1, while the maximum removable surface activity indicates the M&E is Class 3, the M&E should be classified as Class 1.

The improper classification of M&E has serious implications, particularly when it leads to the release of material with residual radioactivity in excess of the AL. For example, if material were mistakenly thought to have a very low potential for having residual radioactivity, the material will be subjected to a survey with lesser scrutiny. This misclassification might result in releasing material that should not be released. The opposing possibility (i.e., when M&E is misclassified as impacted when it is non-impacted) involves the stakeholders expending potentially substantial resources involved in unnecessarily surveying non-impacted M&E.

4.4   Design the Disposition Survey

MARSAME recommends design of disposition surveys that measure 100% of the M&E being investigated whenever practical. This includes survey designs where all of the M&E are physically measured. Survey designs where physical measurements are performed for less than 100% of the M&E may be acceptable if the radioactivity is measurable. Measurable radioactivity is radioactivity that can be quantified and meets the DQOs and MQOs established for the survey. Radioactivity that is quantified using known or predicted relationships developed from process knowledge, historical data, sentinel measurements, or preliminary measurements is considered measurable as long as the relationships are developed and verified as specified in the DQOs and MQOs. An example of such a relationship could be the immobile progeny of the measured radionuclides.

Survey designs that measure 100% of the M&E being investigated reduce the uncertainty in the final decision. Because 100% of the M&E are measured, for practical purposes spatial variability can be ignored. Attention should be given to ensure that all impacted surfaces are measured in 100% scan surveys. Surveys that use known or predicted relationships to estimate radionuclide concentrations or levels of radioactivity need to account for the contribution of spatial variability to total uncertainty.

To make the best use of limited resources, MARSAME places the greatest level of survey effort on M&E that have, or had, the greatest potential for residual radioactivity (i.e., Class 1). This is referred to as a graded approach. As noted in Section 1.3, survey designs that measure 100% of the M&E are often neither practical nor cost-effective, and could drive the user to dispose of any material that is potentially impacted without considering the benefits of reuse or recycle. The use of a graded approach to ensure that a sensible, commensurate balance is achieved between cost and risk reduction should always be incorporated into MARSAME survey designs. The following sections describe the four basic disposition-survey designs:

- Scan-only survey designs (Section 4.4.1),
- In situ survey designs (Section 4.4.2),
- Survey designs that combine scans and static measurements (MARSSIM-type surveys, Section 4.4.3), and
- Method-based survey designs (Section 4.4.4).

Figures 4.5, 4.6, and 4.7 illustrate the process of designing a disposition survey. Classification can be used to provide a graded survey approach to individual survey designs. Information on adjusting the level of survey effort based on classification is provided for each type of survey design. Each survey design can include a variety of survey techniques (Section 5.9).

4.4.1   Scan-Only Survey Designs

Scan-only survey designs use scanning techniques to measure the M&E. The detector is moving at a constant speed relative to the M&E being surveyed while maintaining a constant distance relative to the M&E. In general, scan-only survey designs may be applied to all types of M&E, from small individual items to large quantities of materials to large, complex machines. Scan techniques include hand-held instruments that are moved over the M&E, as well as systems that move the M&E past stationary detectors (e.g., conveyor systems). For example, a scan-only survey may involve the use of a Geiger-Mueller (GM) pancake detector to measure potential surface radioactivity on hand tools. Alternatively, a scan-only survey could involve the use of a conveyorized system that measures large quantities of M&E (e.g., bulk material or laundry). Scan-only surveys generally are applicable to all types of disposition surveys.

Scan-only survey designs often require the least amount of resources to design and implement, and are easy to incorporate into SOPs or project-specific survey designs. In many cases it is not necessary to document the results of individual scanning measurements because it is easy to identify results that exceed some threshold corresponding to the action level. With the real-time feedback available during Class 1 scan-only surveys, the user can implement a "clean as you go" practice by segregating M&E that exceed the threshold for additional investigation. Drawbacks to scan-only surveys include increased measurement uncertainty because of variations in scan speed and source to detector distance making it difficult to detect or quantify radionuclides with action levels close to zero or background.

Scan-only surveys are characterized by large numbers of measurements with relatively short count times. Measurement uncertainty should account for variations in source-to-detector distance, scan speed, and surface efficiency that are commonly associated with scanning measurements.

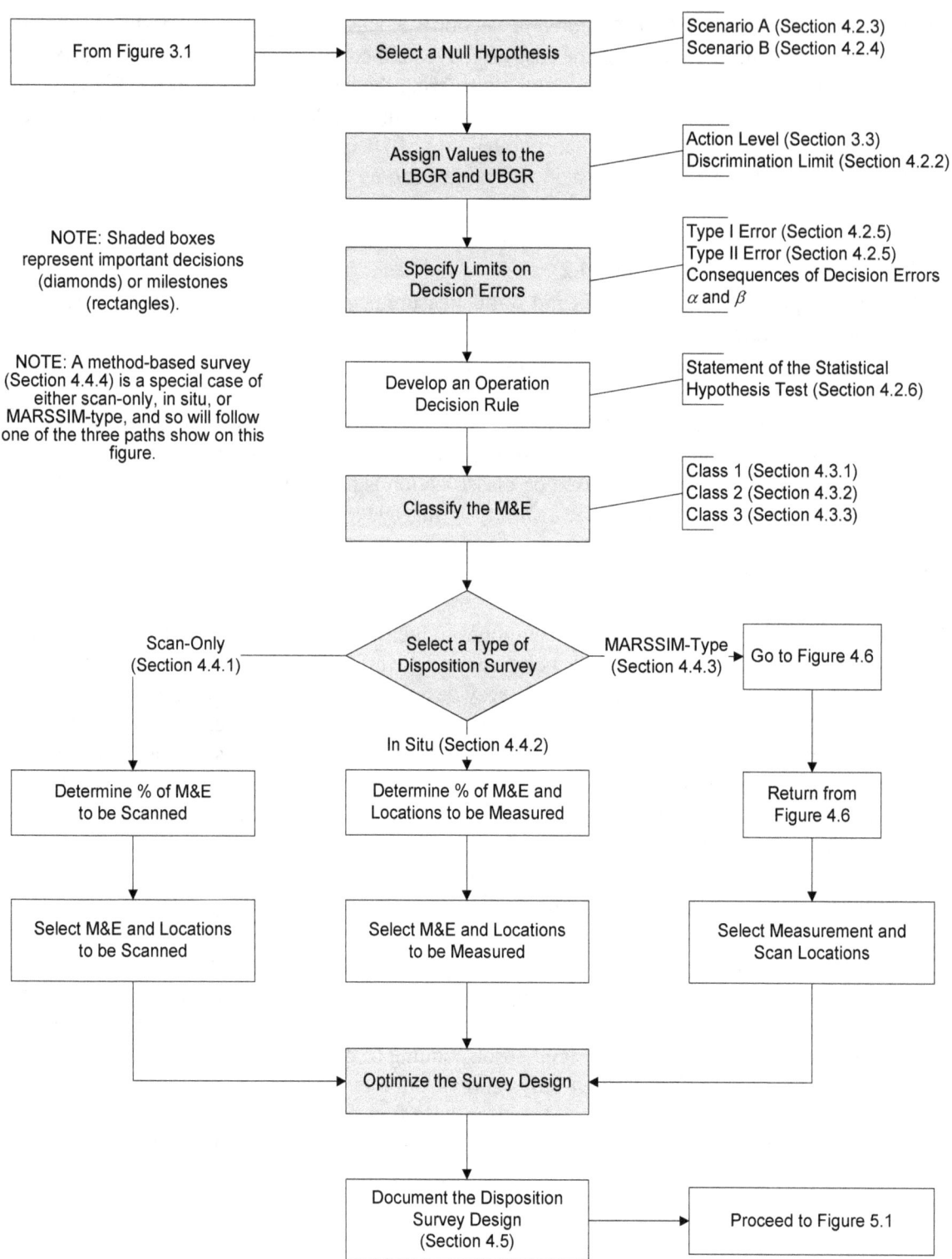

Figure 4.5 Flow Diagram for a Disposition Survey Design

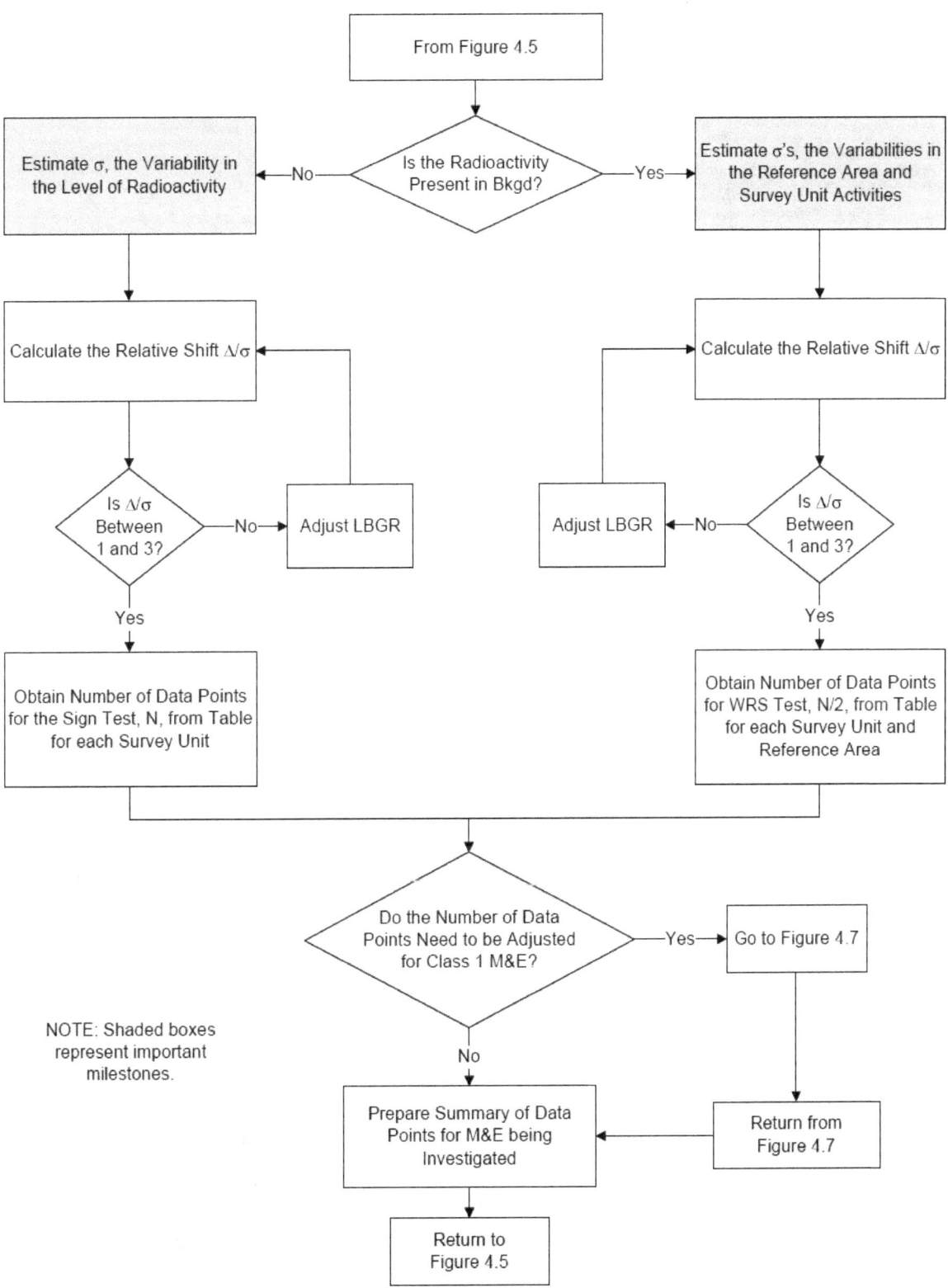

Figure 4.6 Flow Diagram for Identifying the Number of Data Points for a MARSSIM-Type Disposition Survey

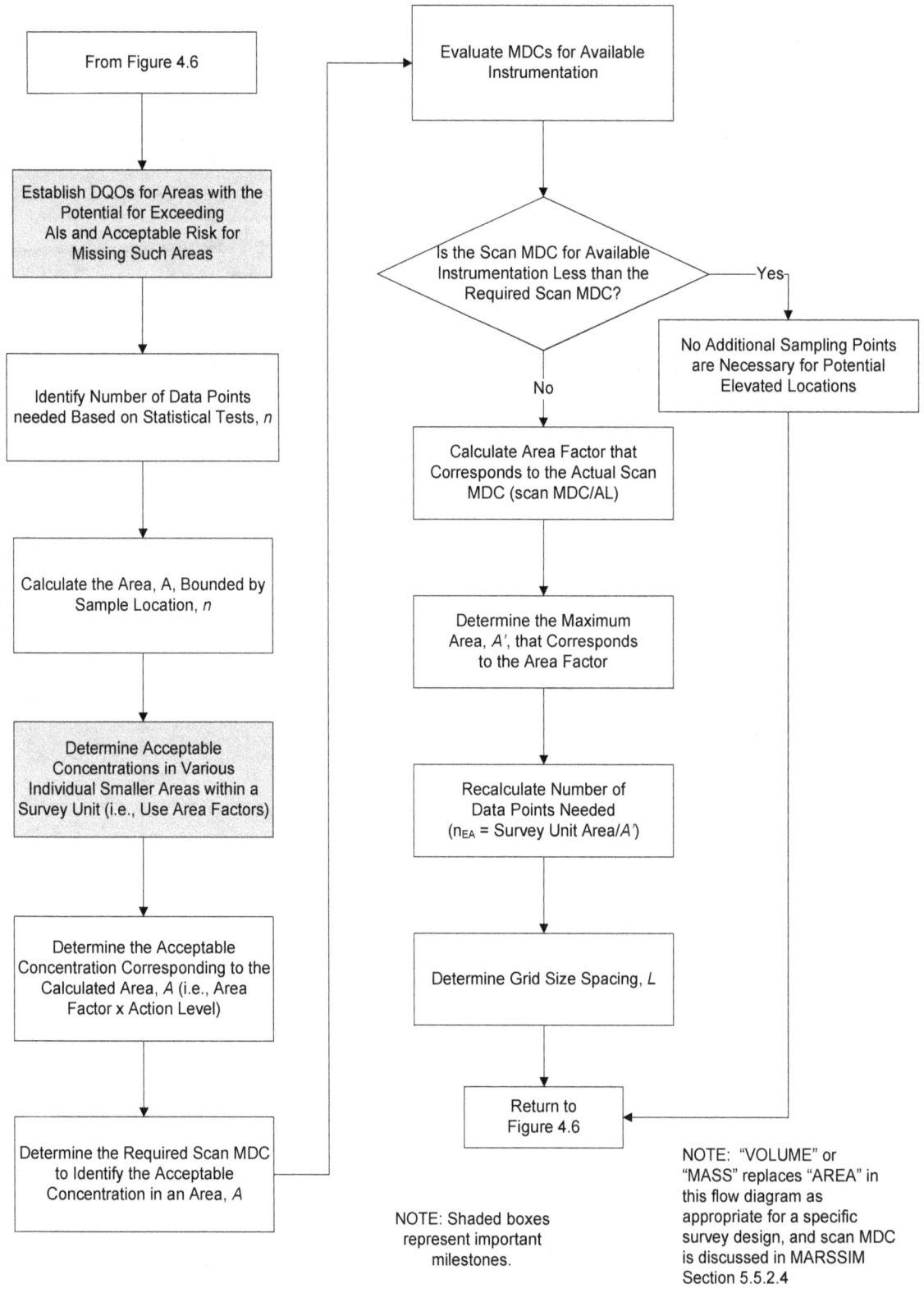

Figure 4.7 Flow Diagram for Identifying Data Needs for Assessment of Potential Areas of Elevated Activity in Class 1 Survey Units for MARSSIM-Type Disposition Surveys

Evaluation of scan-only survey data depends on whether or not individual measurement results are recorded (Section 6.2.5). The decision of whether to record individual measurement results will impact the selection of instrumentation (Section 5.9) and survey documentation requirements (see Sections 4.5, 5.11, and 6.9), and may impact handling of the M&E (Section 5.3).

4.4.1.1   Class 1 Scan-Only Surveys

Class 1 scan-only surveys require that physical measurements be performed for 100% of the M&E being investigated. For individual items this may require scanning both sides of flat items (e.g., sheet metal, boards) and changing the surveyor's grip on the item to ensure all areas are surveyed (e.g., handles). For conveyor systems this may require flipping or rotating the M&E and performing additional measurements. Conveyor systems can also be designed with detectors surrounding the M&E (e.g., above and below a conveyor belt) to provide 100% measurability.

4.4.1.2   Class 2 Scan-Only Surveys

Class 2 scan-only surveys use information about the M&E to reduce the total area surveyed using a graded approach. The amount of the M&E surveyed is calculated based on the relative shift (i.e., Δ/σ). The percent of the M&E to be surveyed is 10%, or the result using Equation 4-1, whichever is larger:

$$\% \text{ Scan} = \frac{\left(10 - \dfrac{\Delta}{\sigma}\right)}{10} \times 100\% \qquad (4\text{-}1)$$

The amount of M&E to be scanned should be rounded up to the next 10 percent, and at least 10% of the M&E must be surveyed. For example, if the percent scan is 51%, then 60% of the M&E will be surveyed. This means that between 10 to 100% of Class 2 M&E would be measured during the disposition survey. Figure 4.8 shows the relationship between the relative shift and the amount of M&E to be scanned.

The scanned percentages need to represent spatially uniform coverage of the survey unit and coincide with the conceptual model for the M&E. Consider spatially uniform coverage when scanning 30% of a desk and 30% of a bucket of bolts. For the desk example, 30% coverage during scanning may be derived from performing scans on the top surface, the legs, inside the drawers, etc., so that essentially 30%

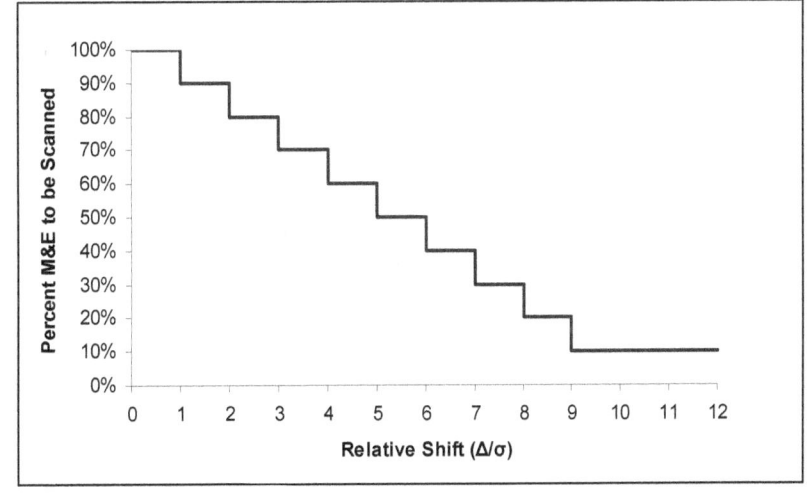

Figure 4.8 Relationship Between the Relative Shift and the Amount of M&E to be Scanned

of each surface is scanned, yielding 30% total coverage of the entire desk. For the bucket of bolts example, 30% scanning coverage means laying out all the bolts and scanning 30% of them as well as 30% of the bucket itself. Alternatively, if the conceptual model for the desk showed a higher potential for contamination on the top, bottoms of legs, and drawer handles, 100% of these areas could be scanned with smaller amounts of the areas with a lower potential for radioactivity scanned to provide a total of 30% coverage for the entire desk. The graded approach should be applied to all aspects of the survey design.

The selection of M&E to survey as part of a Class 2 survey is project specific and is determined based on what is known about the M&E. For example, if all of the M&E is accessible and is expected to have uniform radionuclide concentrations or levels of radioactivity, the M&E to be surveyed should be selected randomly. However, there may be areas that are difficult-to-access with the instrumentation selected to perform the survey. If there is a known and accepted relationship between radionuclides in difficult-to-access areas and radionuclides in accessible areas, the Class 2 measurements may be biased to only accessible areas (i.e., representative of measurements in difficult-to-access areas).

If elevated radionuclide concentrations or levels of radioactivity are restricted to areas that can be readily identified (e.g., discolored areas, corners, cracks, access points) the Class 2 measurements may be designed to concentrate on these biased areas. The Class 2 survey design should include a combination of biased and random areas to check assumptions used to support the survey design.

The selection of M&E to survey may also depend on the physical characteristics of the M&E. For example, surveying 40% of the inside of a railroad car would be different from surveying 40% of a pile of rubblized concrete. Section 5.3 provides information on handling M&E and determining what will be measured during implementation of the survey design.

4.4.1.3   Class 3 Scan-Only Surveys

Class 3 scan-only survey designs are identical to Class 2 scan-only survey designs. The planning team may decide that some Class 3 scan-only disposition surveys require that less than 10% of the M&E will be measured. The decision to design a survey requiring less than 10% of the M&E to be measured should be based on the total uncertainty associated with the disposition decision. The determination of total uncertainty should be based on process knowledge, historical data, and the results of preliminary and disposition surveys.

In addition, some Class 3 scan-only survey designs may be based solely on biased measurements. In other words, random measurement locations are not required for Class 3 scan-only survey designs. However, if biased measurements are reasonable, they should be performed, keeping in mind that Class 3 M&E have very little or no potential for exceeding the AL.

4.4.2   In Situ Survey Designs

In situ survey designs use static measurements to measure 100% of an item. The detector and the item being measured are held in a fixed geometry[5] for a specified count time to meet the MQOs. There are a wide variety of in situ measurement techniques available. Examples include box counters, portal monitors, and in situ gamma spectrometry systems, as well as direct measurements with hand-held instruments (e.g., NaI(Tl), ZnS, GM pancake, and portable gas proportional detectors). In situ surveys generally are applied to situations where scan-only surveys are determined to be unacceptable. For example, variations in source-to-detector distance, scan speed, and surface efficiency that are commonly associated with scanning measurements can often be effectively controlled using an in situ survey design.

In situ surveys are characterized by limited numbers of measurements with long count times (relative to scan-only surveys). Measurement uncertainty will incorporate spatial uncertainty because of the source geometry assumed in the calibration. Thus, special attention needs to be made to the assumptions made in the calibration of in situ systems. Potential deviations from these assumptions need to be propagated through the calibration equation to assess the total measurement uncertainty (see Sections 5.6 and 7.4). Count times are determined by the MQOs rather than the time constant of the measurement system. In situ measurements provide a 100% measurement for some portion of the M&E being investigated. The M&E may be an individual item or piece of equipment, or some fraction of a large quantity of material determined by the solid angle coverage of the detector.

In situ surveys may consist of a single measurement, or a series of measurements. Single measurement surveys typically are performed on individual items or relatively small batches of M&E. A series of in situ measurements may be used to evaluate larger quantities of M&E. In some cases, a series of in situ measurements may be performed of a single item or batch of M&E to provide several estimates of the radionuclide concentrations from different angles. The primary difference between an in situ survey and a MARSSIM-type survey is that an in situ survey measures 100% of an item (using one or several measurements) to determine the average radionuclide concentration for that item. A MARSSIM-type survey uses a statistically based number of measurements (that generally do not measure 100% of the item or group of items being surveyed) to calculate an average radionuclide concentration for that item or group of items.

4.4.2.1   Class 1 In situ Surveys

Class 1 in situ surveys require that physical measurements be performed for 100% of the M&E being investigated. Placing an item inside a 4-π measurement system, performing a series of measurements with overlapping fields of view that incorporate all of the M&E, or rotating the M&E within the field of view of the detector so 100% of the M&E are measured are examples where 100% of the M&E are measured.

---

[5] There are situations where the levels of radioactivity for M&E being measured are expected to be inhomogeneous. Certain measurement systems can rotate the M&E during a measurement to provide an estimate of the average activity. For the purposes of this section, these are considered fixed geometries. Additional discussion on the limitations of these systems is provided in Chapter 5.

4.4.2.2   Class 2 In situ Surveys

Class 2 in situ surveys use information about the M&E to reduce the total area surveyed using a graded approach. The amount of the M&E surveyed is calculated based on the relative shift (i.e., Δ/σ). The percent of the M&E to be surveyed is 10% or the result using Equation 4-2, whichever is larger:

$$\text{\% Measured or \% Solid Angle Coverage} = \frac{\left(10 - \dfrac{\Delta}{\sigma}\right)}{10} \times 100\% \qquad (4\text{-}2)$$

The fraction of the M&E or the solid angle coverage of the M&E to be surveyed should be rounded up to the next 10 percent. If the % coverage is 51%, then 60% of the M&E will be surveyed. This means that 10 to 100% of Class 2 M&E would be measured during the disposition survey. Figure 4.8, on page 4-13, shows the relationship between the relative shift and the amount of M&E to be surveyed.

The selection of M&E to survey as part of a Class 2 survey is project specific and is determined based on what is known about the M&E. For example, if all of the M&E is accessible and is expected to have uniform radionuclide concentrations or levels of radioactivity, the M&E to be surveyed should be selected randomly. However, there may be areas that are difficult-to-access with the instrumentation selected to perform the survey. If there is a known and accepted relationship between radionuclides in difficult-to-access areas and radionuclides in accessible areas, the Class 2 measurements may be biased to only accessible areas (i.e., representative of measurements in difficult-to-access areas). If elevated radionuclide concentrations or levels of radioactivity are restricted to areas that can be readily identified (e.g., discolored areas, corners, cracks, access points) the Class 2 measurements may be designed to concentrate on these biased areas. The Class 2 survey design should include a combination of biased and random areas to check assumptions used to support the survey design.

4.4.2.3   Class 3 In situ Surveys

Class 3 in situ survey designs are identical to Class 2 in situ survey designs. The planning team may decide that some Class 3 in situ disposition surveys require that less than 10% of the M&E will be measured. The decision to design a survey requiring less than 10% of the M&E to be measured should be based on the total uncertainty associated with the decision based on process knowledge, historical data, and the results of preliminary and disposition surveys.

4.4.3   MARSSIM-Type Survey Designs

MARSSIM-type survey designs combine a statistically based number of static measurements to determine average radionuclide concentrations or radioactivity levels with scanning to identify areas of elevated radionuclide concentrations or radioactivity for specified quantities of M&E (i.e., survey units). Identifying survey unit sizes, laying out systematic measurement grids, and calculating project- and item-specific area factors requires a significant effort. Section 5.3 discusses considerations for handling M&E, including locating measurements. The planning team should consider that MARSSIM-type survey designs might be more complex and require more resources than scan-only or in situ survey designs that meet the DQOs. Information on designing MARSSIM-type surveys is found in MARSSIM Section 5.5. In general, MARSSIM-type surveys of M&E are only performed on large, complicated M&E with a high inherent value after scan-only and in-situ surveys have been considered and rejected.

4.4.3.1   Class 1 MARSSIM-Type Surveys

Class 1 MARSSIM-type surveys calculate the required number of measurements in each survey unit based on the shift (i.e., Δ), the variability in the radionuclide concentrations or levels of radioactivity (i.e., σ), and the Type I and Type II decision error rates (i.e., α and β). The number of measurements per survey unit is adjusted to account for small areas of elevated activity using the information in MARSSIM Section 5.5.2.4. In addition, scan measurements are required for 100% of the M&E being investigated.

The development of survey unit boundaries is discussed in Section 3.6. The quantity of M&E in each survey unit should be determined based on the modeling assumptions used to develop the action levels.

The variability in the radionuclide concentrations in each survey unit can be estimated using the standard deviation of preliminary measurements or the uncertainties from individual measurements, whichever is larger. Whenever practical, preliminary data should be used to provide estimates of variability. As a last resort when preliminary data are not available, MARSSIM states that assuming a coefficient of variation on the order of 30% may be reasonable (MARSSIM Section 5.5.2.2, Page 5-26). This 30% is used as a starting point for the DQO process, and should be adjusted iteratively during the development of a final survey design. For M&E, MARSAME recommends using a higher percentage value.

Area factors are specified in a regulation or other guidance, or developed based on the changes in dose or risk associated with changing the area (or volume) of activity to be less than the entire survey unit. For example, DOE Order 5400.5, Figure IV-1 (DOE 1993) allows use of an area factor of up to 3.0 for total surficial radioactivity for all radionuclides, NUREG-1640 (NRC 2003a) is only concerned with average activity and total inventory of radioactivity, which implies that within the survey unit relatively high localized concentrations of radioactivity could exist. This implication does not mean that a large part of the survey unit may be used to intentionally "dilute" high concentrations of radioactivity. Rather, in the course of normal processing there is a non-prescriptive flexibility allowed for inhomogeneity of radionuclide concentrations. Nevertheless, mixing different classes of M&E (Class 1, 2, and 3) is not allowed. The physical characteristics of the M&E combined with potential future exposures based on the selected disposition option mean that area factors (and possibly exposure pathway dose or risk models) need to be developed for each project. In the absence of regulation-specific area factors, assuming an area factor of 1.0 for all radionuclides would be the most conservative approach. Depending on the basis of the action level, an area factor may or may not be applicable. MARSSIM uses completely different scenarios for real property to develop area factors in contrast to those scenarios used for M&E in NUREG-1640 (NRC 2003a). Area factors may be derived on a project-specific basis using project-specific scenarios.

If the radioactivity being measured is present in background, Table 5.3 in MARSSIM provides the number of measurements required in each survey unit as well as in each reference area. MARSSIM Section 5.5.2.2 and NUREG-1505 (NRC 1998a) Sections 9.4 and 9.5 provide information on calculating the number of required measurements when the radioactivity being measured is present in background.

If the radioactivity being measured is not present in background, Table 5.5 in MARSSIM provides the number of measurements required in each survey unit. MARSSIM Section 5.5.2.3 and NUREG-1505 (NRC 1998a) Sections 9.2 and 9.3 provide information on calculating the number of required measurements when the radioactivity being measured is not present in background. For convenience, statistical sample size and critical value tables for the Sign and Wilcoxon Rank Sum (WRS) tests taken from MARSSIM Appendix I are given in MARSAME Appendix A. In addition, Appendix A contains a table of critical values for the Quantile test, taken from NUREG-1505.

Whenever area factors other than 1.0 are used to design the disposition survey, a systematic grid should be used to determine measurement locations. The systematic grid determines the largest area that could be missed by the measurements which is used to determine the required scan MDC. Section 5.3 provides information on handling M&E, including setting up systematic grids.

4.4.3.2   Class 2 MARSSIM-Type Surveys

Class 2 MARSSIM-type surveys are similar to Class 1 MARSSIM-type surveys. The numbers of measurements in each survey unit are determined in the same manner, although the expected radionuclide concentrations or levels of radioactivity and the decision error rates may change. Unlike MARSSIM, the survey unit size remains the same and does not change based on classification. The portion of the survey unit where scan surveys are required is reduced to between 10 and 100%. The information in Section 4.4.1.2 for Class 2 scan-only surveys should be used to determine the areas to be scanned. This recommendation is provided for M&E only, and is not intended to update the guidance in MARSSIM for surface soils and building surfaces.

4.4.3.3   Class 3 MARSSIM-Type Surveys

Class 3 MARSSIM-type surveys are similar to Class 1 MARSSIM-type surveys. The numbers of measurements in each survey unit are determined the same way, although the expected radionuclide concentrations or levels of radioactivity and the decision error rates may change. Unlike MARSSIM, the survey unit size does not change based on classification. The portion of the survey unit where scan surveys are required is reduced to less than 10% and is based on professional judgment. The information in Section 4.4.1 for scan-only surveys should be used to determine the areas to be scanned. This recommendation is provided for M&E only, and is not intended to update the guidance in MARSSIM for surface soils and building surfaces.

4.4.4   Method-Based Survey Designs

The action level selected in Section 3.3 may implicitly or explicitly require using a specific measurement technique (Section 5.9.1) or instrument (Section 5.9.2). A survey design that is based on a required measurement method, or combination of measurement technique and instrumentation, is called a "method-based" survey design.

A method-based survey design is a scan-only, in situ, or MARSSIM-type survey design that incorporates the required measurement method. The survey design will still need to address all of the required components, such as number, type, location, and sensitivity of measurements. Survey components that are not specified as part of the required measurement method should be identified and addressed using the DQO process.

4.4.5   Optimize the Disposition Survey Design

The disposition survey design process described in this supplement could result in the development of multiple potential disposition survey designs. For example, consider the case when simultaneous compliance with more than one action level is required (e.g., DOE 1993). In other cases the decision resulting from one survey may lead to the requirement of another survey, such as failure to demonstrate compliance with the disposition criterion for release resulting in a survey to comply with radioactive waste acceptance criteria. Multiple survey designs could result from selection of multiple potential disposition options, action levels, survey techniques, measurement systems, decision rules, or some combination of these factors. Before the planning team can proceed, all of the potential disposition survey designs need to be reviewed to select a final disposition survey design.

The final step in the DQO process ("Develop the Detailed Plan for Obtaining Data," Step 7) is designed to produce the most resource-efficient survey design that is expected to meet the DQOs. It may be necessary to revisit previous steps in the DQO process and work through this step more than once.

There are five activities included in this step:

1. Review existing data (e.g., historical data, sentinel measurement results, preliminary survey results). Use existing data to support the data collection design. If no existing data are available, consider performing preliminary surveys to acquire estimates of variability to determine numbers of measurements. Evaluate potential problems regarding detection limits or interferences. If new data will be combined with existing data, determine if there are data gaps that need to be filled or deficiencies that can be mitigated prior to implementing the disposition survey design.

2. Evaluate operational decision rules. The theoretical decision rules developed in Section 3.6 were based on the assumption that the true radionuclide concentrations or radioactivity present in the M&E were known. Operational decision rules based on the statistical tests (Chapter 6) should replace the theoretical decision rule (see Sections 3.5 and 4.2.6). Review the parameter of interest (e.g., maximum measured value, mean or median radionuclide concentration) and the possible statistical tests that could be applied to the data to evaluate the operational decision rules.

3. Develop general data collection design alternatives. Sections 4.4.1, 4.4.2, and 4.4.3 provide information on general data collection design alternatives applicable to disposition surveys.

Consider individual instruments and measurements techniques (Section 5.9) combined with general data collection designs to develop alternative survey approaches.

4. Calculate the number of measurements or amount of M&E to be surveyed. Sections 4.4.1, 4.4.2, and 4.4.3 provide general information on determining the level of survey effort for the general data collection design alternatives based on classification. Determine the estimated resources required for each of the alternative survey approached.

5. Select the most resource-effective survey design. Evaluate each of the survey approaches based on the required resources and the ability to meet the DQO and MQO constraints within the tolerable decision error limits. The survey design that provides the best balance between cost and meeting survey objectives while considering the non-technical economic and health factors imposed on the project is usually the most resource-effective. The statistical concept of a power curve (MARSSIM Appendix I.9) is extremely useful in investigating the performance of alternative survey designs.

If none of the alternative survey designs meet the survey objectives within the tolerable decision error limits while considering the budget or other constraints, then the planning team will need to relax one or more of the constraints. Examples include—

- Increasing the budget for implementing the survey;
- Using exposure pathway modeling to develop site-specific action levels;
- Increasing the decision error rates, not forgetting to consider the consequences associated with making an incorrect decision;
- Increasing the width of the gray region for Scenario A surveys by decreasing the average activity associated with the M&E which may require remediation, or negotiating a higher UBGR for Scenario B which may require additional reference area investigations;
- Relaxing other project constraints (e.g., schedule);
- Changing the boundaries—it may be possible to reduce measurement costs by changing or eliminating survey units that will require different decisions;
- Segregating the M&E based on physical or radiological attributes (Section 5.4);
- Evaluating alternative measurement techniques with lower detection limits or lower survey costs;
- Adjusting the list of radionuclides or radiations of concern (Section 3.2); and
- Considering other disposition options that will result in higher action levels.

4.5 Document the Disposition Survey Design

Documentation of the disposition survey design should provide a complete record of the selected survey design. The documentation should include all assumptions used to develop the survey design, a detailed description of the M&E being investigated, along with the DQOs and MQOs for the survey (e.g., MQC, MDC, count time). The regulatory basis for the disposition criterion and calculations showing the derivation of action levels should also be provided. Sufficient data and information should be provided to enable an independent re-creation and evaluation of the disposition survey design. The documentation should provide information on the following topics:

- Information on *who* developed, reviewed, and approved the survey design, as well as training and qualification requirements for such individuals, should be included, along with any requirements for who can implement the survey design.
- Information on *what* M&E were considered when developing the survey design along with a description of M&E to which the survey design applies.
- Information on *when* the survey design was developed along with when the survey design will be implemented including restrictions on time of day, time of year, and count times when applicable.
- Information on *where* the survey design can be applied (including restrictions on local background levels) along with measurement locations including fraction of M&E to be surveyed and locations of direct measurements or samples or methods for selecting locations during implementation,
- Information on *why* a survey should be performed including justification for impacted and non-impacted decisions and assignment of classifications,
- Information on *how* the survey will be performed including measurement techniques and instruments along with instructions for segregating and handling the M&E during the survey.

There are two methods for documenting surveys described in the following sections, based on the type of project—

- Routine or repetitive surveys, and
- Case-specific applications.

4.5.1   Routine Surveys and Standard Operating Procedures

Routine (or repetitive) surveys are disposition surveys that are routinely performed on M&E entering or leaving an operating facility. Examples of routine surveys include–

- Clearance of tools from radiological control areas at a radiation facility,
- Preparation of low-level radioactive waste for disposal, and
- Interdiction of scrap metal entering a recycling facility.

Documenting routine survey designs, for example as SOPs, can be consistent with MARSAME recommendations. SOPs detail the work processes that are conducted or followed within an organization and document the way activities are performed. SOPs that also meet the DQOs for the disposition survey can be used to document routine survey designs. The development and use of SOPs facilitates consistent conformance to technical and quality system requirements. They promote quality through consistent implementation of a process within an organization, even if there are temporary or permanent personnel changes. The benefits of a valid SOP are reduced work effort combined with improved data comparability, credibility, and legal defensibility (EPA 2001). Additional guidance on developing SOPs, including example SOPs, is provided in EPA QA/G-6 (EPA 2001).

4.5.1.1   SOP Process

The organization developing the SOP should have a procedure in place for determining what procedures or processes need to be documented. SOPs documenting these procedures or processes should be written by individuals knowledgeable with the activity and the organization's internal structure. For disposition survey designs, a team approach to writing SOPs is often used. This allows input from subject-matter experts with information critical to the survey process, and promotes acceptance of the SOP once it is completed.

SOPs should be concise and provide step-by-step instructions in an easy-to-read format. They should provide sufficient detail so that a technician with limited experience, but with a basic understanding of the process, can successfully implement the survey design when unsupervised. Disposition survey SOPs should be reviewed and validated by one or more individuals with appropriate training and experience in performing surveys of M&E before they are implemented. It may be helpful to have the draft SOP field tested by someone not directly involved in the development of the SOP. The review process for disposition surveys should include a regulatory review and appropriate stakeholder involvement.

SOPs need to remain current. SOPs should be updated and re-approved whenever survey procedures are changed. SOPs should be systematically reviewed on a periodic basis to ensure that the policies and procedures remain current and appropriate.

Many disposition survey activities use checklists or forms to document completed tasks (e.g., daily instrument checks). Any checklists or forms included as part of the disposition survey should be referenced at the points in the procedure where they are used and attached to the SOP. Remember that the checklist or form is not the SOP, but a part of the SOP.

The organization should have a system for developing, reviewing, approving, controlling, and tracking documents. This process is usually documented in the Quality Management Plan.

4.5.1.2   General Format for Disposition Survey SOPs

In general, disposition survey SOPs consist of five elements:

- Title Page,
- Table of Contents,
- Procedures,
- Quality Assurance and Quality Control, and
- References.

The title page should include a title that clearly identifies the activity, an identification number, date of issue or revision, and the name of the organization to which the SOP applies. The signatures and signature dates of individuals who prepared and approved the SOP also should be included.

The table of contents lists the major section headings and the pages where the information is located. This provides a quick reference for locating the desired information and identifies changes or revisions made to individual sections.

The procedures are specific to the disposition survey design and may include some or all of the following topics:

- Scope and applicability. This section should provide a detailed description of the M&E to which the SOP can be applied. In addition, it is often important to clearly identify M&E to which the SOP does not apply.
- Summary of method. This section briefly describes the overall survey design, identifies the disposition option, lists the action levels, and provides their regulatory basis. The details on the development of the action levels based on the disposition criterion in the regulations is generally referenced or included as an attachment.
- Definitions. This section identifies and defines any acronyms, abbreviations, or specialized terms used in the SOP.
- Health and safety warnings. This section indicates operations that could result in personal injury, loss of life, or uncontrolled release to the environment. Explanations of what could happen if the procedure is not followed or if it is followed incorrectly should appear here as well at the critical steps in the procedure.
- Cautions. This section identifies activities that could result in equipment damage, degradation of data, or possible invalidation of results. Explanations of what could happen if the procedure is not followed or if it is followed incorrectly should appear here as well as the critical steps in the procedure.
- Interferences. This section describes any component of the process that may interfere with the final decision regarding disposition of the M&E.
- Personnel qualifications. This section lists the minimum experience required for individuals implementing the SOP. Any required certifications or training courses should be listed. For many routine surveys the training records of the personnel implementing the survey design are used to document compliance with the SOP.
- Equipment and supplies. This section lists and specifies the equipment, materials, reagents, and standards required to implement the SOP. At a minimum, this section must identify the model number and manufacturer of instruments that will be used to perform the survey.
- Roles and responsibilities of project personnel. This section identifies the decision-maker for the project as well as identifying who is responsible for performing specific tasks. An organizational chart documenting the chain of command and reporting authority (including quality control, health and safety, and any subcontractors) is a useful tool for showing potential interactions between project team members.
- Procedure. This section provides all pertinent steps, in order, and materials needed to implement the survey design. This section should include—
  o Instrument or method calibration and standardization (generally requires a check of the instrument calibration date and lists the appropriate MQOs such as MQC or MDC and references the details for these processes),
  o Type, number, and location of measurements,
  o Data acquisition, calculations, and data reduction requirements,
  o Troubleshooting, and

    o Computer hardware and software.
- Data and records management. This section describes the forms to fill out, reports to be written, and data and record storage information. At a minimum routine survey records should identify the personnel performing measurements and the instruments used to perform the measurements (i.e., model and serial number for all components of the measurement system). These records should show that the personnel performing the survey were properly trained and the instruments used to collect the data were calibrated and operating properly. This section should clearly state whether individual measurement results will be recorded, because this information is not always required.

The QA/QC section describes the activities required to demonstrate the successful performance of the disposition survey. For many organizations the QC activities for individual instruments are provided in separate SOPs describing the proper use of that instrument, so the daily checks of the instruments are included by reference. The QA/QC section should identify QC requirements for the disposition survey such as blanks, replicates, splits, spikes, and performance evaluation checks. The frequency for each QC measurement should be listed along with a discussion of the rationale for decisions. Specific criteria should be provided for evaluating each type of QC measurement, as well as actions required when the results exceed the QC limits. The procedures for reporting and documenting the results of QC measurements should be listed in the QA/QC section. Section 5.10 provides additional information on QC for disposition surveys.

The reference section should list all documents or SOPs that interface with the routine survey SOP. Full references (including SOP versions and dates) should be provided. Published literature and instrument manuals that are not readily available should be attached.

4.5.2 Case-Specific Applications

There are M&E that may require a disposition survey that are not covered by routine surveys. These are collectively referred to as case-specific applications. Case-specific applications include project-specific applications such as decommissioning or cleanup surveys, as well as unique applications involving one-time disposition of special equipment from a facility.

Ideally, documentation of case-specific survey designs involves a comparable level of effort associated with routine surveys. This is obviously the case for large decommissioning or cleanup projects where survey designs are documented as SOPs using a process analogous to routine surveys. The major differences are seen in the requirements for approval and maintenance of SOPs, which generally are less for decommissioning or cleanup projects compared to operating facilities. Disposition survey designs that will be applied during decommissioning or cleanup activities typically are documented as part of the survey design. However, a survey design needs to provide all of the information supporting the development of the disposition survey design, where SOPs typically focus on one aspect of the survey design or implementation. Historical information, process knowledge, description of the M&E, and assumptions used in the disposition survey design need to be included and not referenced.

The assumptions used to develop survey designs for routine surveys cannot be applied to all M&E, so situations will arise where a disposition survey design needs to be developed for special items or unique applications. These types of surveys are often associated with M&E that have a high inherent value (e.g., large quantities of valuable materials, unique or very expensive equipment) to offset the resources required to develop a unique disposition survey design. These special survey designs need to be inclusive, providing all of the information supporting the development of the disposition survey design. Detailed discussions should be provided for all parts of the survey design, including selection of a disposition option, selection and development of action levels, development of MQOs and selection of instruments, and QA/QC requirements for individual measurement systems as well as for the entire disposition survey.

For most applications the disposition survey design is expected to be documented as a stand-alone survey plan or as a series of SOPs. However, the planning team may determine that the survey design documentation can be combined with the results of the survey into a single document. At a minimum, instructions on the type, number, and location of measurements should be documented to provide instructions to the technicians performing the survey. Documenting the entire disposition decision process in a single document is most appropriate for unique applications where there is sufficient historical information or survey precedent such that there is little uncertainty associated with the development of a survey design. The benefit of documenting all of the survey decisions (e.g., design, implementation, and assessment) in one document is the savings in resources to develop multiple documents. The risk associated with not documenting the survey design process until after implementation is that the assessment will identify some problems with the survey design requiring additional data collection which could impact project costs and schedule.

5   IMPLEMENT THE SURVEY DESIGN

5.1   Introduction

This chapter discusses the implementation phase of the data life cycle and focuses on controlling measurement uncertainty and associated MQOs. The information in this chapter describes approaches for safely implementing the final disposition survey design developed in Chapter 4, methods for controlling uncertainty, and techniques to determine whether the measurement results achieve the survey objectives. Figure 5.1 illustrates the implementation phase of the data life cycle.

Similar to MARSSIM, MARSAME excludes specific recommendations for implementing disposition surveys. Instead, MARSAME provides recommendations and information to assist the user in selecting measurement techniques for implementing the survey design. This approach encourages consideration of innovative measurement techniques and emphasizes the flexibility of the information in MARSAME.

Implementation begins with health and safety considerations for the disposition survey (Section 5.2). Section 5.3 provides information on handling M&E, while Section 5.4 discusses segregating M&E based on physical and radiological attributes. Section 5.5 continues the discussion of measurement quality objectives (MQOs) from Chapters 3 and 4. Measurement uncertainty (Section 5.6), detectability (Section 5.7), and quantifiability (Section 5.8), are three MQOs that are described in greater detail. Combining an instrument with a measurement technique to ensure the MQOs are achieved is discussed in Section 5.9. Section 5.10 provides information on quality control (QC), and information on data reporting is provided in Section 5.11.

5.2   Ensure Protection of Health and Safety

Health and safety is emphasized as an issue potentially affecting the implementation of MARSAME disposition surveys. The focus of minimizing hazards is shifted away from environmental hazards (e.g., confined spaces, unstable surfaces, heat and cold stress) and towards scenarios where health and safety issues may affect how a disposition survey is designed and performed. Work areas and procedures that present potential safety hazards must be identified and evaluated to warn personnel of potential hazards. Personnel must be trained with regard to potential physical and chemical safety hazards (e.g., inhalation, adsorption, ingestion, injection/puncturing) and the potential for injury (e.g., slips, trips, falls, burns).

A job safety analysis (JSA) should be performed prior to implementing a disposition survey. The JSA offers an organized approach to the task of locating problem areas for material handling safety (OSHA 2002). The JSA should be used to identify hazards and provide inputs for drafting a health and safety plan (HASP). The HASP will address the potential hazards associated with M&E handling and movement and should be prepared concurrently with the survey design. The HASP identifies methods to minimize the threats posed by the potential hazards. The information in the HASP may influence the selection of a measurement technique and disposition survey procedures. Radiation work permits (RWPs) may be established to control access to

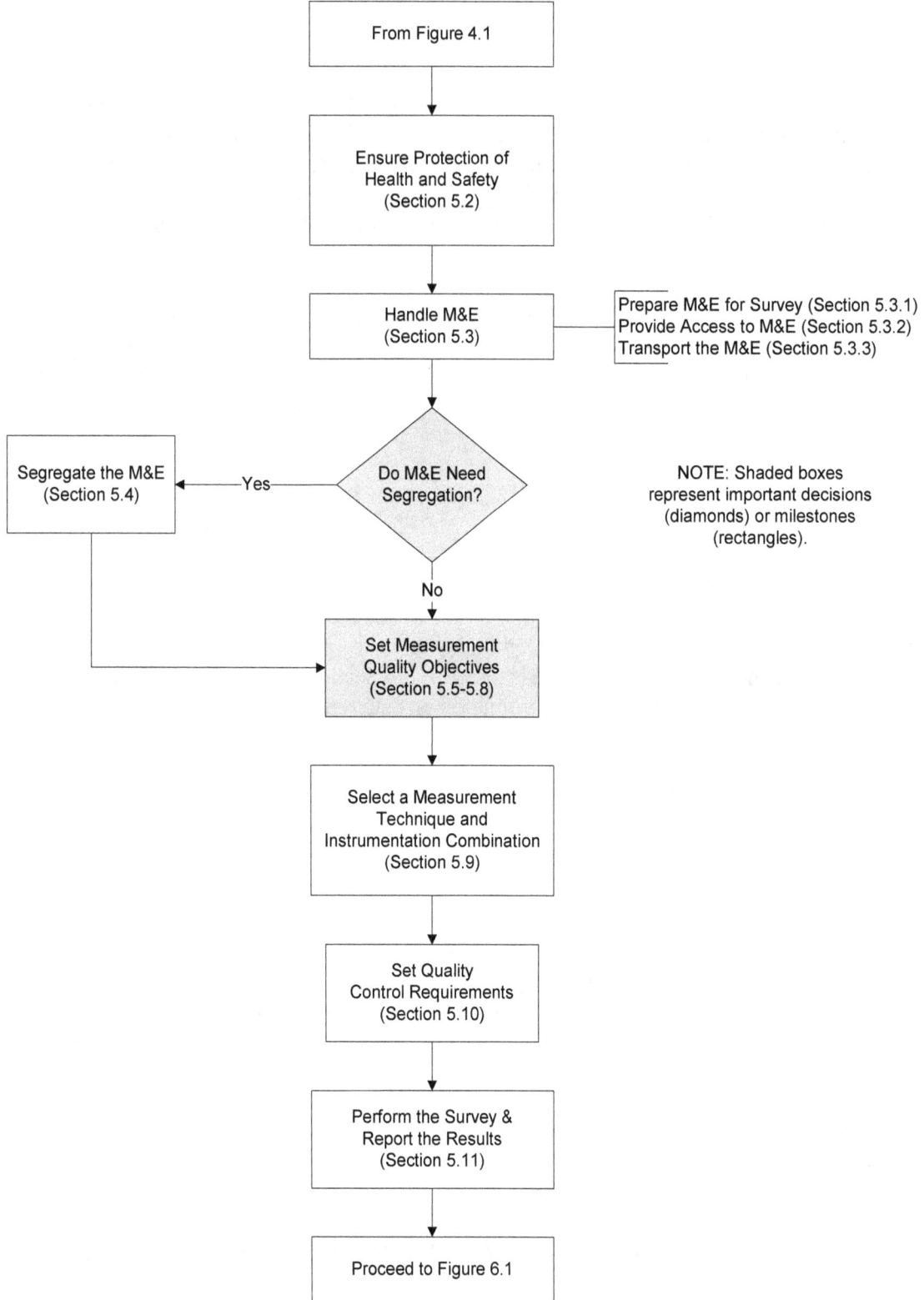

Figure 5.1 Implementation of Disposition Surveys

radiologically controlled areas. RWPs contain requirements from the JSA, such as dosimetry and personal protective equipment (PPE), as well as survey maps illustrating predicted dose rates and related radiological concerns (e.g., removable or airborne radioactivity). Hazard work permits (HWPs) may be used in place of RWPs at sites with primarily physical or chemical hazards. The mineral processing facility concrete rubble example presented in Chapter 8 (see Table 8.9) provides an example of a JSA.

The JSA systematically carries out the basic strategy of accident prevention through the recognition, evaluation, and control of hazards associated with a given job as well as the determination of the safest, most efficient method of performing that job. This process creates a framework for deciding among engineering controls, administrative controls, and PPE for the purpose of controlling or correcting unsafe conditions (Hatch 1978). Examples of these controls include—

- Engineering controls, which are physical changes in processes or machinery (e.g., installing guards to restrict access to moving parts during operation), storage configuration (e.g., using shelves in place of piles or stacks);
- Administrative controls, which are changes in work practices and organization (e.g., restricted areas where it is not safe to eat, drink, smoke, etc.) including the placement of signs to warn personnel of hazards; and
- Personal protective equipment, which are clothing or devices worn by employees to protect against hazards (e.g., gloves, respirator, full-body suits).

Correction measures may incorporate principles of all of the controls listed above. The preferred method of control is through engineering controls, followed by administrative controls, and then personal protective equipment.

Proper handling procedures for hazardous M&E are documented in site-specific health and safety plans. Compliance with all control requirements is mandatory to maintain a safe working environment. Personnel must regard control requirements as a framework to facilitate health and safety, while still taking responsibility for their own well being. Being wary of safety hazards remains an individual responsibility and personnel must be aware of their surroundings at all times in work areas.

5.3   Consider Issues for Handling M&E

Materials and equipment handling is addressed in this document as a process control issue. M&E handling requirements are determined by the final integrated survey design (Section 4.4) and the combination of instrumentation and measurement technique used to perform the survey (Section 5.9). M&E may also require handling to more closely match the assumptions used to develop instrument calibrations used to determine measurement uncertainty (Section 5.6), measurement detectability (see Section 5.7), and measurement quantifiability (Section 5.8). Typically, M&E will be handled to—

- Prepare a measurement grid or arrange M&E to perform a survey,
- Provide access for performing measurements, and

• Transport the M&E to a different location.

5.3.1 Prepare M&E for Survey

Depending on the survey design, or assumptions used to develop the survey design, it may be necessary to prepare the M&E for survey. The amount of preparation required is determined by the DQOs and MQOs, and ranges from identifying measurement locations to adjusting the physical characteristics of the M&E (e.g., disassembly, segregation, physical arrangement).

The performance of a MARSSIM-type survey requires determining the location where the measurements are to be performed. The DQOs will determine the level of effort required to identify, mark, and record measurement locations.

Identifying measurement locations can be problematic because MARSSIM-type surveys recommend samples to be located either randomly (Class 3) or on a systematic grid (Class 1 and Class 2). Class 2 and Class 3 scan-only and in situ surveys do not require 100% of the M&E to be measured, so a method of identifying which portions will be measured is required.

Bulk materials or M&E consisting of many small, regularly shaped objects can be spread out in a uniform layer, and a two-dimensional grid can be superimposed on the surface to identify measurement locations. However, it is virtually impossible to identify random or systematic locations on M&E that consist of relatively few, large, irregularly shaped objects. The reason is that it is virtually impossible to establish a reference grid for these M&E. It is important to note that the objective for random locations is to allow every portion of the survey unit the same opportunity to be measured. Alternatively, the objective of systematic locations is to distribute the measurement locations equally. It is only necessary to establish a reference grid to sufficiently identify the measurement locations to meet the survey objectives.

One way to approximate a reference grid for locating measurements is to establish a grid in the area where the survey will be performed. The M&E to be surveyed are laid out in a single layer within the grid. The grid can then be used to identify measurement locations. Another option for locating measurements involves superimposing a grid on top of the M&E. A net could be laid over the M&E to be surveyed, ropes could be laid over the M&E to form a grid, or lights on a grid could be directed onto the M&E to approximate a grid and identify measurement locations. If measurement locations cannot be identified with a grid, there may be no alternative but to perform biased measurements. Measurements would be performed preferentially in locations more likely to contain radionuclides or radioactivity, based on the results of the initial assessment (IA) (Section 2.5). This process involves professional judgment and may result in overestimating the average radionuclide concentration or level of radioactivity. In all cases, it is important to document the criteria used for identifying measurement locations and to document that these criteria were followed.

Marking measurement locations, once they have been identified, should be done in a way that will not interfere with the measurement. For example, using paint to mark the location of an alpha measurement could end up masking the presence of alpha activity. Using arrows, marking borders, or using an alternate method for marking locations (e.g., encircling with chalk) should be considered for these types of situations.

Recording measurement locations may be required as part of the survey objectives if the measurements may need to be repeated. For example, a large piece of equipment is surveyed prior to use on a decommissioning or cleanup project. If the exact same locations will be surveyed at the completion of the project, it will be necessary to record the measurement locations. Permanent or semi-permanent markings can be used to identify the measurement locations. Video or photographic records of measurement locations can also be used to return to a specific measurement location.

5.3.2   Provide Access

Large pieces of equipment may require special handling considerations. Large, mobile equipment (e.g., front loader, bulldozer, or crane) typically requires a specially trained operator. The operator may need to be available during the disposition survey to provide access to all areas requiring survey (e.g., move the equipment to provide access to the bottom of tires or treads). Other large items may require special equipment (e.g., a crane or lift) to provide access to all areas requiring survey. Special health and safety issues (Section 5.2) may be required to ensure protection of survey personnel from physical hazards (e.g., personnel or items falling from heights, or large items dropping on personnel or equipment). It may be necessary to partially or totally disassemble large pieces of equipment to provide access and ensure measurability.

Piles of M&E may involve special handling precautions. Piles of dispersible M&E (e.g., excavated soil or concrete rubble) may need to be rearranged to match the assumptions used to develop the instrument efficiency. For example, a conical pile of excavated soil may need to be flattened to a uniform thickness to ensure measurability. If the M&E consists of or contains a significant amount of dust, precautions against generating an airborne radiation hazard may be necessary. Because many dust control systems use liquids to prevent the dust from becoming airborne, it may be necessary to account for dust control impacts on measurability of the M&E. For example, adding water to control dust will make it more difficult to measure alpha radioactivity. Piles of scrap may also present other health and safety concerns along with issues related to measurability. Sharp edges, pinch points, and unstable piles are examples of handling problems that may need to be addressed.

Small pieces of M&E may be surveyed individually or combined into groups for survey. Care should be taken when combining items to prevent mixing impacted and non-impacted items, or mixing items with different physical or radiological attributes (see Section 2.2 and Section 5.4). The moving of materials at a given site may require labeling as a quality control measure to ensure M&E movement is tracked and documented. Labeling will help avoid the commingling of impacted and non-impacted materials, and facilitate the staging and storage of impacted and non-impacted M&E in appropriate areas.

5.3.3   Transport the M&E

Identification of impacted and non-impacted areas within a facility will assist in selecting areas for storing, staging, and surveying impacted M&E. In general, impacted M&E should be stored, staged, and surveyed in impacted areas. Care should be taken when moving or handling impacted M&E to prevent the spread of radionuclides to non-impacted areas. M&E in areas with airborne radioactivity issues should be moved to protect the personnel conducting surveys and reduce the possibility of contaminating survey instruments.

Disposition surveys can be performed with the M&E in place, or the M&E can be moved to another location. For example, work areas with high levels of radioactivity may make it difficult or resource intensive to meet the MQOs for measurement detectability (Sections 5.7 and 7.5) or quantifiability (Sections 5.8 and 7.6). Moving the M&E to areas with lower levels of radioactivity will help reduce radiation exposure for personnel conducting surveys and facilitate meeting the survey objectives.

5.4   Segregate the M&E

The purpose of segregation is to separate M&E based on the estimated total measurement uncertainty, ease of handling, and disposition options. Segregation is based on the physical and radiological attributes determined during the IA (Chapter 2), not only on radionuclide concentrations or radiation levels (i.e., classification).

In general, segregation based on measurement uncertainty should consider the physical and radiological attributes that affect efficiency (i.e., geometry and fluence rate). M&E with simple geometries, such as drums (cylinder) and flat surfaces (plane), should be separated from M&E with complex geometries. Fluence rate is affected by location of the radioactivity (i.e., surficial or volumetric) as well as surface effects (e.g., rough or smooth), density of the M&E, and type and energy of radiation. High fluence rates are associated with surface radioactivity with high energy on flat smooth surfaces made from materials with high atomic number (due to increased backscatter). Volumetric activity, shielded surfaces, alpha or low energy or beta radiations, irregular shapes, or rough surfaces can cause lower fluence rates. All of these factors should be considered when segregating M&E.

Segregation of M&E should be performed conservatively. This means that the user should separate M&E when they are not obviously similar. It is always possible to combine M&E but it is not always practical or possible, to separate M&E once they have been combined. For example, consider a facility where all the waste materials (e.g., paper, wood, metal, broken equipment) are combined into a single "trash pile." When the planning team considers different measurement methods and disposition options, they identify an innovative measurement method that only applies to non-ferrous scrap metal. This would allow for recycling of these materials with significant cost recovery as opposed to disposal. If the cost of re-segregating the M&E is not offset by the value of recycling these materials, it may not be practical to segregate the non-ferrous metals.

It is important to note that segregation does not require physical separation. Consider a generic large box geometry, such as an empty shipping container or railroad car. The large, flat sides could be considered separate survey units from the corners. Therefore, separate surveys would be designed for the corners and the sides even though the entire railroad car would remain intact throughout implementation of the disposition survey. Alternatively (or additionally), obvious flaws, corrosion areas, or damaged areas could be segregated from the areas in good condition. Even if the entire object is eventually surveyed using a single in situ measurement (e.g., in situ gamma spectroscopy) it is important to segregate the M&E (at least conceptually) so an adequate evaluation of alternate measurement methods can be performed (Section 5.9).

Handling of M&E during disposition surveys should also be considered during segregation (Section 5.3). Physical characteristics of the M&E should be considered when segregating based on handling requirements. Small, light items are easier to move and gain access to all surfaces than large, massive items. M&E that will require preparation (e.g., disassembly, crushing, chopping) prior to survey should be segregated from M&E that can be surveyed in their present form. Disposition options should also be considered when segregating M&E. M&E that can be reused or recycled should be segregated from M&E that is being considered for disposal. Selection of disposition options is discussed in Section 2.4.

5.5   Set Measurement Quality Objectives

A number of terms with specific statistical meanings are used in this and subsequent sections. These terms are defined in Chapter 7. The concept of Measurement Quality Objectives (MQOs) and in particular the required measurement method uncertainty is introduced in Section 3.8.1. These ideas are discussed in greater detail in the *Multi-Agency Radiological Laboratory Analytical Protocols* manual (MARLAP 2004) Chapter 3 and Appendix C. While MARLAP is focused on radioanalytical procedures, these concepts are applicable on a much broader scale and will be used in MARSAME to guide the selection of measurement methods for disposition surveys for materials and equipment.

Section 4.2 discusses the DQO process for developing statistical hypothesis tests for the implementation of disposition decision rules using measurement data. These concepts are further developed in Chapter 7. This includes formulating the null and alternative hypotheses, defining the gray region using the action level and discrimination limit, and setting the desired limits on potential Type I and Type II decision error probabilities that a decision-maker is willing to accept for project results. Decision errors are possible, at least in part, because measurement results have uncertainties. Because DQOs apply to both sampling and measurement activities, method performance characteristics specifically for the measurement process of a particular project are needed from a measurement perspective. These method performance characteristics (Section 3.8) are the measurement quality objectives (MQOs).

DQOs define the performance criteria that limit the probabilities of making decision errors by—

- Considering the purpose of collecting the data,
- Defining the appropriate type of data needed, and
- Specifying tolerable probabilities of making decision errors.

DQOs apply to both sampling and measurement activities. MQOs can be viewed as the measurement portion of the overall project DQOs (Section 3.8). MQOs are—

- The part of the project DQOs that apply to the measured result and its associated uncertainty,
- Statements of measurement performance objectives or requirements for a particular measurement method performance characteristic (e.g., measurement method uncertainty and detection capability),
- Used initially for the selection and evaluation of measurement methods, and
- Used subsequently for the ongoing and final evaluation of the measurement data.

Measurement method uncertainty refers to the *predicted* uncertainty of a measured value that would be calculated if the method were applied to a hypothetical sample with a specified concentration. Measurement method uncertainty is a characteristic of the measurement method and the measurement process. Measurement uncertainty, as opposed to sampling uncertainty, is a characteristic of an individual measurement.

The true measurement method standard deviation, σ_M, is a theoretical quantity and is never known exactly, but it may be estimated using the methods described in Section 7.4. The estimated value of σ_M will be denoted here by σ_M and called the "measurement method uncertainty." The measurement method uncertainty, when estimated by uncertainty propagation, is the predicted value of the combined standard uncertainty ("one-sigma" uncertainty) of the measurement for material with concentration equal to the upper bound of the gray region (UBGR). Note that the term "measurement method uncertainty" and the symbol u_M actually apply not just to the measurement method but also to the entire measurement process: it should include uncertainties in how the measurement method is actually implemented. This definition of measurement method uncertainty is independent of the null hypothesis and applies to both Scenario A and Scenario B.

The true standard deviation of the measurement method, σ_M, is unknown, but the required measurement method uncertainty, σ_{MR}, is intended to be an upper bound for σ_M. In practice, σ_M is actually used as an upper bound for the method uncertainty, σ_M, which is an estimate of σ_M. Therefore, the estimated value of σ_{MR} will be called the "required measurement method uncertainty" and denoted by u_{MR}. Note that when referring to a theoretical population standard deviation, the symbol σ is used. Estimates of the value of σ in specific cases are denoted by the symbol u, for uncertainty. An uncertainty is not a standard deviation because its evaluation involves concepts from metrology as well as statistics, however, in many cases it is treated mathematically as if it were a standard deviation.

The principal MQOs in any project will be defined by the required measurement method uncertainty, u_{MR}, at and below the UBGR and the relative required measurement method uncertainty, φ_{MR}, at and above the UBGR:

$$\varphi_{MR} = \frac{u_{MR}}{UBGR} \tag{5-1}$$

Section 7.7 provides the rationale and guidance for establishing project-specific MQOs for controlling u_M.

> Note: When making decisions about individual measurement results, u_{MR} usually should be about 0.3Δ, and when making decisions about the mean of several measurement results, u_{MR} usually should be about 0.1Δ, where Δ is the width of the gray region, $\Delta = UBGR - LBGR$. These rules of thumb require certain assumptions as discussed in Chapter 7.

This check of measurement quality against the required measurement method uncertainty relies on having realistic estimates of the measurement uncertainty. Often reported measurement uncertainties are underestimated, particularly if they are confined to the estimated Poisson counting uncertainty (Section 7.8). Tables of results are sometimes presented with a column listing simply "±" without indicating how these numbers were obtained. Often it is found that they simply represent the square root of the number of counts obtained during the measurement. The method for calculating measurement uncertainty, approved by both the International Organization for Standardization (ISO) and the National Institute of Standards and Technology (NIST) is discussed in the next section.

5.6   Determine Measurement Uncertainty

This section discusses the evaluation and reporting of measurement uncertainty. Measurements always involve uncertainty, which must be considered when measurement results are used as part of a basis for making decisions. Every measured and reported result should be accompanied by an explicit uncertainty estimate. One purpose of this section is to give users of data an understanding of the causes of measurement uncertainty and of the meaning of uncertainty statements; another is to describe procedures that can be used to estimate uncertainties. Much of this material is derived from MARLAP Chapter 19.

In 1980, the Environmental Protection Agency published a report entitled "Upgrading Environmental Radiation Data," which was produced by an ad hoc committee of the Health Physics Society (EPA 1980). Two of the recommendations of this report were that–

1. Every reported measurement result (x) should include an estimate of its overall uncertainty (u_x) that is based on as nearly a complete assessment as possible, and
2. The uncertainty assessment should include every significant source of inaccuracy in the result.

The concept of traceability is also defined in terms of uncertainty. Traceability is defined as the "property of the result of a measurement or the value of a standard whereby it can be related to stated references, usually national or international standards, through an unbroken chain of comparisons all having stated uncertainties" (ISO 1996). Thus, to realistically make the claim that a measurement result is "traceable" to a standard, there must be a chain of comparisons (each measurement having its own associated uncertainty) connecting the result of the measurement to that standard.

This section considers only the measurement standard deviation, σ_M. Reducing sampling standard deviation, σ_S, by segregating M&E was discussed in Section 5.4. The sampling standard deviation is often larger than the measurement standard deviation. Although this statement may be true in some cases, this is not an argument for failing to perform a full evaluation of the measurement uncertainty, u_M, to evaluate σ_M. A realistic estimate of the measurement uncertainty is one of the most useful data quality indicators for a result (Section 3.8).

Although the need for reporting uncertainty has been recognized, often it consists of only the estimated component due to Poisson counting statistics. This is done because it is easier than a full uncertainty analysis, but it can be misleading because it is at best only a lower bound on the uncertainty and may lead to incorrect decisions based on overconfidence in the measurement. Software is available to perform the mathematical operations for uncertainty evaluation and propagation, eliminating much of the difficulty in implementing the mathematics of uncertainty calculations. There are several examples of such software (McCroan 2006, GUM Workbench 2006, Kragten 1994, Vetter 2006).

The methods, terms, and symbols recommended by MARSAME for evaluating and expressing measurement uncertainty are described in the *Guide to the Expression of Uncertainty in Measurement*, or GUM, which was published by ISO (ISO 1995). The ISO methodology is summarized in the NIST Technical Note TN-1297 (NIST 1994). The details of applying this methodology are given in Section 7.4 and 7.8.

5.7   Determine Measurement Detectability

This section is a summary of issues related to measurement detection capabilities. Much of this material is derived from the MARLAP Chapter 20. More detail may be found in Section 7.9. Radioactivity measurements may involve material with very small amounts of the radionuclide of interest. Measurement uncertainty often makes it difficult to distinguish such small amounts from zero. Therefore, an important MQO of a measurement process is its detection capability, which is usually expressed as the smallest concentration of radioactivity that can be reliably distinguished from zero. Effective project planning requires knowledge of the detection capabilities of the measurement method that will be or could be used. This section explains an MQO called the minimum detectable concentration (MDC) and describes radioactivity detection capabilities, as well as methods for calculating it.

The method most often used to make a detection decision about radiation or radioactivity involves the principles of statistical hypothesis testing. It is a specific example of a Scenario B hypothesis testing procedure described in Section 7.2.4. To "detect" the radiation or radioactivity requires a decision on the basis of the measurement data that the radioactivity is present. The detection decision involves a choice between the null hypothesis (H_0): There is no radiation or radioactivity present (above background), and the alternative hypothesis (H_1): There is radiation or radioactivity present (above background). Making the choice between these hypotheses requires the calculation of a critical value. If the measurement result exceeds this critical value, the null hypothesis is rejected and the decision is that radiation or radioactivity is present. If the null hypothesis is rejected when it is true, a Type I decision error is made. In this case, a sample with no additional radiation or radioactivity above background is deemed to actually contain such. The rate at which this decision error occurs is denoted by α. The critical value depends directly on the value of α. The planning team has to make a choice about the establishment of the acceptable rate for mischaracterizing a background count for a real detection count, i.e., establish a Type I error rate, α, for mistakenly deciding a background measurement is really a detection of additional radiation or radioactivity.

Radioactivity measurements are often recorded as counts or count rates. Radiation exposure measurements are often expressed in different terms, e.g., ionization current. The term "instrument signal" is used in the following so that all types of measurement are included.[1]

The relationship between the critical value of the net instrument signal (or count), S_C, and the minimum detectable net instrument signal, S_D, is shown in Figure 5.2. More detail on the calculation of the minimum detectable value of the net instrument signal (or count), S_D, is given in Section 7.9. The net instrument signal obtained for a blank sample will usually be distributed around zero as shown. Occasionally, a net instrument signal above S_C may be obtained by chance. The probability that this happens is controlled by the value of α, the Type I decision error rate, shown as the lightly shaded area in Figure 5.2. Smaller values of α result in larger values of S_C and vice versa. The minimum detectable value of the net instrument signal S_D is that value of the mean net instrument signal that results in a detection decision with probability $1 - \beta$. That is, there is only a probability β, the Type II decision error rate shown as the more darkly shaded area in Figure 5.2, of yielding an observed instrument signal less than S_C. Smaller values of β result in larger values of S_D and vice versa. The planning team has to decide what an acceptable value of β should be, i.e. when additional radiation or radioactivity is present, at what rate is it acceptable to mistakenly attribute the measurement result to only background. Note that S_D depends on the values of both α and β.

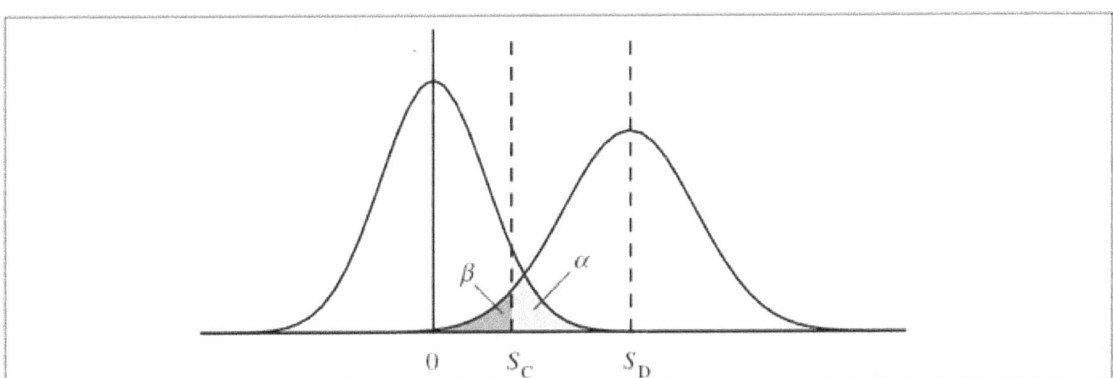

Figure 5.2 The Critical Value (S_C) and the Minimum Detectable Value (S_D) of the Net Instrument Signal (or Count)

The MDC is usually obtained from the minimum detectable value of the net instrument signal (or count), S_D. The MDC is by definition an estimate of the true concentration of the radiation or radioactivity required to give a specified high probability that the measured response will be

---

[1] "Net instrument signal," is used here as a general term, because many radiation-detection instruments may have output other than "counts" (e.g., current for ionization chambers). In cases where the instrument output *is* in counts, the term "net counts" can be substituted for the term "net instrument signal."

greater than S_C. The common practice of comparing a measured concentration to the MDC, instead of S_C, to make a detection decision is incorrect.

To calculate the MDC, the minimum detectable value of the net instrument signal, S_D, must first be converted to the detectable value of the net instrument signal rate (often a count rate), S_D/t_S (s^{-1}), where t_S is the duration of the measurement in seconds. This in turn must be divided by the instrument efficiency, ε (s^{-1}/Bq) to get the minimum detectable activity, y_D. Finally, the minimum detectable activity can be divided by the sample volume or mass to obtain the MDC. At each stage in this process, additional uncertainty may be introduced by the uncertainties in time, efficiency, volume, mass, etc. Prudently conservative values of these factors should be used so that the desired detection power, $1 - \beta$, at the MDC is maintained. Another approach would be to recognize that y_D itself has an uncertainty which can be calculated using the methods of Section 7.8. Thus, any input quantity that is used to convert from S_D to y_D that has significant uncertainty, can be incorporated to assess the overall uncertainty in the MDC.

MARSAME recommends that when a detection decision is required, it generally should be made by comparing the net instrument signal to its corresponding critical value. Expressions for S_C and S_D should be chosen that are appropriate for the structure and statistics of the measurement process. An appropriate background should be used to predict the instrument signal produced when there is no radioactivity present in the sample. The MDC should be used only as a MQO for the measurement method. To make a detection decision, a measurement result should be compared to S_C and never to the MDC. Finally, additional discussion of the calculation of the MDCs is given in Section 7.9.

5.8   Determine Measurement Quantifiability

This section discusses issues related to measurement quantifiability. Much of this material is derived from the MARLAP Chapter 20.

Action levels are frequently stated in terms of a quantity or concentration of radioactivity, rather than in simply in terms of whether radioactivity is detected. In these cases, project planners may need to know the quantification capability of a measurement method, or its capability for precise measurement. The quantification capability is expressed as the smallest concentration of radiation or radioactivity that can be measured with a specified relative measurement standard deviation. This section explains an MQO called the *minimum quantifiable concentration* (MQC), which may be used to describe quantification capabilities.

The MQC, y_Q, is defined as the concentration at which the measurement process gives results with a specified relative standard deviation, $1/k_Q$, where k_Q is usually chosen to be 10 for comparability. Thus, the MQC is generally the concentration at which the relative measurement uncertainty is 10%.

Historically much attention has been given to the detection capabilities of radiation and radioactivity measurement processes, but less attention has been given to quantification capabilities. For some projects, quantification capability may be a more relevant issue. For example, suppose the purpose of a project is to determine whether the ^{226}Ra concentration on material at a site is below an action level. Because ^{226}Ra can be found in almost any type of naturally occurring material, it may be assumed to be present in every sample, making detection decisions unnecessary. The MDC of the measurement process obviously should be less than the action level, but a more important question is whether the MQC is less than the action level.

A common practice in the past has been to select a measurement method based on the MDC, which is defined in Section 5.7 and Section 7.5. For example, MARSSIM says:

> During survey design, it is generally considered good practice to select a measurement system with an MDC between 10-50% of the DCGL [action level].

Such guidance implicitly recognizes that for cases when the decision to be made concerns the mean of a population that is represented by multiple measurements, criteria based on the MDC may not be sufficient and a somewhat more stringent requirement is needed. The requirement that the MDC (approximately 3–5 times σ_M) be 10% to 50% of the action level is tantamount to requiring that σ_M be 0.02 to 0.17 times the action level. In other words, the relative measurement standard deviation should be approximately 10% at the action level. However, the concentration at which the relative measurement standard deviation is 10% of the MQC when k_Q assumes its conventional value of 10. Thus, a requirement that is often stated in terms of the MDC may be more naturally expressed in terms of the MQC (e.g., by saying that the MQC should not exceed the action level). Further details on calculating the MQC can be found in Section 7.10.

5.9   Select a Measurement Technique and Instrumentation Combination

The combination of a measurement technique with instrumentation, or measurement method, is selected to implement a disposition survey design based on the ability to meet the MQOs (see Sections 3.3.2 and 5.5). Note that measurement techniques are separate from survey designs. The relationship between the two is explained in Sections 5.9.1.1, 5.9.1.2, and 5.9.1.3. A realistic determination of the measurement method uncertainty (Section 5.6) is critical to demonstrating a method meets the MQOs. Other considerations when selecting a measurement method include—

- Health and safety concerns (Section 5.2),
- M&E handling issues (Section 5.3),
- Segregation (Section 5.4),
- Measurement detectability (Section 5.7), and
- Measurement quantifiability (Section 5.8).

The measurement techniques discussed in Section 5.9.1 all can be classified as scanning measurements (constant motion involved in the surveying procedure) or fixed measurements (surveying discrete locations without motion). Fixed measurements consist of in situ measurements (the detection instrument moves to the M&E or measures the M&E in its entirety), and sampling (removing part of the M&E for separate analysis).

Instrumentation for performing radiological measurements is varied and constantly being improved. Section 5.9.2 provides an overview of some commonly used types of instruments and how they might be applied to disposition surveys. The purpose of the discussions on instrumentation is not to provide an exhaustive list of acceptable instruments, but to provide examples of how instrumentation and measurement techniques can be combined to meet the survey objectives. Additional information on instrumentation is found in Appendix D. Section 5.9.3 provides information on selecting a combination of measurement technique and instrumentation to provide a measurement method. It is necessary that the selected measurement method meet the MQOs established during survey design (Section 3.8). Selection of instrumentation can be an iterative process. The appropriate MQO (e.g., MDC, MQC) may not be attainable with some measurement methods. In some cases selection of a different instrument may be all that is necessary, while in other cases a different measurement technique or an entirely different measurement method will need to be considered.

5.9.1   Select a Measurement Technique

A measurement technique describes how a measurement is performed. The detector can be moved relative to the M&E (i.e., scanning), used to perform static measurements of the M&E in place (i.e., in situ or direct measurements), or some representative portion of the M&E can be taken to a different location for analysis (i.e., sampling). These three measurement techniques are described in Sections 5.9.1.1, 5.9.1.2, and 5.9.1.3, respectively. Smears are a type of sampling, where a portion of the removable radioactivity is collected (Section 5.9.1.4).

5.9.1.1   Scanning Techniques

Scanning techniques generally consist of moving portable radiation detectors at a specified distance above the physical surface of a survey unit at some specified speed to meet the MQOs. Alternatively, the M&E can be moved past a stationary instrument at a specified distance and speed (e.g., conveyorized systems or certain portal monitors). Scanning techniques can be used alone to demonstrate compliance with a disposition criterion (i.e., scan-only surveys, Section 4.4.1), or combined with sampling in a MARSSIM-type survey design (Section 4.4.3). Scanning is used in MARSSIM-type surveys to locate radiation anomalies by searching for variations in readings, indicating gross radioactivity levels that may require further investigation or action. Scanning techniques can more readily provide thorough coverage of a given survey unit and are often relatively quick and inexpensive to perform. Scanning often represents the simplest and most practical approach for performing MARSAME disposition surveys.

Maintaining the specified distance and speed during scanning can be difficult, especially with hand-held instruments and irregularly shaped M&E. Variations in source-to-detector distance and scan speed can result in increased total measurement method uncertainty. Determining a calibration function for situations other than surficial radionuclides uniformly distributed on a plane can be complicated, and may also contribute to the total measurement method uncertainty.

5.9.1.2   In Situ Measurements

In situ measurements are taken by placing the instrument in a fixed position at a specified distance[2] from the surface of a given survey unit of M&E and taking a discrete measurement for

---

[2]Measurements at several distances may be needed. Near-surface or surface measurements provide the best indication of the size of the area of elevated radionuclide concentrations or radioactivity, and are useful for model

a pre-determined time interval. Single in situ measurements can be performed on individual objects or groups of M&E. Multiple in situ measurements can be combined to provide several different views of the same object, or used to provide measurements for a specified fraction of the M&E. In situ measurements can also be performed at random or systematic locations, combined with scanning measurements, in a MARSSIM-type survey design. In situ measurements are used generally to provide an estimate of the average radionuclide concentration or level of radioactivity over a certain area or volume defined by the calibration function.

Determining a calibration function for situations other than radionuclides uniformly distributed on a plane or through a regularly shaped volume (e.g., a disk or cylinder) can be complicated and may contribute to the total measurement method uncertainty. In situ techniques are not typically used to identify small areas or volumes of elevated radionuclide concentration or activity.

5.9.1.3   Sampling

Sampling consists of removing a portion of the M&E for separate laboratory analysis. This measurement technique, when combined with laboratory analysis, surpasses the detection capabilities of measurement techniques that may be implemented with the M&E left in place. This facilitates the analysis of complicated radioisotope mixtures, difficult-to-measure radionuclides, and extremely low concentrations of residual radioactivity. Sampling is used to provide an estimate of the average radionuclide concentration or level of radioactivity for a specified area or volume. The sample locations may be located using a random or systematic grid, depending on the objectives of the survey. Sampling is typically combined with scanning in a MARSSIM-type survey design, where sampling is used to evaluate the average concentration or activity and scanning is used to identify small areas or volumes with elevated radionuclide concentrations or radioactivity. Sampling may also be used to validate data collected using other measurement techniques.

Sampling (combined with laboratory analysis) typically requires the most time for data generation of all the surveying techniques discussed in this chapter and is often the most expensive. Sampling is not an effective technique for identifying small areas or volumes of elevated radionuclide concentrations or levels of radioactivity.

5.9.1.4   Smears

Smears are used to provide an estimate of removable surface radioactivity. Smears are also referred to as smear tests, swipes, or wipes. Smears are a type of sample where a filter paper or other substance is used to wipe a specified area of a surface. The filter paper or other substance is then tested for the presence of radioactivity.

Individual smear results collected by hand usually have a high uncertainty because the fraction of surface radioactivity transferred to the smear is unknown and variable and the surface area

---

implementation. Gamma measurements at one meter provide a good estimate of potential direct external exposure (MARSSIM 2002).

covered by the smear is variable. In addition, the results may vary with time due to environmental factors or interactions of surface activity with the surface itself. Action levels for removable activity based on smear measurements may include assumptions about the fraction of surface radioactivity transferred to a single smear or specify a surface area to be smeared. For example, DOT surface contamination guidelines assume that 10% of the surface radioactivity is transferred to a single smear. Also, DOE Order 5400.5 Figure IV-1 (DOE 1993) provides instructions for using smears to measure removable radioactivity. These instructions specify wiping an area of 100 cm^2 with a dry filter or soft absorbent paper while applying moderate pressure. The instructions also discuss how to account for minor variations from the procedure.

Determining a collection or removal fraction for smears can be complicated. The uncertainty and variability in the removal fraction estimate and surface area smeared can result in increased total measurement method uncertainty. Using a template or cutout with a known area can help control the variability in the area covered by a smear. Using a tool that applies consistent pressure while collecting smears can reduce the variability in the fraction of radioactivity removed. Implementing a protocol for preparing surfaces and sorting materials prior to survey can reduce variability in surface textures and conditions resulting in lower variability in smear collection conditions.

5.9.2   Select Instrumentation

This section briefly describes the typical types of instrumentation that may be used to conduct MARSAME disposition surveys. More detailed information relevant to each type of instrument and measurement method is provided in Appendix D.

5.9.2.1   Hand-Held Instruments

Hand-held instruments typically are composed of a detection probe (utilizing a single detector) and an electronic instrument to provide power to the detector and to interpret data from the detector to provide a measurement display. They may be used to perform scanning surveys or in situ measurements. Hand-held measurements also allow the user the flexibility to constantly vary the source-to-detector geometry for obtaining data from difficult-to-measure areas.

5.9.2.2   Volumetric Counters (Drum, Box, Barrel, 4-π Counters)

Box counting systems typically consist of a counting chamber, an array of detectors configured to provide 4-π counting geometry, and microprocessor-controlled electronics that allow programming of system parameters and data-logging. Volumetric counters are used to perform in situ measurements on entire pieces of small M&E.

5.9.2.3   Conveyorized Survey Monitoring Systems

Conveyorized survey monitoring systems automate the routine scanning of M&E. Conveyorized survey monitoring systems typically perform scanning surveys by moving M&E through a detector array on a conveyor belt. Conveyorized survey monitoring systems may be utilized to take in situ measurements by halting the conveyor and continuing the measurement to improve the detection efficiency.

5.9.2.4   In Situ Gamma Spectroscopy

Some in situ gamma spectroscopy (ISGS) systems consist of a small hand-held unit that incorporates the detector and counting electronics into a single package. Other ISGS systems consist of a semiconductor detector, a cryostat, a multi-channel analyzer (MCA) electronics package that provides amplification and analysis of the energy pulse heights, and a computer system for data collection and analysis. ISGS systems typically are applied to perform in situ measurements, but they may be incorporated into innovative detection equipment set-ups to perform scanning surveys.

5.9.2.5   Portal Monitors

Portal monitors utilize a fixed detector array through which M&E are passed to typically perform scanning surveys (objects may also remain stationary within the detector array to perform in situ measurements). Portal monitors typically are used to perform scanning surveys of vehicles.[3] In situ measurements may be utilized with portal monitors by taking motionless measurements to improve the detection efficiency.

5.9.2.6   Laboratory Analysis

Laboratory analysis consists of analyzing a portion or sample of the M&E. The laboratory will generally have recommendations or requirements concerning the amount and types of samples that can be analyzed for radionuclides or radiations. Communications should be established between the field team collecting the samples and the laboratory analyzing the samples. More information on sampling is provided in Section 5.9.1.3. Laboratory analyses can be developed for any radionuclide with any material, given sufficient resources. Laboratory analyses typically require more time to complete than field analyses. The laboratory may be located onsite or offsite. The quality of laboratory data typically is greater than data collected in the field because the laboratory is better able to control sources of measurement method uncertainty. The planning team should consider the resources available for laboratory analysis (e.g., time, money), the sample collection requirements or recommendations, and the requirements for data quality (e.g., MDC, MQC) during discussions with the laboratory.

5.9.3   Select a Measurement Method

Table 5.1 and Table 5.2 illustrate the potential applications and associated size restrictions for combinations of the instrument and measurement techniques discussed in Sections 5.9.1 and 5.9.2, respectively. Sampling followed by laboratory analysis is not included in these tables, but is considered "GOOD" for all applications. Please note the following qualifiers:

---

[3] Specialized vehicle monitors are available that monitor rates of change in ambient background to account for differences in vehicles being scanned to improve measurement detectability.

GOOD	The measurement technique is well-suited for performing this application
FAIR	The measurement technique can adequately perform this application
POOR	The measurement technique is poorly suited for performing this application
NA	The measurement technique cannot perform this application
Few	A relatively small number, usually three or less
Many	A relatively large number, usually more than three

Table 5.1 illustrates that most measurement techniques can be applied to almost any M&E and type of radioactivity. The quantity of M&E to be surveyed becomes a major factor for the selection of measurement instruments and techniques described in this chapter. Hand-held measurements and techniques generally are the most efficient technique for surveying small quantities of M&E.

Table 5.1 Potential Applications for Instrumentation and Measurement Technique Combinations

Radiation Type	Hand-Held Instruments	Volumetric Counters	Portal Monitors	In Situ Gamma Spectroscopy	Conveyorized Survey Monitoring Systems
In Situ Measurements					
Alpha	FAIR	FAIR	POOR	NA	FAIR
Beta	GOOD	FAIR	FAIR	NA	GOOD
Photon	GOOD	GOOD	GOOD	GOOD	GOOD
Neutron	GOOD	FAIR	GOOD	NA	GOOD
Scanning Surveys					
Alpha	POOR	NA	POOR	NA	POOR
Beta	GOOD	NA	FAIR	NA	FAIR
Photon	GOOD	NA	GOOD	GOOD	GOOD
Neutron	FAIR	NA	FAIR	NA	FAIR

Table 5.2 Survey Unit Size and Quantity Restrictions for Instrumentation and Measurement Technique Combinations

Size of Items	Number of Survey Units or Items	Hand-Held Instruments	Volumetric Counters	Portal Monitors	In Situ Gamma Spectroscopy	Conveyorized Survey Monitoring Systems
In Situ Measurements						
> 10 m³	Few	GOOD	NA	FAIR	GOOD	POOR
	Many	POOR	NA	FAIR	GOOD	POOR
1 to 10 m³	Few	GOOD	FAIR	FAIR	GOOD	FAIR
	Many	POOR	FAIR	FAIR	GOOD	FAIR
< 1 m³	Few	GOOD	GOOD	POOR	GOOD	GOOD
	Many	FAIR	GOOD	POOR	GOOD	GOOD

Table 5.2 Survey Unit Size and Quantity Restrictions for Instrumentation and Measurement Technique Combinations (Continued)

Size of Items	Number of Survey Units or Items	Hand-Held Instruments	Volumetric Counters	Portal Monitors	In Situ Gamma Spectroscopy	Conveyorized Survey Monitoring Systems
Scanning Surveys						
> 10 m³	Few	GOOD	NA	GOOD	FAIR	POOR
	Many	FAIR	NA	GOOD	FAIR	POOR
1 to 10 m³	Few	GOOD	NA	FAIR	FAIR	FAIR
	Many	FAIR	NA	FAIR	FAIR	FAIR
< 1 m³	Few	GOOD	NA	POOR	FAIR	GOOD
	Many	GOOD	NA	POOR	FAIR	GOOD

Facilities that conduct routine surveys on substantial quantities of specific types of M&E may benefit financially from investing in measurement instruments and techniques that require less manual labor to conduct disposition surveys. For example, it will require significantly more time for a health physics technician to survey a toolbox of tools and equipment used in a radiologically controlled area using hand-held surveying techniques and instruments than the time to complete the surveying using a box counting system. Use of such automated systems will also reduce the potential for ergonomic injuries, and attendant costs, associated with routine, repetitive surveys performed using hand-held instruments.

Hand-held surveying remains the more economical choice for a small quantity of tools and toolboxes, but as the quantity of tools and toolboxes increases, the cost of a box counting system becomes an increasingly worthwhile investment to reduce manual labor costs associated with surveying. Note that some M&E have no survey design options that are described as "GOOD" in these two tables (e.g., a large quantity of M&E impacted with residual alpha radioactivity with survey unit sizes greater than 10 m³). The planning team should revisit earlier DQO selections to see if a different approach is more acceptable (e.g., review selection of disposition options in Section 2.4).

Each type of measurement technique has associated advantages and disadvantages, some of which are summarized in Table 5.3. All the measurement techniques described in this table include source-to-detector geometry and sampling variability as common disadvantages.

Table 5.3 Advantages and Disadvantages of Instrumentation and Measurement Technique Combinations

Instrument	Measurement Technique	Advantages	Disadvantages
Hand-Held Instruments	In Situ	• Generally allows flexibility in media to be measured • Detection equipment is usually portable • Detectors are available to efficiently measure alpha, beta, gamma, x-ray, and neutron radiation • Generally acceptable for performing measurements in difficult-to-measure areas • Measurement equipment is relatively low cost • May provide a good option for small quantities of M&E	• Requires a relatively large amount of manual labor as a surveying technique; may make surveying large quantities of M&E labor-intensive • Detector windows may be fragile • Most do not provide nuclide identification
Hand-Held Instruments	Scanning	• Generally allows flexibility in media to be measured • Detection equipment is usually portable • Detectors are available to efficiently measure beta, gamma, x-ray, and neutron radiation • Generally good for performing measurements in difficult-to-measure areas • Measurement equipment is relatively low cost • May provide a good option for small quantities of M&E	• Requires a relatively large amount of manual labor as a surveying technique; may make surveying large quantities of M&E labor-intensive • Detector windows may be fragile • Most do not provide nuclide identification • Incorporates more potential sources of uncertainty than most instrument and measurement technique combinations • Potential ergonomic injuries and attendant costs associated with repetitive surveys.
Hand-Held Instruments	Smear	• Only measurement technique for assessing removable radioactivity • Removable radioactivity can be transferred and assessed in a low background counting area.	• Instrument background may not be sufficiently low. • Detectors with counting sensitive region larger than the smear surface area may require counting adjustments to account for inherent backgrounds associated with other media located under the detector sensitive region.

Table 5.3 Advantages and Disadvantages of Instrumentation and Measurement Technique Combinations (Continued)

Instrument	Measurement Technique	Advantages	Disadvantages
Volumetric Counters	In situ	• Able to measure small items • Designs are available to efficiently measure gamma, x-ray, and alpha radiation • Requires relatively small amount of labor • May be cost-effective for measuring large quantities of M&E	• May not be suited for measuring radioactivity in difficult-to-measure areas • Size of instrumentation may discourage portability
Portal Monitors	In situ	• Able to measure large objects • Designs are available to efficiently measure gamma, x-ray, and neutron radiation • Requires relatively small amount of labor • May be cost-effective for measuring large quantities of M&E	• Not ideal for measuring alpha or beta radioactivity • May not be ideal for measuring radioactivity in difficult-to-measure areas • Size of detection equipment may discourage portability
Portal Monitors	Scanning	• Able to measure large objects • Efficient designs available for gamma, x-ray, and neutron radiation • Residence times generally are short • May not require objects to remain stationary during counting • Requires relatively small amount of labor • May be cost-effective for measuring large quantities of M&E	• Not ideal for measuring alpha or beta radioactivity • Source geometry is an important consideration • May not be ideal for measuring radioactivity in difficult-to-measure areas • Size of detection equipment may discourage portability
In Situ Gamma Spectroscopy (ISGS)	In situ	• Provides quantitative measurements with flexible calibration • Generally requires a moderate amount of labor • May be cost-effective for measuring large quantities of M&E	• Instrumentation may be expensive and difficult to set up and maintain • May require liquid nitrogen supply (with ISGS semiconductor systems) • Size of detection equipment may discourage portability
In Situ Gamma Spectroscopy (ISGS)	Scanning	• Provides quantitative measurements with flexible calibration • Generally requires a moderate amount of labor • May be cost-effective for measuring large quantities of M&E	• Instrumentation may be expensive and difficult to set up and maintain • May require liquid nitrogen supply (with ISGS semiconductor systems) • Size of detection equipment may discourage portability

Table 5.3 Advantages and Disadvantages of Instrumentation and Measurement Technique Combinations (Continued)

Instrument	Measurement Technique	Advantages	Disadvantages
Conveyorized Survey Monitoring Systems	In situ	• Requires relatively small amount of labor after initial set up • May be cost-effective for measuring large quantities of M&E	• Instrumentation may be expensive and difficult to set up and maintain • May not be ideal for assessing radioactivity in difficult-to-measure areas • Size of detection equipment may discourage portability • Typically does not provide nuclide identification
Conveyorized Survey Monitoring Systems	Scanning	• Requires relatively small amount of labor after initial set up • May be cost-effective for measuring large quantities of M&E	• Instrumentation may be expensive and difficult to set up and maintain • May not be ideal for assessing radioactivity in difficult-to-measure areas • Size of detection equipment may discourage portability • Typically does not provide nuclide identification
Laboratory Analysis	Sampling	• Generally provides the lowest MDCs and MQCs, even for difficult-to-measure radionuclides • Allows positive identification of radionuclides without gammas	• Most costly and time-consuming measurement technique • May incur increased overhead costs while personnel are waiting for analytical results • Great care must be taken to ensure samples are representative • Detector windows may be fragile
Laboratory Analysis	Smear	• Only measurement technique for assessing removable radioactivity • Removable radioactivity can be transferred and assessed in a low background counting area.	• Instrument background may not be sufficiently low. • Detectors with counting sensitive region larger than the smear surface area may require counting adjustments to account for inherent backgrounds associated with other media located under the detector sensitive region.

5.9.4   Measurement Performance Indicators

Measurement performance indicators are used to evaluate the performance of the measurement method. These indicators describe how the measurement method is performing to ensure the survey results are of sufficient quality to meet the survey objectives.

5.9.4.1   Blanks

Blanks are measurements of materials with little or no radioactivity and none of the radionuclide(s) of concern present, and performed to determine whether the measurement process introduces any increase in instrument signal rate that could impact the measurement method detection capability. Blanks should be representative of all measurements performed using a specific method (i.e., combination of instrumentation and measurement technique). When practical, the blank should consist of the same or equivalent material(s) as the M&E being surveyed.

Blanks typically are performed before and after a series of measurements to demonstrate the measurement method was performing adequately throughout the survey. At a minimum, blanks should be performed at the beginning and end of each shift. When large quantities of data are collected (e.g., scanning measurements) or there is an increased potential for radionuclide contamination of the instrument (e.g., removable or airborne radionuclides), blanks may be performed more frequently. In general, a blank should be collected whenever enough measurements have been performed such that it is not practical to repeat those measurements if a problem is identified.

A sudden change in a blank result indicates a condition requiring immediate attention. Sudden changes are caused by the introduction of a radionuclide, a change in ambient background, or instrument instability. Gradual changes in blank values indicate a need to inspect all survey areas for sources of radionuclides or radioactivity. Gradual build up of removable radionuclides over time or instrument drift and deterioration can result in slowly increasing blank values. High variability in blank values can result from instrument instability or improper classification (i.e., high activity and low activity M&E combined into a single survey unit. It is important to correct any problems with blanks to ensure measurement detectability (see Sections 5.7 and 7.5) is not compromised.

5.9.4.2   Replicate Measurements

Replicate measurements are two or more measurements performed on the same M&E, and performed primarily to provide an estimate of precision for the measurement method. The reproducibility of measurement results should be evaluated by replicates to establish this component of measurement uncertainty (see Sections 5.6 and 7.4).

Replicates typically are performed at specified intervals during a survey (e.g., 5% of all measurements or once per day), and should be used to evaluate each batch of data used to support a disposition decision (e.g., one replicate per survey unit). For single measurement surveys or scan-only surveys where decisions are made based on every measurement, typically 5% of all measurements are replicated.

Precision exhibits a range of values and depends in part on the material being measured and the activity level. Small changes in precision are expected, and the acceptable range of variability should be established prior to initiating data collection activities. The main causes for lack of precision include problems with repeating measurements on irregularly shaped M&E, the material being measured, counting statistics when the activity levels are low, and instrument contamination.

5.9.4.3 Spikes and Standards

Spikes and standards are materials with known composition and radioactivity, used to evaluate bias in the measurement method, and typically performed periodically during a survey (e.g., 5% of all measurements or once per day). When spikes and standards are available, they should be used to evaluate each batch of data used to support a disposition decision (i.e., at least one spike or standard per survey unit).

M&E cover a broad range of physical forms and materials that can change a measurement method's expected bias. Tracking results of measurements with known activity can provide an indication of the magnitude of bias. In general, activity levels near the action levels (or discrimination limits in Scenario B) will provide adequate information on the performance of the measurement system.

5.9.5 Instrument Performance Indicators

Instrument performance indicators provide information on how an instrument is performing. Evaluation of these indicators provides information on the operation of the instruments.

5.9.5.1 Performance Tests

Performance tests should be performed periodically and after maintenance to ensure that the instruments continue to meet performance requirements for measurements. An example of a performance test is a test for response time. Performance requirements should be met as specified in the applicable sections of ANSI N323A (ANSI 1997), ANSI N42.17A (ANSI 2003b), and ANSI N42.17C (ANSI 1989). These tests may be conducted as part of the calibration procedure.

5.9.5.2 Functional Tests

Functional tests should be performed prior to initial use of an instrument. These functional tests should include—

- General condition,
- Battery condition,

- Verification of current calibration (i.e., check to see that the date due for calibration has not passed),
- Source and background response checks (and other tests as applicable to the instrument), and
- Constancy check.

The effects of environmental conditions (temperature, humidity, etc.) and interfering radiation on an instrument should be established prior to use. The performance of functional tests should be appropriately documented. This may be as simple as a checklist on a survey sheet, or may include more detailed statistical evaluation such as a chi-square test.

5.9.5.3 Instrument Background

All radiation detection instruments have a background response, even in the absence of a sample or radiation source (Section 3.4.2). Inappropriate background correction will result in measurement error and increase the uncertainty of data interpretation.

5.9.5.4 Efficiency Calibrations

Detector efficiency is critical for converting the instrument response to activity (MARSAME Section 7.8.2.2, MARSSIM Section 6.5.4, MARLAP Chapter 16). Routine performance checks may be used to demonstrate the system's operational parameters are within acceptable limits, and these measurements typically are included in the assessment of bias. The system's operational parameters may be tracked using control charts.

5.9.5.5 Energy Calibrations (Spectrometry Systems)

Spectrometry systems identify radionuclides based on the energy of the detected radiations. A correct energy calibration is critical to accurately identify radionuclides. An incorrect energy calibration may result in misidentification of peaks, or failure to identify radionuclides present in the M&E being investigated.

5.9.5.6 Peak Resolution and Tailing (Spectrometry Systems)

The shape of the full energy peak is important for identifying radionuclides and quantifying their activity with spectrometry systems. Poor peak resolution and peak tailing may result in larger measurement uncertainty, or in failure to identify the presence of peaks based on shape. Consistent problems with peak resolution indicate the presence of an analytical bias.

5.9.5.7 Voltage Plateaus (Gas Proportional Systems)

The accuracy of results using a gas proportional system can be affected if the system is not operated with its detector high voltage adjusted such that it is on a stable portion of the operating plateau.

5.9.5.8   Self Absorption, Backscatter, and Crosstalk

Alpha and beta measurement results can be affected by the M&E through self-absorption and backscatter. Measurement systems simultaneously detecting alpha and beta particles using an electronic discriminator (e.g., gas flow proportional detectors) can be affected by crosstalk (i.e., identification of alpha particles as beta particles and vice versa). Accurate differentiation between alpha and beta activity depends on the assessment and maintenance of information on self-absorption and crosstalk.

5.10   Report the Results

Once the instruments have been checked to ensure proper operation, the data should be collected in a manner consistent with the survey design. Any field changes and deviations from survey design should be documented and described in sufficient detail to enable an independent recreation and evaluation at some future time.

The reported measurements should comprise raw data that includes background radioactivity (i.e., gross measurement data). Electronic instruments with data logging capabilities should be used when applicable. Electronic data should be exported and backed up periodically to minimize the chance of losing data and the need for re-surveying.

Use of a measurement identification system should be considered. If required by the objectives of the survey, the identification system should be developed and used such that each measurement is assigned and labeled with a unique (preferably sequential) identifying number, the collection date and time, the measurement location, and any applicable comments.

While MARSAME does not make specific recommendations with regard to approved media formats for storing documentation, some users of MARSAME (e.g., private industry nuclear power plants) may be required to retain documentation in media formats prescribed by State and Federal rules of evidence. Similarly, State and Federal rules of evidence may specify retention periods for documentation that exceed internal facility requirements. Compliance with State and Federal rules of evidence is intrinsic to maintaining legally defensible records for insurance and litigation-related purposes.

Projects at large, complex facilities often occur over relatively long time frames (e.g., years or decades). In many cases the project is divided into smaller sub-projects that are performed as resources and information become available. Retention of records, data compatibility, data accessibility, and transfer of data between sub-projects should be considered during the performance of individual surveys.

Documentation of the survey measurements should provide a complete and unambiguous record of the data collected. Documentation should also include descriptions of variability and other conditions pertaining to the M&E that may have affected the measurement capabilities of the survey procedure, and photographs where applicable. The documentation itself should be clear, legible, retained, retrievable, and to the level of detail required.

Negative results (net activity below zero) can be obtained when an instrument background is subtracted from the measurement of a low activity sample. In the case where the activity is close to zero, the measurement uncertainty will result in a distribution of results where approximately one-half are less than zero and one-half are greater than zero. As long as the magnitude of negative values is comparable to the estimated measurement uncertainties and there is no discernible negative bias, negative results should be accepted as legitimate estimates of radionuclide concentrations or levels of radioactivity associated with the M&E. A preponderance of negative results, even if they are close to zero may indicate a bias or systematic error.

The inclusion of the information described above is important in creating comprehensive documentation to make disposition surveys technically and legally defensible. The collection of all necessary data prepares the MARSAME user to assess the results of the disposition survey, which is discussed in Chapter 6.

6   EVALUATE THE SURVEY RESULTS

6.1   Introduction

The assessment phase of the data life cycle involves the interpretation of survey results. Interpretation of survey results is very straightforward when all of the data are below or all of the data are above the action level, and the correct decision regarding disposition of the M&E is obvious. In these cases very little data interpretation is required. However, formal statistical tests provide a valuable tool when the survey results are neither clearly above nor entirely below the action level. In either case, statistical tests always can be used to support the survey design in helping to ensure the quantity and quality of data meet the data quality objectives (DQOs) and measurement quality objectives (MQOs). Figure 6.1 illustrates the assessment phase of the data life cycle.

6.2   Conduct Data Quality Assessment

Data quality assessment (DQA) is a scientific and statistical evaluation that determines whether data are the right type, quality, and quantity to support their intended use (EPA 2006b). There are five steps in the DQA process:

1. Review the DQOs and survey design.
2. Conduct a preliminary data review.
3. Select the statistical test.
4. Verify the assumptions of the statistical test.
5. Draw conclusions from the data.

The effort applied to DQA should be consistent with the graded approach used to develop the survey design. More information on DQA can be found in *Data Quality Assessment: A User's Guide* (EPA QA/G-9R, EPA 2006b) and *Data Quality Assessment: Statistical Tools for Practitioners* (EPA QA/G-9S, EPA 2006c). Data should be verified and validated as described in the quality assurance project plan (QAPP). Guidance on data verification and validation can be found in MARSSIM Section 9.3 and MARLAP Chapter 8. Guidance on developing a QAPP is available in EPA QA/G-5 (EPA 2002a) and MARLAP Chapter 4.

6.2.1   Review the Data Quality Objectives and Survey Design

The first step in the DQA process is a review of the DQO outputs used to develop the survey design to ensure they are still applicable. The review of the DQOs and survey design should also include the MQOs (e.g., measurement uncertainty, detectability, quantifiability). For example, if the data show the measurement uncertainty exceeds the estimate used to design the survey, the DQOs and MQOs should be revisited.

The survey design should be reviewed for consistency with the DQOs. For example, the review should verify that the appropriate number or amount of measurements were performed in the correct locations and were analyzed using measurement methods with adequate sensitivity.

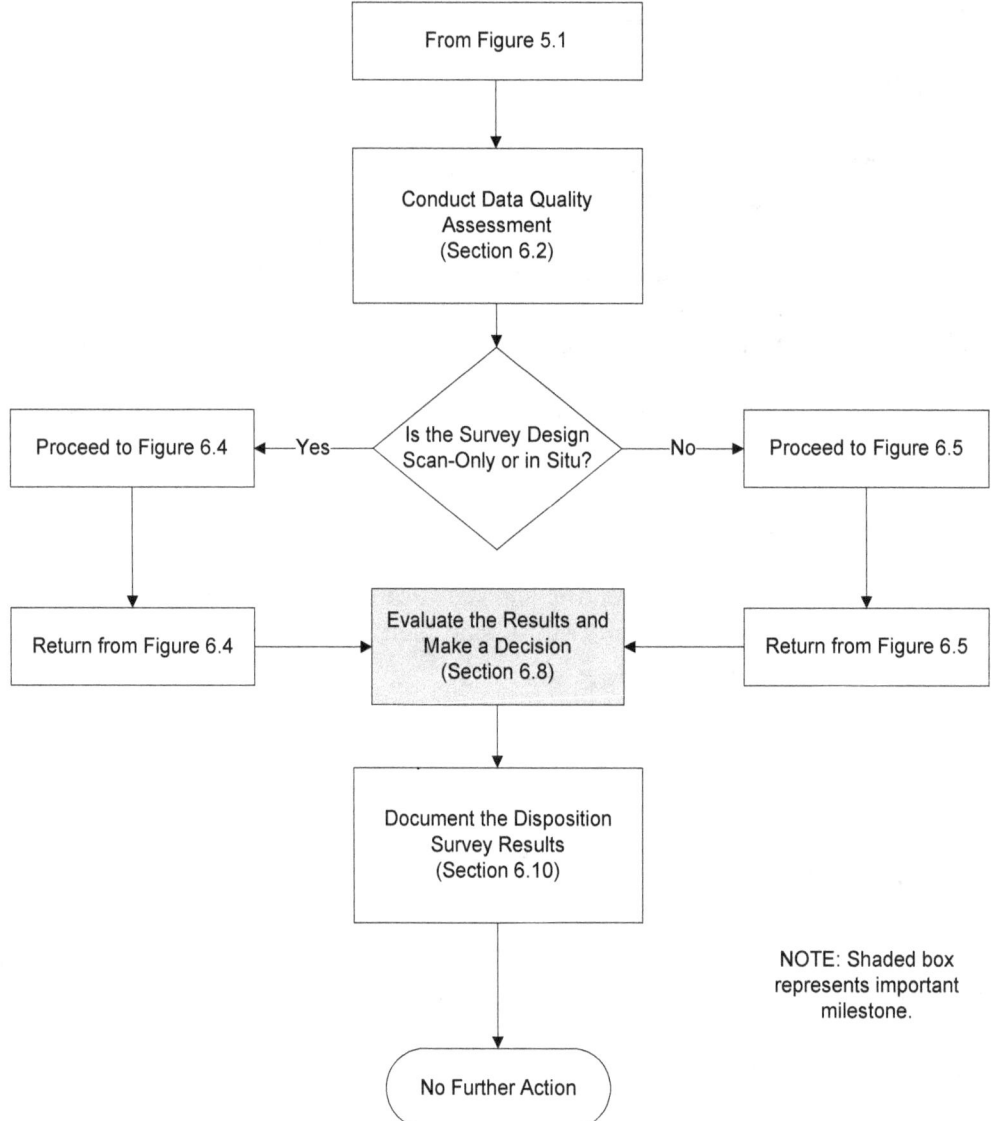

Figure 6.1 The Assessment Phase of the Data Life Cycle

In cases where the survey did not involve taking discrete measurements or samples (i.e., scan-only, conveyor systems, or in situ surveys), it is imperative that the minimum detectable concentrations (MDCs) be calculated realistically and they truly reflect at least 95% probability that concentrations at or about the MDC were detected. Clearly, MDCs must be capable of detecting radionuclide concentrations or levels of radioactivity at or below the upper bound of the gray region (UBGR). When detection decisions are made for individual items (i.e., Scenario B) the MDC should be less than or equal to the UBGR.

The minimum quantifiable concentration (MQC) is defined as the radionuclide concentration or level of radioactivity at which the measurement method gives results with a specified relative standard deviation $1/k_Q$, where k_Q is usually chosen to be 10 (see Section 5.8, MARLAP Section

19.4.5, MARLAP Section 19.7.3). MARSAME recommends that the MQC should be no larger than the upper bound of the gray region (UBGR) when making quantitative comparisons of the mean survey data to the action level (i.e., Scenario A). This is an expression of the fact that the MQC, unlike the MDC used for a simple detection decision, addresses the relative uncertainty of the data value obtained. If the objective of the disposition survey is to quantify radionuclide concentrations near the UBGR, the MQC should be no larger than the UBGR.[1]

For MARSSIM-type surveys (Section 4.4.3) it is important to collect sufficient data to support a disposition decision. This is particularly important in cases where the radionuclide concentrations are near the action level. This can be done prospectively during survey design to test the efficacy of a proposed survey design (see Chapter 4), or retrospectively during interpretation of survey results to demonstrate the objectives of the survey design have been achieved. The procedure for generating power curves for the Sign test and the Wilcoxon Rank Sum test are provided in Appendix I of MARSSIM. Note that the accuracy of a prospective power curve depends on estimates of data variability and the planned number of measurements. After the data are analyzed, the sample standard deviation provides an estimate of data variability and the actual number of valid measurements are known, and these two parameters are used to generate a retrospective power curve (see MARSSIM Appendix I). The consequence of inadequate power is an increased Type II decision error rate. For Scenario A, this means M&E that actually meet the release criteria have a higher probability of being incorrectly determined not to meet the release criterion. For Scenario B, this means M&E that actually do not meet the release criterion have a higher probability of being incorrectly determined to meet the release criterion.

6.2.2   Conduct a Preliminary Data Review

A preliminary data review is performed to learn more about the structure of the data by identifying patterns, relationships, or potential anomalies. The preliminary data review includes reviewing quality assurance (QA) and quality control (QC) reports, performing a graphical data review, and calculating basic statistical quantities.

6.2.2.1   Review Quality Assurance and Quality Control Reports

Quality assurance reports describing data collection and reporting processes provide valuable information about potential problems with or anomalies in the data. EPA QA/G-9R (EPA 2006b) recommends a review of (1) data validation reports that document the data collection, handling, analysis, reduction, and reporting procedures; (2) QC reports from laboratories or field stations that document measurement system performance including data from blanks, replicates, spikes, standards, and certified reference materials, or other internal QC measures; and (3) technical systems reviews, performance evaluation audits, and audits of data quality including data from performance evaluation measurements. EPA QA/G-9R (EPA 2006b) also suggests paying particular attention to information that can be used to check assumptions made during survey design using the DQO process, especially any anomalies in recorded data, missing values, deviations from SOPs, or the use of nonstandard data collection methods (e.g., new, emerging, or "cutting edge" technology). Verification of instrument calibrations and review of MQOs are

---

[1] The UBGR is either the action level for Scenario A or the discrimination limit for Scenario B (see Section 4.2).

particularly important to disposition surveys. Periodic measurements must be made to ensure the measurement systems remain within acceptable calibration and control limits.

Quality control measurements are performed during implementation of the survey design to monitor performance of the measurement methods, identify problems, and initiate corrective actions when necessary. The evaluation of QC measurements used to control measurement methods is distinct from the evaluation from survey results. MARLAP Section 18.3 ("Evaluation of Performance Indicators"), Attachment 18A ("Control Charts"), and Attachment 18B ("Statistical Tests for QC Results") provide information on the evaluation of quality control measurements.

Reviewing QA and QC reports is the only preliminary data review performed for surveys where individual measurements are not recorded (e.g., scan-only surveys with hand-held instruments). This increases the importance of the QA and QC reports and should be considered during survey planning to ensure data quality is adequate to meet the survey objectives.

6.2.2.2   Perform a Graphical Data Review

Preparing and evaluating graphs and other visual depictions of the data may identify trends in the data that go unnoticed using purely numerical methods. The graphical data review may include posting plots, frequency plots, quantile plots, or other methods for visually interpreting data. General guidance on performing a graphical data review and exploratory data analysis is provided in EPA QA/G-9R (EPA 2006b) and by the National Institute of Science and Technology (NIST 2006). A graphical data review cannot be performed unless the measurement results are recorded. Surveys where recording individual measurement results is not required (e.g., scan-only surveys with hand-held instruments) do not receive a graphical data review.

A posting plot is simply a map of the survey unit with the data values entered at the measurement locations. This type of plot potentially reveals heterogeneities in the data, especially possible clusters of elevated radionuclide concentrations. For a reference material survey a posting plot can reveal spatial trends in background data that might affect the results of the statistical tests. If the posting plot reveals systematic spatial trends in the M&E, the cause of the trends should be investigated. In some cases the trends could be attributable to residual radioactivity, but they may also be caused by inhomogeneities in the ambient background in the area the survey is performed. EPA QA/G-9S (EPA 2006c) provides additional diagnostic tools for examining spatial trends. The role of a posting plot for a conveyorized system would be a time series display of the data showing any trends between adjacent batches of M&E conveyed past the detector.

The geometric configuration of most M&E survey units composed of a few large irregularly shaped pieces of M&E is transitory. The arrangement of tools and piles of scrap metal, for example, changed as volumes of material were moved, or even as individual pieces were handled during the survey (Section 5.3). In these cases some identifying marks, numbers, or bar-code labels should have been used to identify and track where measurements were made, at least until it is determined that the M&E meet the disposition criteria. Such marking and labeling need not be permanent, but may be made with materials such as chalk or removable labels.

A frequency plot, or histogram, is a useful tool for examining the general shape of a distribution. This plot is a bar chart of the number of data points within a certain range of values. A frequency plot reveals any obvious departures from symmetry, such as skewness or bimodality (two peaks), in the data distributions for the M&E or reference material.

The presence of two peaks in the M&E data set frequency plot may indicate the presence of small areas of elevated activity. In some cases it may be possible to identify an appropriate background distribution within the M&E data set. This type of data interpretation generally depends on site-specific considerations and should only be pursued after consultation with the responsible regulatory agency.

The presence of two peaks in the M&E or reference material frequency plots may also indicate a mixture of materials with different intrinsic radiation backgrounds. The greater variability in the data caused by the presence of such a mixture reduces the power of the statistical tests. These situations should be avoided whenever possible through segregation of M&E (see Section 5.4) and carefully matching the reference materials to the M&E being surveyed.

When data are obtained from scan-only surveys incorporating data loggers, large quantities of data are usually recorded. In essence, 100% of Class 1 M&E are measured. While the survey coverage may be less than 100% for Class 2 and Class 3 M&E, the number of data points is still likely to be large. As long as there was no bias in the selection of areas that were scanned, the frequency plot will be close to the population distribution of radioactivity levels in the M&E. The mean and standard deviation calculated from these logged values should be very close to the corresponding population values.

For conveyorized survey monitors, the data may be interpreted batch-by-batch as it is scanned. In this case, the data treatment would be most similar to a single in situ measurement used to evaluate all of the M&E. If, on the other hand, the data were logged continuously the data treatment would be similar to a scan-only survey using data loggers.

6.2.2.3   Calculate Basic Statistical Quantities

Radiological survey data are usually obtained in units (e.g., counts per unit time) that have no intrinsic meaning relative to the action levels. For comparison of survey data to action levels, survey data from laboratory and field analyses are converted into action level units. MARSSIM Section 6.6 provides guidance on data conversion. Any uncertainty associated with data conversion should be included in the estimate of measurement uncertainty (Section 5.6). For surveys where individual results are not recorded (e.g., scan-only surveys with hand-held instruments) the uncertainty is associated with converting the action level into the units provided by the instrument in the field. Because individual results are not recorded, no statistical quantities can be calculated.

Basic statistical quantities that should be calculated for the sample data set include the mean, standard deviation, and the median. Other statistical quantities may be calculated based on the survey objectives.

Example 1: Suppose the following 10 measurement results are obtained from a disposition survey:

$$9.1, 10.7, 13.6, 3.4, 13.3, 7.9, 4.5, 7.7, 8.3, 10.4$$

The mean of the data (μ) is 8.89 and the standard deviation (σ) is 3.3231.
The next 10 measurement results are from an appropriate matching reference material:

$$6.2, 13.8, 15.2, 9.3, 6.7, 4.9, 7.1, 3.6, 8.8, 8.9$$

The mean of the reference data (μ) is 8.45 and the standard deviation (σ) is 3.6713.

The means of the two data sets can be compared to provide a preliminary indication of the survey unit status.[2] The difference is 0.44, with the M&E being investigated having a higher mean concentration. If the mean for the M&E exceeds the mean for the reference material by more than the action level, the M&E clearly do not meet the disposition criterion. On the other hand, if the difference between the largest M&E measurement (13.6 for this example) and the smallest reference material measurement (3.6 for this example) is below the action level, the M&E will pass the Wilcoxon Rank Sum (WRS) test (Section 6.6), but will have to meet other criteria as well.

The value of the sample standard deviation is especially important. If the standard deviation is too large compared to what was assumed for variability during development of the survey design, this may indicate an insufficient number of samples were collected to achieve the desired power for the statistical test. As previously mentioned, inadequate power can lead to an increase in the Type II decision error rate.

The median is the middle value of the data set when the number of data points is odd or the mean of the two middle values when the number of data points is even. A large difference between the mean and the median indicate a potential skew in the data. This would also be evident in a histogram of the data.

Examining other statistical quantities such as the maximum, minimum, and range may provide additional useful information. When there are 30 or fewer data points, range values greater than 4 or 5 standard deviations would be unusual.

Example 2: For the example M&E data set the minimum is 3.4 and the maximum is 13.6. The range is $13.6 - 3.4 = 10.2$. The range is equal to 3.1 standard deviations (i.e., 10.2/3.3). Thus, the range for this example data set is not unusually large. The range may be greater for larger data sets.

---

[2] Note the use of significant digits in this example. Because all of the numbers in the text are interim values in calculating the difference between two means, they are not rounded. If the mean and standard deviation values were to be reported as results they would be rounded to two significant digits because the original data is a mixture of numbers with two and three significant digits. If the data were rounded after each calculation, the difference in the rounded means appears to be 0.4 (i.e., 8.9 minus 8.5), but the actual difference is 0.44 based on the un-rounded means (i.e., 8.89 minus 8.45). This is an example of how rounding numbers too early in the process can result in additional uncertainty.

6.2.3   Select the Statistical Tests

In most cases the selection of a statistical test is determined by the survey design used to collect the data. The most appropriate procedure for summarizing and analyzing the data is chosen based on the preliminary data review. If the preliminary data review indicates that the assumptions used to develop the survey design are valid, the statistical tests and evaluation methods determined should then be applied. If the assumptions used to develop the survey design are determined to be invalid, it may be necessary to consult a statistician to determine the most appropriate statistical test for evaluating the survey results.

6.2.3.1   Scan-Only Surveys

Scan-only surveys generate large amounts of data. Class 1 surveys measure all of the M&E. When less than 100 percent of the M&E are measured (i.e., Class 2 or Class 3 surveys) the areas that are measured are assumed representative of the areas that are not measured. This assumption should be checked during the preliminary data review (Section 6.2.2). The radionuclide concentrations or radioactivity in the areas that are not measured can be inferred based on the measurement results in the areas that are measured. Data indicating this inference may not be reasonable should result in re-evaluation of the survey design. For example, suppose the survey design specifies that ^{137}Cs is the radionuclide of concern and scanning 50% of the M&E is appropriate based on the expected distribution of radionuclide concentrations, expected levels of radioactivity, and the beta-gamma emissions from the radionuclide of concern. If additional historical data is found showing ^{239}Pu is also a radionuclide of concern, the survey design should be re-evaluated based on the presence of an alpha emitting radionuclide as well.

If disposition decisions will be made for individual items or based on individual measurement results, all of the results should be compared to the action level. Comparison to the action level based on a detection decision or measurement (Section 5.7) is discussed in Section 6.3. Individual measurement results can be recorded for scan-only surveys. The benefit of logging individual measurement results is the ability to statistically evaluate the data (e.g., calculate a mean and an upper confidence limit). If disposition decisions will be made based on the mean of logged data, an upper confidence limit for the mean is calculated and compared to the UBGR. This means that compliance with the disposition criterion can be demonstrated for the entire survey unit, even if some of the results exceed the UBGR. Evaluations using the upper confidence limit are discussed in Section 6.4. When less than 100% of the M&E are measured (i.e., Class 2 and Class 3 surveys), the total uncertainty includes both spatial and measurement uncertainty. Measuring 100% of the M&E (i.e., Class 1 survey) accounts for spatial variability, but there is still an uncertainty component resulting from variability in the measurement process.

Conveyorized systems that continually log the survey results also generate large amounts of data. An upper confidence limit for the mean can be used for the evaluation of data from these types of systems (see Section 6.4) in the same manner as logged scan data. Conveyorized systems that operate in a batch mode are essentially treated as single in situ measurements of small batches of M&E. The results generated by these types of systems are evaluated as a series of comparisons to the UBGR; using detection decisions based on the MDC (Section 6.3).

6.2.3.2   In Situ Surveys

In situ surveys may consist of a series of isolated measurements covering all or part of the M&E, a series of measurements with overlapping fields of view incorporating all (Class 1) or a portion (Class 2 or Class 3) of the M&E, or a single measurement incorporating all of the M&E (Section 4.4.2).

Similar to scan-only surveys, if disposition decisions will be made for individual items or based on individual measurement results, all of the results should be compared to the action level. Comparison to the action level based on a detection decision (Section 5.7) is discussed in Section 6.3. Unlike scan-only surveys, in situ surveys are likely based on a limited number of data points. To perform in situ measurements, assumptions were made about the distribution of radioactivity within the volume of M&E being measured. These assumptions are inherent in the calibration of in situ measurement systems and the validity of these assumptions determines the appropriateness of the measurement. It is important to account for uncertainty in these assumptions when calculating the MDC and to evaluate these assumptions using QC measurements performed during the survey. If there is uncertainty about the true MDC or critical value, use conservative values for the efficiency as described in Section 7.5.2.

6.2.3.3   MARSSIM-Type Survey Designs

MARSSIM-type survey designs generally are used when instrumentation for scan-only or in situ measurement surveys do not provide sufficient sensitivity (e.g., the MDC is greater than the UBGR). A statistically based number of measurements is used to provide an estimate of the mean activity in each survey unit, and scanning is used to identify small areas of elevated activity between sample locations.

The number of measurements is determined by the statistical test. In most cases the statistical tests used in MARSSIM are appropriate for Scenario A. The criteria for choosing between the Sign test and the WRS test are described in MARSSIM Section 8.2.3. In general, when the radionuclide is not present in background (or its background concentration is negligible compared to the action level) and radionuclide-specific measurements are made, the Sign test (Section 6.5) is used. Otherwise, the WRS (Section 6.6) test should be used. The Sign test is designed to detect whether there is radioactivity in the M&E above the action level. The WRS test is used to compare measurements of the M&E to measurements performed on the reference material.

When Scenario B is used, the statistical tests described in NUREG-1505 (NRC 1998a) generally are used. The Sign test and the WRS test are still used, but the application of the test is adjusted to account for the difference in the null hypothesis. When using Scenario B, there is a potential for the WRS test to miss non-uniform radioactivity (i.e., slightly elevated radionuclide concentrations or levels of radioactivity over a portion of the survey unit). Randomization of the M&E through mixing or homogenization can eliminate this possibility. If randomization is not practical, the Quantile test (Section 6.7) should be used to evaluate survey units when the WRS test fails to reject the null hypothesis.

The results of scanning measurements performed as part of a MARSSIM-type survey are evaluated using the elevated measurement comparison (EMC). The EMC is simply a comparison to an action level (see Section 6.3). The action level used for the EMC is the action level for small areas of elevated activity. If there is no action level for elevated activity, the scanning results are compared to the action level for the mean activity in the survey unit. Additional information on the EMC is available in MARSSIM Section 8.5.1 and NUREG-1505 Chapter 8 (NRC 1998a).

6.2.4   Verify the Assumptions of the Tests

An evaluation to determine the data are consistent with the underlying assumptions of the statistical tests helps to validate the use of a particular test. One may also determine that certain departures from these assumptions are acceptable when given the actual data and other information about the project. The nonparametric tests described in this chapter assume that the data from the M&E or the reference material consist of independent measurements from each distribution. The primary issue associated with the evaluation of scan-only and single in situ measurement survey data is the MDC or MQC as discussed in Section 6.2.1.

Asymmetry in the data can be identified using a histogram or a Quantile plot. Information on histograms and Quantile plots is provided in MARSSIM Appendix I and NUREG-1505 Section 4.2.2 (NRC 1998a). As discussed in Section 6.2.2.3, data transformations can sometimes be used to minimize the effects of asymmetry.

One of the primary advantages to using the nonparametric tests is that they involve fewer assumptions about the data than their parametric counterparts. If parametric tests are used (e.g., Student's t test) any additional assumptions made in using these tests should be verified (e.g., testing for normality). These issues are discussed in detail in EPA QA/G-9S (EPA 2006c).

One of the more important assumptions made in the survey design is that the number of measurements is sufficient to achieve the DQOs set for the Type I (α) and Type II (β) decision error rates. Verification of the power of the statistical tests ($1-\beta$) may be of particular interest. Methods for assessing power are discussed in Appendix I.9 of MARSSIM. If there is not reasonable assurance the DQOs have been achieved, additional investigations including repeating the survey may be needed. The planning team can develop survey designs cautiously to avoid unnecessary and potentially costly decision errors by—

- Estimating the potential data variability conservatively,
- Taking more measurements than suggested by the DQO process, and
- Estimating the MDCs conservatively.

In the absence of other data, each of these estimates could be multiplied by a safety factor of 1.2 (i.e., increase the estimate by 20%). Examples of assumptions and possible methods for evaluating and verifying these assumptions are summarized in Table 6.1.

Table 6.1 Issues and Assumptions Underlying the Evaluation Method

Evaluation Method	Issue	Verification Method	Survey Type
Compare single measurements to a limit (Section 6.3)	Verify the MDC and Measurement Uncertainty	Review the MDC Review QA/QC Reports Review IA and DQOs	Scan-Only In situ
Compare an upper confidence limit for the mean to a limit (Section 6.4)	Verify the MQC and Measurement Uncertainty	Review the Measurement Uncertainty Review QA/QC Reports Review IA and DQOs	Scan Only In situ
Statistical Tests (Sections 6.5, 6.6, 6.7)	Verify the Assumptions of the Statistical Test (e.g., spatial independence, symmetry, data variance, power)	Preliminary Data Review (e.g., posting plot, histogram, summary statistics, power curve)	MARSSIM-Type Survey

Verification of scan-only and in situ survey results focuses on the estimates of the MDC and MQC values used to design the survey. If the assumptions used to estimate these values are incorrect, the survey design may be invalid.

The first step in evaluating the MDC and MQC is to review the assumptions used to develop these values. In general, the key assumptions are made in determining the source and detector efficiencies. QA and QC reports should be reviewed to evaluate measurement performance (e.g., scan speed, source geometry, distance from M&E to the detector, non-uniform response of large area detectors). The description of the M&E from the IA should be compared to the assumptions used to develop the efficiency.

In some cases it may be possible to compare the survey results of multiple measurement techniques. For example, if there are multiple radiations associated with the M&E it may be possible to compare gamma measurement results to alpha or beta measurement results to verify the survey results. Direct measurements may provide more quantitative results for areas of elevated activity identified during scan-only surveys.

It may be possible to use an entirely different survey method to provide information to support verification of assumptions used to design a survey. For example, smears or surface scrapings can be used to verify the presence of radionuclides or radioactivity on the surface.[3]

In situ measurements or sample collection and analysis may be used to verify the results of scan-only survey designs. Care must be taken to ensure comparability of survey methods before evaluating the results to avoid generating conflicting results. For example, consider an in situ survey used to demonstrate the mean activity is less than the action level. A scan-only survey method is used to verify the results and identifies an area of elevated activity. This discrepancy in results warrants additional investigation of the small area of elevated activity. The additional investigation should determine if the activity in this area actually causes the mean activity to exceed the disposition criterion.

---

[3] This smear procedure does not rule out additional volumetric activity.

6.2.5   Draw Conclusions from the Data

The types of measurements performed on M&E are—

- Scans,
- In situ or direct measurements at discrete locations, and
- Samples collected at discrete locations.

Specific details for conducting the Sign test and the WRS test are provided in Sections 6.5 and 6.6, respectively. When the data clearly show that the M&E meets or exceeds the disposition criterion, the result is often obvious without performing the formal statistical analysis. This is the expected outcome for Class 2 and Class 3 surveys. Table 6.2 summarizes examples of circumstances leading to specific conclusions based on a simple examination of the data.

Table 6.2 Summary of Evaluation Methods and Statistical Tests

Evaluation Method or Statistical Test	Survey Result	Conclusion
Comparison to a Limit (AL=0) – Scenario B only – Results may or may not be recorded – Scan-only or In situ surveys	All measurements less than the critical value corresponding to the MDC (e.g., does not exceed alarm set point)	M&E meet the disposition criterion
	Any measurement exceeds the critical value corresponding to the MDC	M&E do not meet the disposition criterion
Comparison to a Limit (AL≠0) – Scenario A or B – Results not recorded – Scan-only or In situ surveys	All measurements less than the critical value corresponding to the UBGR	M&E meet the disposition criterion
	Any measurement exceeds the critical value corresponding to the UBGR	M&E do not meet the disposition criterion
Comparison to Upper Confidence Limit – Scenario A or B – Results must be recorded – Scan-only or In situ surveys	Upper confidence limit less than UBGR	M&E meet the disposition criterion
	Upper confidence limit greater than UBGR	M&E do not meet the disposition criterion
Sign Test – Radionuclide not in background – Nuclide-specific measurements – Scenario A or B – MARSSIM-type surveys	All measurements less than the action level	M&E meet the disposition criterion
	Mean greater then the action level	M&E do not meet the disposition criterion
	Any measurement greater than the action level and the mean less than the action level	Conduct Sign test (and elevated measurement comparison, if necessary)
Wilcoxon Rank Sum Test – Radionuclide in background – Nuclide non-specific measurements – Scenario A or B – MARSSIM-type surveys	Difference between maximum survey unit measurement and minimum reference area measurement is less than the UBGR	M&E meet the disposition criterion
	Difference of survey unit mean and reference area mean is greater than the action level	M&E do not meet the disposition criterion
	Difference between any survey unit measurement and any reference area measurement greater than the action level or the difference of survey unit mean and reference area mean is less than the action level	Conduct WRS test (and elevated measurement comparison, if necessary)

Table 6.3 Summary of Evaluation Methods and Statistical Tests (Continued)

Evaluation Method or Statistical Test	Survey Result	Conclusion
Quantile Test – Test for non-uniform radioactivity – Combine with WRS test – Scenario B only – MARSSIM-type surveys	Difference between maximum survey unit measurement and minimum reference area measurement is less than the UBGR	M&E meet the disposition criterion
	Difference of survey unit mean and reference area mean is greater than the action level	M&E do not meet the disposition criterion
	Difference between any survey unit measurement and any reference area measurement greater than the action level or the difference of survey unit mean and reference area mean is less than the action level	Conduct Quantile test (and elevated measurement comparison, if necessary)

6.3   Compare Results to the UBGR

When disposition decisions will be made about individual items, or decisions will be based on individual measurement results, each result (plus or minus a multiple of its combined standard uncertainty) will be compared to the action level (see MARLAP Appendix C.4). In practice, this means that any result that exceeds the critical value (S_C, see Section 5.7 and Section 7.5.1) when the minimum detectable level (S_D, see Section 5.7 and Section 7.5.2) equals the UBGR provides evidence that the result exceeds the UBGR.

For Scenario A, if all the results are less than the action level, then the mean and the maximum activity must also be below the action level. Thus, the radionuclide concentrations or levels of radioactivity associated with the M&E demonstrate compliance with the disposition criterion. For Scenario B when the action level is not zero or background, all of the results must be below the critical value corresponding to the MDC set equal to the UBGR. If the action level is zero or background, Scenario B must be used and any indication of the presence of radionuclide concentrations or radioactivity above background (i.e., above the discrimination level) would result in rejecting the null hypothesis. For this situation, any measurement result exceeding the critical value corresponding to the required MDC indicates the potential presence of radionuclides or radioactivity above background. This applies to single in situ measurements as well as series of in situ measurements.

If there is an action level based on small areas of elevated activity or the maximum allowable value, the individual results can be compared directly to the action level. This applies primarily to the evaluation of scanning results for MARSSIM-type surveys (i.e., the EMC), but may be applied to scan-only survey data as well.

6.4   Compare Results Using an Upper Confidence Limit

The use of the upper confidence limit (UCL) can apply to both Scenario A and B for scan-only or in situ surveys where individual results are recorded. When disposition decisions are made about the estimated mean of a sampled population, the assessment of the survey results is accomplished by comparing a UCL for the mean to the UBGR. For scan-only surveys where there are a large number of data points, a simple comparison of the mean activity to the UBGR may be sufficient.

If individual scan-only survey results are recorded, a non-parametric confidence interval can be used to evaluate the results of the disposition survey. Similarly, a confidence interval can be used to evaluate a series of in situ measurements with overlapping fields of view. A one-tailed version of Chebyshev's inequality or software (e.g., EPA's ProUCL software) can be used to evaluate the probability of exceeding the UBGR (i.e., using a UCL). The use of a UCL applies to both Scenario A (where the UBGR equals the action level) and Scenario B (where the UBGR equals the discrimination limit).[4]

6.4.1   Calculate the Upper Confidence Limit

Chebyshev's inequality calculates the probability that the absolute value of the difference of the true but unknown mean of the population and a random number from the data set is at least a specified value. That is, given a specified positive number (n), a mean (μ), and a random number from the data set (r), then the probability that $|\mu\text{-}r|$ is greater than or equal to n is equal to α. In addition, a one-tailed version of the inequality can be used to calculate a UCL for a data set that is independent of the data distribution (i.e., there is no requirement to verify the data are from a normal, lognormal, or any other specified kind of distribution) by letting the inequality equal the UCL, as described in the following steps:

1. Calculate the mean (μ) and standard deviation (σ) of the number of results (n) in the data set.
2. For Scenario A, retrieve the Type I error rate (α) used to design the survey.
3. Using Chebyshev's inequality, calculate the maximum UCL using equation 6-1:

$$\text{UCL} = \mu + \sqrt{\frac{\sigma^2}{n\alpha} - \frac{\sigma^2}{n}}$$

(6-1)

4. For Scenario B, substitute the Type II error rate (β) used to design the survey for α in Equation 6-1.
5. If the maximum UCL is less than the UBGR, the survey demonstrates compliance with the disposition criterion (i.e., reject the null hypothesis for Scenario A or fail to reject the null hypothesis for Scenario B).

Chebyshev's inequality must be used with caution when there are very few points in the data set. This is because the population mean and standard deviation in the Chebyshev formula are being estimated by the sample mean and sample standard deviation. In a small data set from a highly skewed distribution, the sample mean and sample standard deviation may be underestimated if the high concentration but low probability portion of the distribution is not captured in the sample data set. EPA has issued guidance on calculating UCLs for exposure point concentrations (EPA 2002b).[5] Software for implementing EPA's guidance is available (EPA 2006d).

---

[4] In the case of Scenario B, if the action level is zero and the radionuclide of concern does not appear in background, any positive radionuclide-specific detection would result in a rejection of the null hypothesis that there is zero activity.

[5] In MARSAME, "exposure point concentration" is used to mean a conservative estimate of the mean radionuclide concentration(s) in or on M&E.

6.4.2   Upper Confidence Limit Example: Class 1 Concrete Rubble

This example illustrates the survey design for concrete rubble using 3 inch × 3 inch NaI(Tl) detectors mounted on a conveyorized survey system to measure ^{137}Cs. A pile of concrete rubble was loaded on the conveyor and passed beneath the detectors at a pre-determined speed. Each one-second count recorded by a detector corresponds to approximately 9,800 cm^3 of concrete rubble (i.e., a 5-cm thick disk with a 50-cm diameter). The following information was used to design the survey:

- The selected disposition option was clearance, using Scenario A with the null hypothesis that the residual radioactivity exceeds the action level.
- The IA indicated the concrete was potentially volumetrically contaminated prior to being converted to rubble.
- The concrete rubble had a maximum particle dimension of less than 0.5 cm.
- The average background count rate was estimated to be 38,000 cpm based on preliminary surveys of non-impacted concrete, and was used for the LBGR.
- The action level was set at 20,000 cpm above the average background count rate, so the UBGR was set at 58,000 cpm.
- The estimated standard deviation of background count rate is 2,500 cpm based on preliminary survey data.
- The Type I decision error rate was set at 0.10, or 10%.

The survey consisted of 9,616 independent, one-second measurements that were recorded using a data logger. The mean count rate for the survey was 39,252 cpm, with a standard deviation (σ) of 5,465 cpm. The standard deviation of the mean, σ_N, was calculated using the following equation:

$$\sigma_N = \frac{\sigma}{\sqrt{N}} = \frac{5,465\ cpm}{\sqrt{9,616}} = 55.7\ cpm \tag{6-2}$$

As noted earlier, with such a large data set, one can expect that the sample mean and standard deviation should be fairly close to their population values. The minimum count rate was 30,080 cpm, and the maximum count rate was 72,805 cpm. Note that although the mean concentration is well below the action level, there are data points that exceed the action level. Thus, a test against an UCL for the mean is warranted. Figure 6.2 shows a frequency plot of the survey results.

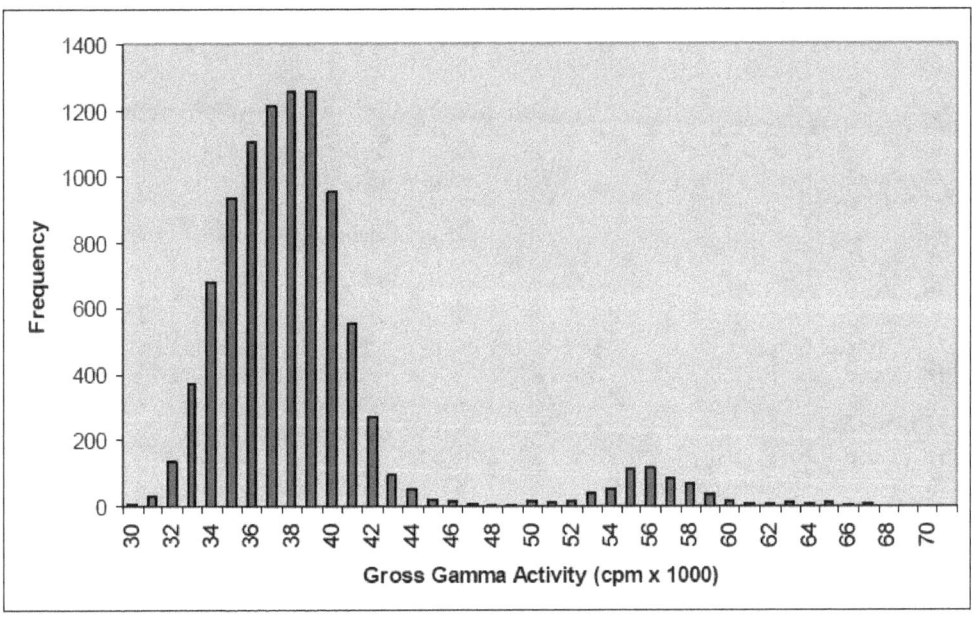

Figure 6.2 Frequency Plot of Concrete Rubble Data

If the sample size were small, however, the upper part of the bimodal distribution could be missed and the Chebyshev UCL could be underestimated. In this case, with a sample size of 9,616, the UCL was calculated using Equation 6-1 in Section 6.4.1.

$$\text{UCL} = 39{,}252 + \sqrt{\frac{(5{,}465)^2}{(0.10)(9{,}616)} + \frac{(5{,}465)^2}{(9{,}616)}} = 39{,}474 \,\text{cpm} \tag{6-3}$$

The UCL of 39,474 cpm is much less than the action level of 58,000 cpm. The null hypothesis that the level of radioactivity exceeds the disposition criterion is rejected.

The EPA ProUCL software was also applied to these data and the results are shown in Figure 6.3. The software has failed to find a good fit to the data for normal, lognormal or gamma distributions, which is hardly surprising given the bimodal nature of the data. The recommendation is that either a Student's *t* or a modified Student's *t* 95% UCL be used. These are both listed as about 39,343. These are lower than the 90% Chebyshev UCL of 39,474 used above, but that would not change the conclusion. A 95% Chebyshev UCL calculated according to Section 6.4.1 would have been 39,574. Note that the 95% Chebyshev UCL calculated by ProUCL, rounded to the nearest count, is slightly different, 39,495, because of the way that the sample mean and standard deviation are estimated before entering them in the Chebyshev formula. The ProUCL User's Manual can be consulted for details. However, with the number of data points at hand, there is little difference among any of the methods for computing an UCL.

	A	B	C	D	E	F	G	H	I
UCL Statistics for CPM									
1	Data File				Variable:	CPM			
2									
3	Raw Statistics				Normal Distribution Test				
4	Number of Valid Samples			9616	Lilliefors Test Statisitic				0.2044466
5	Number of Unique Samples			6441	Lilliefors 5% Critical Value				0.0090352
6	Minimum			30080	Data not normal at 5% significance level				
7	Maximum			72805					
8	Mean			39251.847	95% UCL (Assuming Normal Distribution)				
9	Median			38267	Student's-t UCL				39343.497
10	Standard Deviation			5465.0563					
11	Variance			29866840	Gamma Distribution Test				
12	Coefficient of Variation			0.1392306	A-D Test Statistic				585.96505
13	Skewness			2.4504964	A-D 5% Critical Value				0.7522512
14					K-S Test Statistic				0.1788147
15	Gamma Statistics				K-S 5% Critical Value				0.01814
16	k hat			61.605609	Data do not follow gamma distribution				
17	k star (bias corrected)			61.586458	at 5% significance level				
18	Theta hat			637.14729					
19	Theta star			637.34542	95% UCLs (Assuming Gamma Distribution)				
20	nu hat			1184799.1	Approximate Gamma UCL				39335.904
21	nu star			1184430.8	Adjusted Gamma UCL				39335.917
22	Approx.Chi Square Value (.05)			1181899.7					
23	Adjusted Level of Significance			0.049975	Lognormal Distribution Test				
24	Adjusted Chi Square Value			1181899.4	Lilliefors Test Statisitic				0.1656286
25					Lilliefors 5% Critical Value				0.0090352
26	Log-transformed Statistics				Data not lognormal at 5% significance level				
27	Minimum of log data			10.311616					
28	Maximum of log data			11.19554	95% UCLs (Assuming Lognormal Distribution)				
29	Mean of log data			10.569616	95% H-UCL				N/A
30	Standard Deviation of log data			0.1224578	95% Chebyshev (MVUE) UCL				39441.02
31	Variance of log data			0.0149959	97.5% Chebyshev (MVUE) UCL				39533.758
32					99% Chebyshev (MVUE) UCL				39715.923
33									
34					95% Non-parametric UCLs				
35					CLT UCL				39343.516
36					Adj-CLT UCL (Adjusted for skewness)				39345.004
37					Mod-t UCL (Adjusted for skewness)				39343.729
38					Jackknife UCL				39343.497
39					Standard Bootstrap UCL				39341.993
40					Bootstrap-t UCL				39342.895
41	RECOMMENDATION				Hall's Bootstrap UCL				39343.322
42	Data are Non-parametric (0.05)				Percentile Bootstrap UCL				39337.588
43					BCA Bootstrap UCL				39344.539
44	Use Student's-t UCL				95% Chebyshev (Mean, Sd) UCL				39494.773
45	or Modified-t UCL				97.5% Chebyshev (Mean, Sd) UCL				39599.887
46					99% Chebyshev (Mean, Sd) UCL				39806.364

Figure 6.3 Screen Capture of Output from ProUCL Software for the Sample Data Set

6.5   Conduct the Sign Test

The Sign test is used to compare the measurement results from each survey unit with the applicable disposition criterion. The Sign test can be applied to either Scenario A or Scenario B. The Sign test should only be used if the radionuclide being measured is not present in background or if the radionuclide being measured is present at such a small fraction of the action level as to be considered insignificant. Otherwise, the WRS test described in Section 6.6 should be applied. Additional information on the Sign test can be found in Section 8.3 of MARSSIM and Chapter 5 of NUREG-1505 (NRC 1998a).

6.5.1   Apply the Sign Test to Scenario A

The Sign test is applied to Scenario A by counting the number of measurements from each survey unit that are less than the action level (i.e., UBGR). Each result is subtracted from the action level ($AL - X_i$), and the number of positive values is summed. The result is the test statistic $S+$. Discard any measurement that is exactly equal to the action level and reduce the sample size, N, by the number of such measurements. The value of $S+$ is compared to the critical values in A.3. If $S+$ is greater than the critical value (q) in the table, the null hypothesis is rejected.

6.5.2   Apply the Sign Test to Scenario B

The Sign test is applied to Scenario B in a manner similar to that used for Scenario A. However, for Scenario B the action level (i.e., LBGR) is subtracted from each result ($X_i - AL$), and the number of positive values is summed. The result is the test statistic $S+$. Discard any measurement that is exactly equal to the action level and reduce the sample size, N, by the number of such measurements. The value of $S+$ is compared to the critical values in Table A.3. If $S+$ is greater than the critical value (q) in the table, the null hypothesis is rejected.

6.5.3   Sign Test Example: Class 1 Copper Pipes

This example illustrates the disposition survey design for copper pipe sections using a gas-flow proportional counter to measure ^{239}Pu. Because the alpha background on the copper material is essentially zero, it was decided the Sign test would be used to determine whether the material meets the disposition criterion. The sample size was determined using the DQO Process and inputs such as the disposition option, action level, expected standard deviation of the measurement results, and the acceptable probability of making Type I and Type II decision errors.

The following inputs were used to develop the survey design–

- The selected disposition option was clearance.
- The survey was designed using Scenario A, with the null hypothesis that the residual radioactivity exceeds the action level.
- The IA indicated that the inside surfaces of the pipes potentially came in contact with liquids containing ^{239}Pu, but the outside surfaces were non-impacted.
- The gross activity action level was 100 dpm/100 cm^2. When converted to cpm the gross activity action level was 10 cpm (i.e., total efficiency = 0.10 counts per disintegration).
- The LBGR (i.e., the DL) was set at the expected activity level on the copper pipe sections (i.e., 5 net cpm, the same as the gross mean for an alpha background of 0).
- The standard deviation for the measurements was estimated at 2 cpm.

- The relative shift was calculated as $(10-5)/2 = 2.5$.
- The Type I and Type II decision error rates were both set at 0.05.

Table A.2a shows the number of measurements estimated to be needed for the Sign test, N, is 15 ($\alpha=0.05$, $\beta=0.05$, and $\Delta/\sigma=2.5$). Therefore, 15 surface activity measurements were randomly collected from the inside surfaces of the copper pipe sections. Survey results are shown in Table 6.3.

Table 6.3 Sign Test Example Data

Surface Concentration (cpm/100 cm²)	Surface Concentration (dpm/100 cm²)	< Action Level?
4	40	Yes
3	30	Yes
11	110	No
1	10	Yes
1	10	Yes
4	40	Yes
6	60	Yes
3	30	Yes
9	90	Yes
6	60	Yes
14	140	No
1	10	Yes
4	40	Yes
10	100	No
2	20	Yes
Number of measurements less than the action level ($S+$) = 12		

The surface activity values in Table 6.3 are determined by dividing the measured cpm by the total efficiency (0.10). No probe area correction is necessary. The mean count rate is 5 cpm, compared to the estimate of 5 cpm used for the LBGR, and the median is 4 cpm. The standard deviation is 4 cpm, which is higher than the value of 2 used to develop the survey design.[6] Thus, the power of the test is lower than planned. With the actual value of the relative shift $(10-5)/4=1.2$, 23 measurements should be collected.

With the 15 measurements collected, the actual Type II decision error rate is between 0.10 and 0.25 (the closest entries in Appendix A, Table A.2a are for $\alpha=0.05$, $\beta=0.10$, and $\Delta/\sigma=1.2$ with $N=18$, and $\alpha=0.05$, $\beta=0.25$, and $\Delta/\sigma=1.2$ with $N=12$). Three measurements exceed the action level. The portion of the material associated with these measurements merits further investigation using the elevated measurement comparison described in MARSSIM Section 8.5.1.

---

[6] Values are reported to one significant figure based on the data in Table 6.3. Interim calculations generally carry extra figures, so rounding to the appropriate number of significant figures only occurs for the final calculation. Rounding results too soon in the calculation may result in unnecessarily deleting individual results (i.e., when the result is exactly equal to the UBGR) resulting in lower statistical power.

The value of $S+$, 12, was compared to the appropriate critical value, q, in Appendix A, Table A.3. In this case, for $N=15$ and $\alpha=0.05$, the critical value is 11. Because $S+$ exceeds q, reject the null hypothesis that the survey unit exceeds the action level. In this case, the slight loss of power attributable to underestimating the standard deviation did not affect the result. Pending the outcome of the investigation of the three elevated measurements, this survey unit has satisfied the disposition criteria established for clearance.

6.6   Conduct the Wilcoxon Rank Sum Test

The WRS test is used to compare each material survey unit with an appropriately chosen reference material. Each reference material should be selected on the basis of its similarity to the survey unit material, as discussed in Section 3.9. The WRS test can be applied to either Scenario A or Scenario B. Further information on the WRS test can be found in Section 8.4 of MARSSIM and Chapter 6 of NUREG- 1505 (NRC1998a).

6.6.1   Apply the WRS Test to Scenario A

The WRS test is applied to Scenario A as outlined in the following steps and further illustrated by the example in Section 6.6.2.

1. Obtain the adjusted reference material measurements, Z_i, by adding the action level to each reference material measurement, X_i. $Z_i = X_i + $ AL.
2. The m adjusted reference sample measurements, Z_i, from the reference material and the n sample measurements, Y_i, from the survey unit are pooled and ranked in order of increasing size from 1 to N, where $N = m + n$.
3. If several measurements are tied (i.e., have the same value), they are all assigned the mean rank of that group of tied measurements.
4. If there are t "less than" values, they are all given the mean of the ranks from 1 to t. Therefore, they are all assigned the rank $t(t+1)/(2\,t) = (t+1)/2$, which is the mean of the first t integers. If there is more than one MDC,[7] all observations below the largest MDC should be treated as "less than" values. If more than 40% of the data from either the reference material or the survey unit are reported as less than detectable, the WRS test *cannot* be used.
5. The sum of all the ranks, which is the sum of the first N positive integers, is $N(N+1)/2$, which equals W_r added to W_s. Thus, one needs only to sum the ranks of the either the adjusted reference measurements (W_r) or the sum of the ranks of the sample measurements (W_s).
6. Compare W_r with the critical value (q) given in Table A.4 for the appropriate values of n, m, and α. If W_r is greater than the tabulated value for q, reject the hypothesis that the survey unit exceeds the disposition criterion.

6.6.2   Apply the WRS Test to Scenario B

The WRS test is applied to Scenario B as outlined in the following steps:

---

[7] Examples of situations where there could be more than one MDC include using multiple laboratories to perform sample analyses and using different instruments with different backgrounds and different efficiencies to perform measurements.

1. Obtain the adjusted survey unit measurements, Z_i, by subtracting the LBGR from each survey unit measurement, Y_i. $Z_i = Y_i -$ LBGR.
2. The n adjusted survey unit measurements, Z_i, and the m reference material measurements, X_i, are pooled and ranked in order of increasing size from 1 to N, where $N = m + n$.
3. If several measurements are tied (i.e., have the same value), they are all assigned the mean rank of that group of tied measurements.
4. If there are t "less than" values, they are all given the mean of the ranks from 1 to t. Therefore, they are all assigned the rank $t(t+1)/(2\ t) = (t+1)/2$, which is the mean of the first t integers. If there is more than one MDC, all observations below the largest MDC should be treated as "less than" values. If more than 40% of the data from either the reference material or the survey unit are reported as less than detectable, the WRS test *cannot* be used.
5. Sum the ranks of the adjusted measurements from the survey unit, W_s. The sum of all the ranks, which is the sum of the first N positive integers, is $N(N+1)/2$, which equals W_r added to W_s. Thus, one needs only to sum the ranks of the either the adjusted reference measurements (W_r) or the sum of the ranks of the sample measurements (W_s).
6. Compare W_s with the critical value (q) given in Table A.4 for the appropriate values of n, m, and α. (Note that when using this table for Scenario B, the roles of m and n are reversed. If the Quantile test is being used in addition to the WRS test, then $\alpha/2$ should be used rather than α.) If W_s is greater than the tabulated value for q, reject the hypothesis that the difference in the median concentration between the survey unit and the reference area is less than the LBGR.

6.6.3   WRS Test Scenario A Example: Class 2 Metal Ductwork

This example illustrates the use of the WRS test for releasing Class 2 metal ductwork. Assume that a gas-flow proportional detector was used to make gross (non-radionuclide-specific) surface activity measurements.

The DQOs from this survey unit include $\alpha = 0.05$ and $\beta = 0.05$, and the action level converted to units of gross cpm is 2,300 cpm, which is the UBGR. In this case, the WRS test is used because the estimated background level (2,100 cpm) was large compared to the action level. The estimated standard deviation of the measurements, σ, is 375 cpm. The estimated added activity level is 800 cpm; the LBGR is set at this value, and represents the DL. The relative shift is calculated as Δ/σ, which is (action level – LBGR)/σ, which equals 4.

The sample size needed for the WRS test can be found in Table A.2b for these DQOs. The result is nine measurements in each survey unit and nine in each reference material $\alpha = 0.05$, and $\beta = 0.05$, and $\Delta/\sigma = 4$). The ductwork was laid flat onto a prepared grid, and the 9 measurements needed in the survey unit were made using a random-start triangular grid pattern. For the reference materials, the measurement locations were chosen randomly on a suitable batch of material. Table 6.4 lists the gross count rate data obtained.

Table 6.4 Scenario A WRS Test Example Data

Data (cpm)	Area	Adjusted Data	Ranks	Reference Material Ranks
2180	R	4480	15	15
2398	R	4698	16	16
2779	R	5079	18	18
1427	R	3727	10	10
2738	R	5038	17	17
2024	R	4324	13	13
1561	R	3861	11	11
1991	R	4291	12	12
2073	R	4373	14	14
2039	S	2039	3	0
3061	S	3061	8	0
3243	S	3243	9	0
2456	S	2456	7	0
2115	S	2115	4	0
1874	S	1874	2	0
1703	S	1703	1	0
2388	S	2388	6	0
2159	S	2159	5	0
		Sum =	171	126

In the "Area" column, the code "R" denotes a reference material measurement and "S" denotes a survey unit measurement. The adjusted data were obtained by adding the action level to the reference material measurements (see Section 6.6.1, Step 1). The ranks of the data range from 1 to 18, because there are a total of 9+9 measurements (see Section 6.6.1, Step 2). Note that the sum of all of the ranks is still $18(18+1)/2 = 171$. Checking this value with the formula in Step 5 of Section 6.6.1 is recommended to guard against errors in the rankings.

The total of the ranks belonging to the reference material measurements is 126. This is compared with the entry for the critical value of 104 in Table A.4 for $\alpha = 0.05$, with $n = 9$ and $m = 9$. Because the sum of the reference material ranks is greater than the critical value, the null hypothesis (i.e., that the mean survey unit concentration exceeds the action level) is rejected, and the ductwork is released.

This conclusion can be reached quickly by noting the difference between the largest survey unit measurement (3,243 cpm) and the smallest reference area measurement (1,427 cpm). This difference (3,243 – 1,427 = 1,816 cpm) is less than the action level of 2,300 cpm. Because the largest possible difference is less than the action level, the mean difference must also be less than the action level.

6.6.4 WRS Test Scenario B Example: Class 2 Metal Ductwork

This example illustrates the use of the Scenario B WRS test for releasing Class 2 metal ductwork, using the same data as in Section 6.6.3. The null hypothesis for Scenario B is that there is no detectable radioactivity above background.

In this case, the action level is set at no radioactivity detectable above the estimated background level (2,100 cpm). The LBGR is equal to the action level, and is set to zero. The regulator specified that the survey be able to detect an average excess of even 1,500 cpm being released. This value is the DL. The UBGR is set equal to the DL (i.e., 1,500 cpm), with $\beta = 0.025$. The owner of the ductwork felt that there was very little if any radioactivity above background present, and was willing to set $\alpha = 0.20$. The estimated standard deviation of the measurements, σ, was 375 cpm. The relative shift is $\Delta/\sigma = (UBGR - LBGR)/\sigma = (1,500 - 0)/375 = 4$.

The sample size needed for the WRS test can be found in Table A.2b. The result is 9 measurements in each survey unit and 9 in each reference material $\alpha/2 = 0.10$, and $\beta = 0.025$, and $\Delta/\sigma = 4$. The data were obtained as in Section 6.6.3. Table 6.4 (on the previous page) lists the gross count rate data obtained. These data were reanalyzed using Scenario B and the results are shown in Table 6.5.

Table 6.5 Scenario B WRS Test Example Data

Data (cpm)	Area	Adjusted Data	Ranks	Survey Unit Ranks
2180	R	2180	11	0
2398	R	2398	13	0
2779	R	2779	16	0
1427	R	1427	1	0
2738	R	2738	15	0
2024	R	2024	6	0
1561	R	1561	2	0
1991	R	1991	5	0
2073	R	2073	8	0
2039	S	2039	7	7
3061	S	3061	17	17
3243	S	3243	18	18
2456	S	2456	14	14
2115	S	2115	9	9
1874	S	1874	4	4
1703	S	1703	3	3
2388	S	2388	12	12
2159	S	2159	10	10
		Sum =	171	94

In the "Area" column, the code "R" denotes a reference material measurement and "S" denotes a survey unit measurement. The adjusted data would be obtained by subtracting the LBGR from the survey unit measurements (see Section 6.6.2, Step 1), but because the LBGR is zero, no adjustment is needed. The ranks of the adjusted data range from 1 to 18, because there are a total of 9+9 measurements (see Section 6.6.2, Step 2). Note that the sum of all of the ranks is still $18(18+1)/2 = 171$. Checking this value with the formula in Step 5 of Section 6.6.2 is recommended to guard against errors in the rankings. The total of the ranks belonging to the survey unit measurements is 94. This is compared with the entry for the critical value of 100 in Table A.4 for $\alpha = 0.10$, with $n = 9$ and $m = 9$. Because the sum of the reference material ranks is less than the critical value, the null hypothesis (i.e., that there is no detectable radioactivity above background) is not rejected, and the ductwork may be released if the Quantile test is passed.

6.7   Conduct the Quantile Test

The Quantile test was developed to detect differences between the surveyed M&E and the reference material that consist of a shift to higher values in only a fraction of the surveyed M&E. The Quantile test is only performed when Scenario B is used, and only if the null hypothesis is not rejected for the WRS test. Using the Quantile test, in tandem with the WRS test, results in higher power to identify M&E that do not meet the disposition criterion than either test by itself. Apply the Quantile test as follows:

1. Calculate α_Q ($\alpha_Q = \alpha/2$).
2. Obtain the adjusted survey unit measurements, Z_i, by subtracting the LBGR from each survey unit measurement, Y_i. $Z_i = Y_i$ - LBGR.
3. The n adjusted survey unit measurements, Z_i, and the m reference material measurements, X_i, are pooled and ranked in order of increasing size from 1 to N, where $N = m + n$.
4. If several measurements are tied (i.e., have the same value), they are all assigned the mean rank of that group of tied measurements.
5. Look up the values for r and q in Table A.5 based on the number of measurements in the survey unit (n), the number of measurements in the reference area (m), and α_Q. The operational decision described in the next step is made using the values for r and q.
6. If q or more of the r largest measurements in the combined ranked data set are from the survey unit, the null hypothesis is rejected.

This form of the Quantile test gives only approximate results, Because Table A.5 provides a limited number of combinations of n, m, and α_Q. It is recommended that several combinations of n, m, and α_Q be considered when interpreting the results of the Quantile test. Sections 7.2 and 7.3 of NUREG-1505 (NRC 1998a) provide additional guidance on interpreting the results of the Quantile test.

As an example, the Quantile test can be applied to the Class 2 Metal Ductwork example of section 6.6.4. Using $n = 9$, $m = 9$, and $\alpha_Q = 0.10$, the nearest entry in Table A.5d has for $r = 3$ $q = 3$ with $\alpha_Q = 0.105$ when $n = 10$ and $m = 10$. This means that all three of the highest measurement would have to be from the survey unit in order to reject the null hypothesis. From Table 6.5, one can see that the two largest measurements are from the survey unit, but the third largest is from the reference area. Because the ductwork has passed both the WRS and the Quantile test in the Scenario B example, one would conclude that it could be released from radiological controls.

6.8   Evaluate the Results: The Decision

Once the data and results of the tests are obtained, the specific steps required to make a disposition decision depends on the procedures approved by the regulator. The following considerations are suggested for the interpretation of the test results with respect to the disposition criteria. Note that the tests need not be performed in any particular order.

The interpretation of results from the data evaluation or statistical test is the decision to reject or not to reject the null hypothesis. For some of the survey designs the decision is straightforward, while for other designs the interpretation is more complex. Figures 6.4 and 6.4 summarize the interpretation of results.

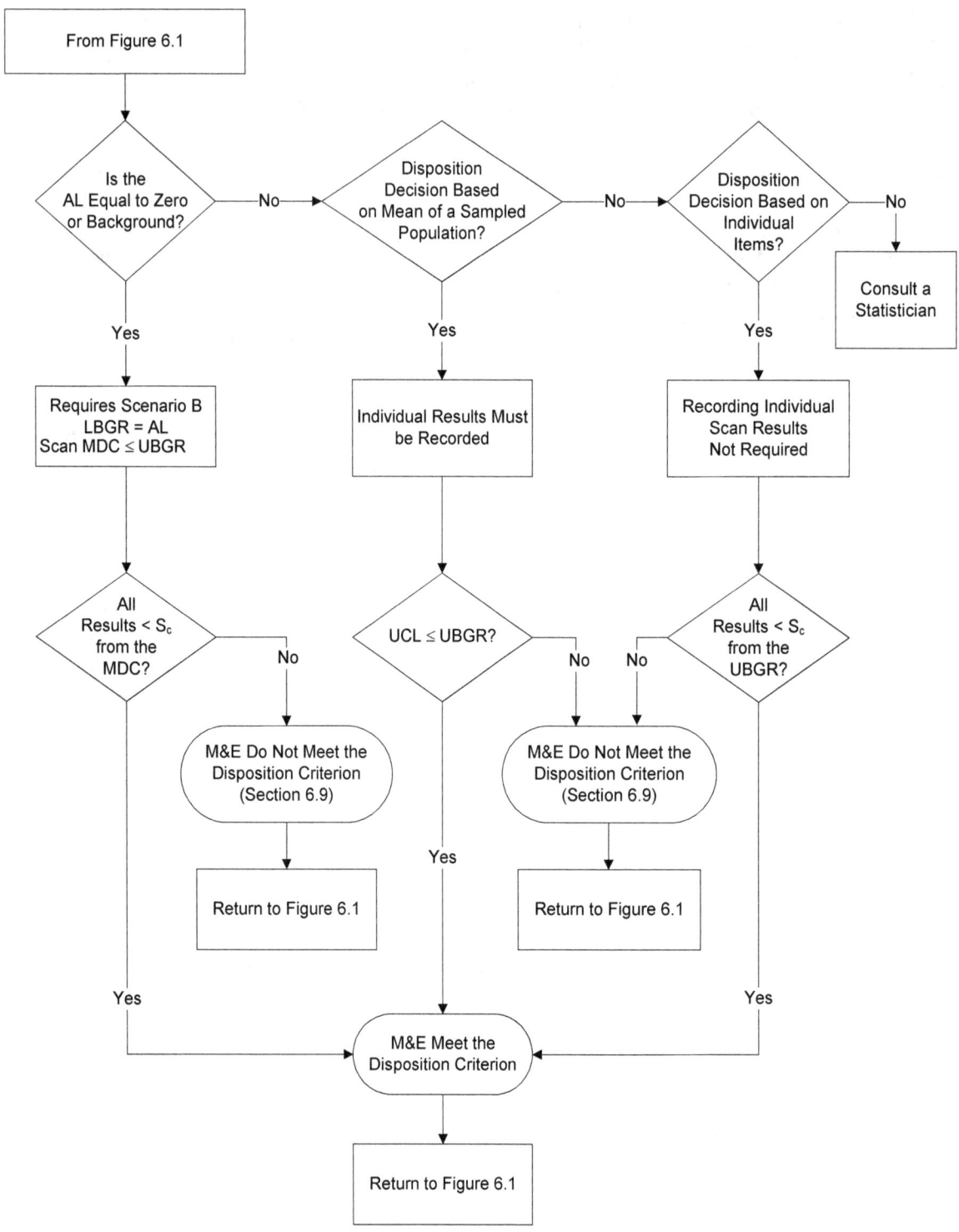

Figure 6.4 Interpretation of Survey Results for Scan-Only and In Situ Surveys

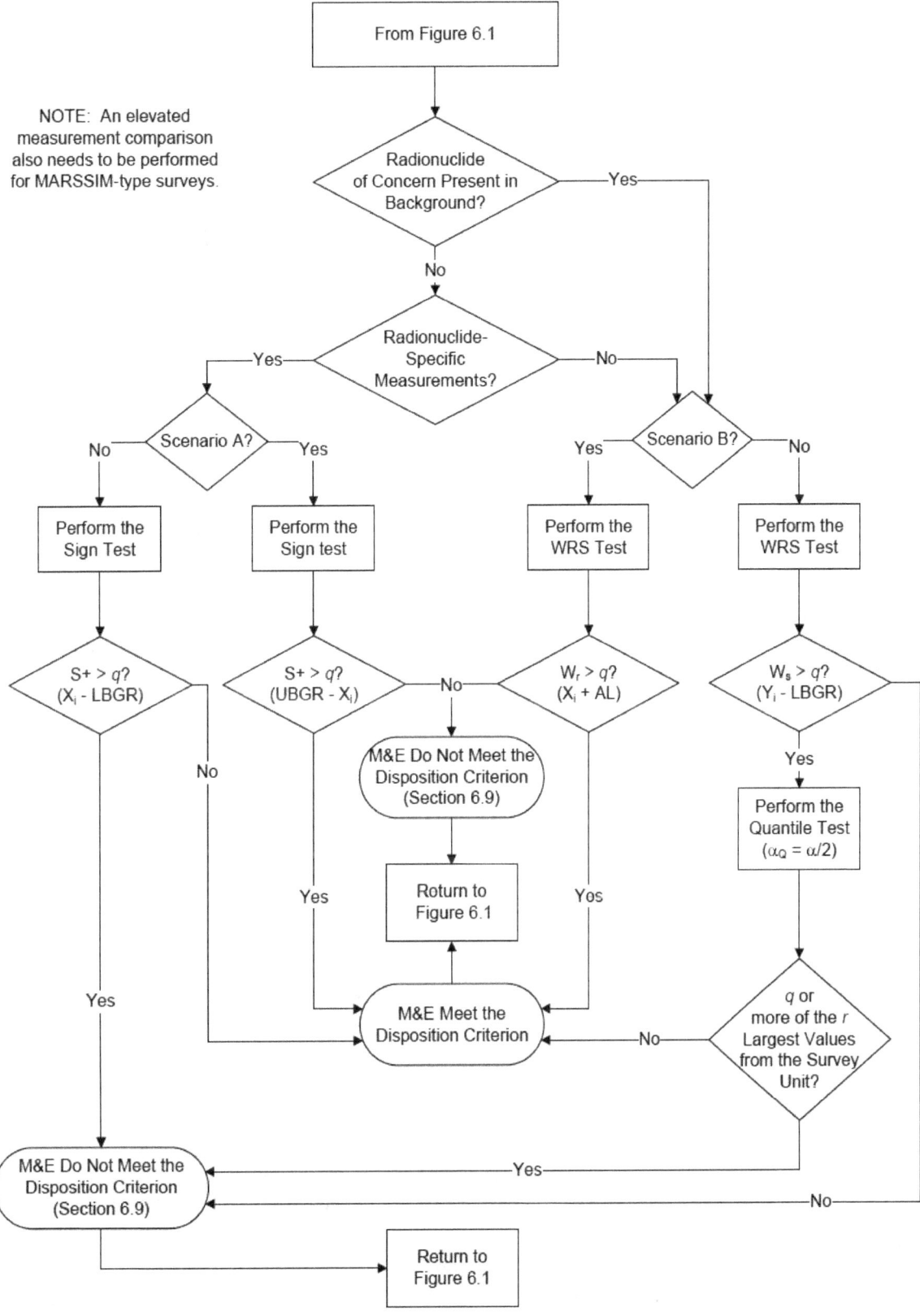

Figure 6.5 Statistical Interpretation of Results for MARSSIM-Type Surveys

6.8.1 Compare Results to the UBGR

The process for interpreting results compared to the UBGR depends on the action level used to develop the survey design. Refer to Table 6.1 for issues and assumptions underlying this evaluation method.

If the action level is zero or background, Scenario B must be used:

- Compare every measurement result to the critical value corresponding to the required scan MDC.
- If all results are below the critical value, the M&E demonstrate compliance with the disposition criterion.
- Any results that exceed the critical value provide evidence of radionuclide concentrations or radioactivity levels exceeding the disposition criteria, so the M&E do not demonstrate compliance with the release criterion.

If the action level is not zero or background—

- Compare every measurement result to the critical value corresponding to the UBGR.
- If all results are below the critical value, the M&E demonstrate compliance with the disposition criterion.
- Any results that exceed the critical value provide evidence of radionuclide concentrations or radioactivity levels exceeding the disposition criteria, so the M&E do not demonstrate compliance with the release criterion.

Scan-only results are usually available as the data are collected. This real-time availability of results allows the surveyor to make decisions as the data are collected. M&E that exceed the action level can be identified and segregated during implementation of the survey. This "clean as you go" approach to surveys is only applicable for Class 1 surveys where there is high confidence in the quality and accuracy of detection decisions around the UBGR. Extensive documentation of the measurement process, previous applications of the process to the same or similar M&E, and verification of MDCs and MQCs is generally necessary to implement a "clean as you go" survey design.

6.8.2 Compare Results Using an Upper Confidence Limit

When decisions are made based on the mean of a sampled population, the survey results should be evaluated by comparison to a UCL (refer to Table 6.1 for issues and assumptions underlying this evaluation method):

- Compare every measurement result to the critical value corresponding to the UBGR.
- If all results are below the critical value, the M&E demonstrate compliance with the disposition criterion.
- If any results are above the critical value, calculate the UCL (Section 6.4.1).
- If the UCL is less than the UBGR, the M&E demonstrate compliance with the disposition criterion.

- If the UCL exceeds the UBGR, the M&E do not demonstrate compliance with the disposition criterion.
- Investigate measurements exceeding the UBGR.
- Results above the UBGR trigger a reevaluation of classification as Class 2.
- Results above the MDC trigger a reevaluation of classification as Class 3.

6.8.3 Compare Results for MARSSIM-Type Surveys

The process for evaluating MARSSIM-type survey results is more complicated. This process is explained in more detail in MARSSIM Section 8.5 (refer to Table 6.1 for issues and assumptions underlying this evaluation method):

- Calculate the test statistics (see Section 6.5.1, 6.6.1, 6.6.2, and 6.7).
- Look up the critical value in the appropriate statistical table in Appendix A.
- Evaluate the results of the statistical test as described in Figures 6.3 and 6.4.
- Evaluate individual results using the elevated measurement comparison (EMC).
- M&E must pass the statistical test and the EMC (if applicable) to demonstrate compliance.

If the null hypothesis is rejected under Scenario A, there is sufficient evidence to show the median radionuclide concentrations or radiation levels are below the disposition criterion. Under Scenario B, failing to reject the null hypothesis means there is insufficient evidence to overturn the initial assumption the M&E demonstrate compliance with the disposition criterion.

If the null hypothesis is rejected under Scenario B, additional investigations are required to determine the final disposition of the M&E (see Section 6.8.2). Failure to reject the null hypothesis under Scenario A also requires additional investigations.

6.9 Investigate Causes for Survey Unit Failures

When M&E fail to demonstrate compliance with the disposition criterion, the first step is to review and confirm the data that led to the decision. Once this is done, the DQO process can be used to evaluate potential problem areas leading to failure.

If the level of radioactivity on or in some Class 1 M&E exceeds the UBGR, the simplest solution might be to segregate those items for a different disposition decision. The concept of "clean as you go" for Class 1 M&E was discussed in Section 6.8.1 where individual objects or sample locations were identified during implementation of the survey design. A simple modification to this approach is to physically segregate the objects exceeding the action level as they are identified, or after reanalysis shows the cleaning was not effective. The segregated M&E can then be evaluated for a different disposition option (e.g., reuse, disposal).

Sometimes activity in excess of background can be removed from the M&E, or remediated, followed by re-evaluation or re-survey of the M&E. This approach may include evaluation of alternatives for remediation and a remedial action support survey prior to performing another final disposition survey.

If the radionuclides of concern have short half-lives, storage of the M&E until the radionuclides have decayed to acceptable levels, or "decay in place," may be an option. The planning team should consider the intrinsic value of the M&E along with storage and disposal costs when considering this option. When multiple radionuclides are present with significantly different half-lives (e.g., order of magnitude) radionuclide-specific measurements may be required to fully evaluate the acceptability of this option.

In other cases, a different disposition option (e.g., reuse, disposal) may be selected. If such a situation were encountered in evaluating Class 2 or Class 3 M&E, the classification would be questioned and the M&E would be reclassified and surveyed as Class 1 M&E. This may also bring other classification decisions into question.

As a general rule, it may be useful to anticipate possible modes of failure. These can be formulated as the problem to be solved using the DQO Process. Once the problem has been stated, the decision concerning the failing survey unit can be developed into a decision rule. For example, decide whether to attempt to remove the radioactivity or simply segregate certain types of M&E for low-level waste disposal. Next, determine the additional data, if any, needed to document that a survey unit where pieces with elevated measurements have been removed or areas of added activity removed demonstrates compliance with the disposition criterion. Alternatives to resolving the decision rule should be developed for each type of M&E that may fail the surveys. These alternatives can be evaluated against the DQOs, and a disposition survey design that meets the objectives of the project can be selected.

6.10 Document the Disposition Survey Results

Documentation of survey results is an important part of the disposition survey process. The form of this documentation can vary greatly depending on the survey objectives and regulatory or administrative requirements. Documentation of disposition survey results should be considered during survey design to ensure adequate records are provided during implementation. Generally, survey documentation requirements are provided as part of the documented survey design. Documented items may include—

- A description of the final disposition, such as disposal in a landfill, return to manufacture for refurbishment, sold as salvage, recycled as ferrous metal, etc.;
- A release statement to the transport carrier and recipient of the material indicating that the M&E described in the bill of laden meet(s) applicable state and federal regulations; and
- Results of QC measurements made during the conduct of release surveys and confirmation of compliance with facility SOPs and action levels.

In both routine and non-routine surveys, the documentation should comply with all applicable regulatory requirements. Development of survey documentation should allow for any necessary or required reviews.

If the disposition survey is a routine survey, then the survey will be documented as specified in the SOP. For example, routine surveys performed to clear M&E from a facility may require documentation that the instruments were calibrated and functioning properly and that trained personnel were on duty to perform the surveys. Quality assurance reviews and audits would be performed periodically (typically under a separate SOP) to document that the clearance surveys were being performed properly and that no M&E were cleared without first being surveyed. These records would document that properly trained personnel had adequately surveyed all M&E leaving the facility using properly functioning instruments. Documentation of individual measurement results may not be required or necessary.

If the survey is not routine, significantly more documentation may be required. This documentation should provide a complete and unambiguous record of the radiological status of the M&E relative to the selected action levels. In addition, sufficient data and information should be provided to enable an independent evaluation of the survey results, including repeating measurements at some future time Additional information on documentation is provided in Section 2.5, Section 3.6, Section 4.5, MARSSIM Sections 3.8 and 8.6, and MARSSIM Chapter 5.

7   STATISTICAL BASIS FOR MARSAME SURVEYS

The statistically rigorous quantitative application of measurement quality objectives (MQOs) plays a central role in the MARSAME process. MQOs did not appear explicitly in *Multi-Agency Radiation Survey and Site Investigation Manual* (MARSSIM 2002), but were subsequently developed for radioanalytical chemistry measurements as part of the *Multi-Agency Radiological Laboratory Analytical Protocols* (MARLAP) manual. However, these concepts apply equally well to field measurements of radiation and radioactivity. The MARSAME process incorporates these ideas and extends them to these measurements.

A major development since the publication of MARSSIM was the publication of the *Guide to the Expression of Uncertainty in Measurement*, or "GUM" (ISO 1995). The procedures described in this document have become a de facto standard for estimating the uncertainty associated with measurements of any type. The GUM methodology is essential for the assessment of measurement uncertainty, but was not previously treated in MARSSIM.

Data quality objectives (DQO) form the backbone of the MARSAME process, and are discussed in detail in Chapters 2 and 3. A number of terms with specific statistical meanings are used in this and subsequent sections. The concept of measurement quality objectives (MQOs) and in particular the required measurement method uncertainty was introduced in Section 3.8. These ideas are discussed in greater detail in MARLAP Chapter 3 and Appendix C. While MARLAP is focused on radioanalytical procedures, these concepts are applicable on a much broader scale and are used in MARSAME in Sections 5.5 through 5.8 to guide the selection of measurement methods for disposition surveys for materials and equipment.

In Section 7.1 the general concepts of statistical survey design and hypothesis testing are discussed, with more detail in Section 7.2. In Sections 7.3, 7.4, 7.5, and 7.6, calculation of measurement quality objectives (particularly the required method uncertainty), measurement uncertainty, minimum detectable concentrations (MDCs) and minimum quantifiable concentrations (MQCs), respectively, are introduced. Further details and examples of these topics for the interested reader are then given in Sections 7.7, 7.8, 7.9, and 7.10. This advanced material is optional on initial reading, and may be referred to later as needed. Section 7.11 shows a detailed calculation of a scan MDC, which is used in Chapter 8. This process was described and used in MARSSIM, but a systematic example was constructed for M&E. These calculations are detailed, and are also optional on first reading.

In developing the results in this chapter, a number of new and sometimes only subtly different definitions and symbols are used. For the convenience of the reader, many of these are summarized in the tables below. Table 7.1 provides a summary of notation used for DQOs and MQOs, used primarily in Sections 7.1 and 7.2. Table 7.2 contains notation used for setting MQOs for required method uncertainties (Sections 7.3 and 7.7) and in uncertainty calculations (Sections 7.4 and 7.8). MDC calculations (Sections 7.5 and 7.9) and MQC calculations (Sections 7.6 and 7.10) use the notation added in Table 7.3 and Table 7.4, respectively. Symbols may not have an entry for both formula or reference and type.

Table 7.1 Notation for DQOs and MQOs

Symbol	Definition	Formula or reference	Type
α	Probability of a Type I decision error		Chosen during DQO process
β	The probability of a Type II decision error		Chosen during DQO process
Δ	Width of the gray region	(UBGR-LBGR)	Chosen during DQO process
φ_{MR}	Required relative method uncertainty above the UBGR	u_{MR}/UBGR	Chosen during DQO process
S_C	The critical value of the net instrument signal (e.g., net count)	Calculation of S_C requires the choice of a significance level for the test. The significance level is a specified upper bound for the probability, α, of a Type I error. The significance level is usually chosen to be 0.05.	If a measured value exceeds the critical value, a decision is made that radiation or radioactivity has been detected
$\hat{S}$	net signal		Experimental
σ	The total standard deviation of the data	$(\sigma_S^2 + \sigma_M^2)^{\frac{1}{2}}$	Theoretical population parameter
σ_N	The standard deviation of the mean of N independent measurements	$\sigma_N = \sigma/\sqrt{N}$	
σ_S	Standard deviation due to sampling		Theoretical population parameter
σ_M	Standard deviation of the measurement method		Theoretical population parameter
σ_{MR}	Required method standard deviation at and below the UBGR	Upper bound to the value of σ_M	Theoretical population parameter
u_{MR}	Required method uncertainty at and below the UBGR	Upper bound to the value of u_M	Chosen during DQO process
$u_c^2(y)$	Combined variance of y	Uncertainty propagation	Calculated
$u_c(y)$	Combined standard uncertainty of y	Uncertainty propagation	Calculated
$z_{1-\alpha}$ $(z_{1-\beta})$	$1-\alpha$ (or $1-\beta$) quantile of a standard normal distribution function	Table of standard normal distribution	Theoretical

Table 7.2 Notation for Uncertainty Calculations

Symbol	Definition	Formula or reference	Type
a	Half-width of a bounded probability distribution	Type B evaluation of uncertainty	Estimated
c_i	Sensitivity coefficient	$\partial f / \partial x_i$, the partial derivative of f with respect to x_i	Evaluated at the measured values $x_1, x_2, \ldots, x_N$
$f(x_1, x_2, \ldots, x_N)$	The calculated value of the output quantity from measurable input quantities for a particular measurement	$y = f(x_1, x_2, \ldots, x_N)$	Experimental
$f(X_1, X_2, \ldots, X_N)$	Model equation expressing the mathematical relationship, between the measurand, Y and the input quantities X_i	$Y = f(X_1, X_2, \ldots, X_N)$	Theoretical
k	Coverage factor for expanded uncertainty	Numerical factor used as a multiplier of the combined standard uncertainty in order to obtain an expanded uncertainty	Chosen during DQO process
p	Coverage probability for expanded uncertainty	Probability that the interval surrounding the result of a measurement determined by the expanded uncertainty will contain the value of the measurand	Chosen during DQO process
$r(x_i, x_j)$	Correlation coefficient for two input estimates, x_i and x_j	$u(x_i, x_j) / (u(x_i)\, u(x_j))$	Experimental
$s(x_i)$	Sample standard deviation of the input estimate x_i	$s(x_i) = \sqrt{\dfrac{1}{(n-1)}\sum_{k=1}^{n}(x_{i,k} - \overline{x_i})^2}$	Experimental
$u(x_i)$	Type B standard uncertainty of the input estimate x_i		Estimated
$u_i(y)$	Component of the combined standard uncertainty $u_c(y)$ generated by the standard uncertainty of the input estimate x_i, $u(x_i)$	$u_i(y) = c_i\, u(x_i)$	Estimated
$u_c(y)$	Combined standard uncertainty of y	Uncertainty propagation	Calculated
$u_c^2(y)$	Combined variance of y	Uncertainty propagation	Calculated
U	Expanded uncertainty	"Defining an interval about the result of a measurement that may be expected to encompass a large fraction of values that could reasonably be attributed to the measurand" (GUM)	Calculated
$u(x_i, x_j)$	Covariance of two input estimates, x_i and x_j		Experimental

Table 7.2 Notation for Uncertainty Calculations (Continued)

Symbol	Definition	Formula or reference	Type
$u_c(y)/y$	Relative combined standard uncertainty of the output quantity for a particular measurement		Experimental
$u(x_i)/x_i$	Relative standard uncertainty of a nonzero input estimate x_i for a particular measurement		Experimental
$w_1, w_2, ..., w_N$	Input quantities appearing in the numerator of $y = f(x_1, x_2, ..., x_N)$	See "$z_1, z_2, ..., z_N$" below	
$X_1, X_2, ..., X_N$	Measurable input quantities		Theoretical
$x_1, x_2, ..., x_N$	Estimates of the measurable input quantities for a particular measurement		Experimental
Y	The output quantity or measurand		Theoretical
y	Estimate of the output quantity for a particular measurement		Experimental
$z_1, z_2, ..., z_N$	Input quantities appearing in the denominator of $y = f(x_1, x_2, ..., x_N)$	$N = n+m$	Experimental

Table 7.3 Notation for MDC Calculations

Symbol	Definition	Formula or reference	Type
N_B	Background count		Experimental
N_S	Gross sample count		Experimental
t_S	Count time for the test source or sample		Experimental
t_B	Count time for the background		Experimental
R_B	Mean count rate of the blank	$R_B = \dfrac{N_B}{t_B}$	
d	Parameter in the Stapleton equation for the critical value of the net instrument signal	Usually has the value 0.4	
ε	Efficiency	Calibration	Experimental or Theoretical
F	Calibration function	$X = F(Y)$	
F^{-1}	Evaluation function	$Y = F^{-1}(X)$, closely related to the mathematical model $Y = f(X_1, X_2, ..., X_N)$	

Table 7.3 Notation for MDC Calculations (Continued)

Symbol	Definition	Formula or reference	Type
S_C	Critical value of the net instrument signal	Net instrument signal is calculated from the gross signal by subtracting the estimated background and any interferences	
S_D	Minimum detectable value of the net instrument signal	Net instrument signal that gives a specified probability, $1-\beta$, of yielding an observed signal greater than its critical value S_C	
X	Observable response variable, measurable signal		Experimental
x_C	The critical value of the response variable	Calculation of y_C requires the choice of a significance level for the test. The significance level is a specified upper bound for the probability, α, of a Type I error. The significance level is usually chosen to be 0.05.	If a measured value exceeds the critical value, a decision is made that radiation or radioactivity has been detected
Y	State variable, measurand	Uncertainty propagation	
y_C	Critical value of the concentration	$y_C = F^{-1}(x_C)$	
y_D	Minimum detectable concentration (MDC)	$y_D = \dfrac{S_D}{\varepsilon}$	
Δ_B	Relative systematic error in the background determination		Experimental
Δ_A	Relative systematic error in the sensitivity		Experimental

Table 7.4 Notation for MQC Calculations

Symbol	Definition	Formula or reference	Type
k_Q	Multiple of the standard deviation defining y_Q, usually chosen to be 10	$k_Q = \dfrac{\sqrt{\sigma^2(y\,\vert\,Y = y_Q)}}{y_Q}$	Chosen during DQO process
$\sigma^2(y\,\vert\,Y = y_Q)$	The variance of y given the true concentration Y equals y_Q		Theoretical
y_Q	Minimum quantifiable concentration (MQC)	The concentration at which the measurement process gives results with a specified relative standard deviation $1/k_Q$, where k_Q is usually chosen to be 10	Theoretical
R_I	Mean interference count rate		Experimental

Table 7.4 Notation for MQC Calculations (Continued)

Symbol	Definition	Formula or reference	Type
$\sigma(\hat{R}_I)$	Standard deviation of the measured interference count rate		Experimental
$\phi_{\hat{\varepsilon}}^2$	Relative variance of the measured efficiency, $\hat{\varepsilon}$		Experimental

7.1   Overview of Statistical Survey Design and Hypothesis Testing

Designing a MARSAME survey involves the following key statistical parameters:

(1) The uncertainty in the measurement method. The measurement method uncertainty can be affected by changes to the measurement method, such as changing counting times, or performing repeated measurements. Generally, the measurement method uncertainty is characterized by its standard deviation, σ_M. This value may be a constant, meaning that all measurements will have the same standard deviation. Alternatively, this value may vary with the level of radionuclide concentration or radioactivity, such that the standard deviation increases with increasing radionuclide concentration or radioactivity.

(2) The uncertainty in the distribution of radionuclide concentrations or radioactivity in the population of materials and equipment (M&E) to be measured. This variation of radionuclide concentrations or radioactivity in space and time can be characterized by the sampling standard deviation, σ_S.

(3) The number of samples, N, from the population of radionuclide concentrations or radioactivity that comprises the survey unit.

(4) The null (H_0) and alternative (H_1) hypotheses to be examined. The symbol Δ represents the detectable difference between the null hypothesis concentration value (the action level, or AL), and the alternative hypothesis concentration value (the discrimination limit, or DL). The range of concentrations between the AL and the DL is referred to as the gray region.

(5) The values of α and β that quantify acceptable limits for Type I and Type II decision errors, respectively. A Type I decision error occurs when the null hypothesis is rejected when it is actually true. A Type II decision error occurs when the null hypothesis is not rejected but should have been rejected. The value of $1-\beta$ is termed the power, or the ability of the statistical test to reject the null hypothesis, when appropriate. For a specific survey design, the power ($1-\beta$) of the survey can be compared at different values of α, since the power is the probability of rejecting the null hypothesis at a given value of α.

Note: Designing a survey involves collecting a number of measurements, N, that will yield the desired α and power ($1-\beta$), given a detectable difference Δ, the σ_M for the measurement method selected and the σ_S for the distribution of radionuclide concentrations or radioactivity in the population of materials and equipment (M&E) to be measured. The relationships between these parameters are complex and interrelated. The choice or determination of one parameter affects the choice or determination of the other parameters.

When a single measurement is taken, the variance of that measurement will equal:

$$\sigma^2 = \sigma^2_M + \sigma^2_S \tag{7-1}$$

In some cases, the distribution of radionuclide concentrations or radioactivity in the population of M&E to be measured and thus σ_S may not be important to a MARSAME survey, e.g., in cases where there is no sampling variability. It then becomes important how the measurement method uncertainty changes when repeated measurements of the same sampling unit are taken. It may be reasonable to assume that the mean of N independent measurements of the same sampling unit will have a standard deviation:

$$\sigma_N = \sigma_M / \sqrt{N} \tag{7-2}$$

When variability in the distribution of radionuclide concentrations or radioactivity in the population of M&E to be measured occurs over time and space, then σ_S is not equal to zero, and must be included in the MARSAME survey design. The variance of the mean of a random sample of N measurements will fall in a range between

$$\sigma_N^2 = [\sigma^2_M + \sigma^2_S]/N \tag{7-3}$$

and

$$\sigma_N^2 = \sigma^2_M + \sigma^2_S/N \tag{7-4}$$

Equation 7.3 corresponds to measurement method uncertainties that are completely uncorrelated, and equation 7.4 corresponds to measurement method uncertainties that are completely correlated, due to common parameters with the same uncertainty. Generally, as more measurements are taken, the contribution of the sampling variance, σ^2_S, to the overall variance of the mean tends to disappear, whereas some or all of the measurement method variance, σ^2_M, may remain. The special case where 100% of the M&E is measured may be regarded as the limit when N approaches infinity. Some or all of the measurement method variance may still remain.

Once σ is estimated, the power $(1-\beta)$ of a study will depend upon:

1. The Type I decision error rate (α),
2. The size of the gray region (Δ), and
3. The number of measurements made (N).

The gray region Δ is the range of radionuclide concentrations or quantities between the DL and the AL. In other words, differences between the DL and the AL less than Δ will be detected with power less than the required $1-\beta$ and therefore are uncertain, or "gray." If the AL is defined as the upper bound of the gray region (UBGR), then the lower bound of the gray region (LBGR) is the DL, and is determined by subtracting Δ from the AL.

All of these factors are interdependent. Generally, the process begins with a known AL, and a DL based on process knowledge. With an estimate of σ, an appropriate number of measurements, N, is found to fulfill the desired limits on decision error rates α and β. If any of these are changed, it will affect the others.

In MARSAME, the null and alternative hypotheses concern the true difference in the M&E between containing radionuclide concentrations or radioactivity in excess of the AL above the appropriate background reference M&E.[1] Scenario A uses a null hypothesis that assumes the radionuclide concentration or radioactivity associated with the M&E exceeds the AL. Scenario A is sometimes referred to as "presumed not to comply" or "presumed not clean." Scenario B uses a null hypothesis that assumes the radionuclide concentration or radioactivity associated with the M&E is less than or equal to the AL. Scenario B is sometimes referred to as "indistinguishable from background" (when the AL is zero) or "presumed clean."

> **Note**: Under Scenario A, the M&E are only deemed suitable for release if the null hypothesis is rejected, whereas under Scenario B, the M&E are suitable for release only if the null hypothesis is not rejected.

For example, under Scenario A, if the true, but unknown, value of the radionuclide concentration or radioactivity in excess of background is less than or equal to the DL, then the hypothesis test upon which the survey is designed will have power $1-\beta$ to reject the null hypothesis that the true, but unknown, value is greater than or equal to the AL at Type I error rate α. Under Scenario B, if the true, but unknown, value of radionuclide concentration or radioactivity in excess of background is greater than the DL (AL + Δ), then the hypothesis test upon which the survey is designed will once again have power $1-\beta$ to reject this null hypothesis at Type I error rate α.

For a given α and $1-\beta$, Δ depends on σ, so it is important that the measurement method (and sampling fraction, where appropriate) selected is sensitive enough to provide a small enough σ, in order that Δ meets survey design requirements for the DL. This ensures that the DL is not set too low in Scenario A or too high in Scenario B. For normally distributed measurements.

$$\Delta/\sigma = (z_{1-\beta} + z_{1-\alpha}) \tag{7-5}$$

Segregation according to likely radionuclide concentrations or radioactivity or a measurement method with a longer counting time may improve σ and therefore Δ. Hypothesis testing (i.e., accepting or rejecting the null hypothesis) consists of comparing an estimate of the radionuclide concentration or radioactivity to a "critical value," S_C. The result indicates whether the observed estimate is consistent with the null value for a given Type I error rate α, after taking account of the uncertainty σ of the measurement. For Scenario A, the critical value is

$$S_C = AL - z_{1-\alpha}\,\sigma \tag{7-6}$$

And for Scenario B the critical value is

$$S_C = AL + z_{1-\alpha}\sigma \tag{7-7}$$

---

[1] Note that the radionuclides of concern may not be contained in the background reference M&E. If radionuclide specific measurements are made, background reference data will be unnecessary.

Where $z_{1-\alpha}$ is the $1-\alpha$ quantile of the standard normal distribution. In situations where the distribution of the estimate may not be normally distributed, more specialized statistical analysis may be needed. By definition, the power $1-\beta$ is the probability as computed under the alternate hypothesis of rejecting the null hypothesis, or that the probability that the observed estimate is less than the critical value S_C for Scenario A, and greater than S_C for Scenario B.

7.2   Statistical Decision-Making

In Section 4.2, MARSAME recommends the planning team complete the following steps:

- Select a null hypothesis,
- Choose a discrimination limit,
- Define Type I and Type II decision errors,
- Set a tolerable Type I decision error rate at the action level, and
- Set a tolerable Type II decision error rate at the discrimination limit.

7.2.1   Null Hypothesis

In hypothesis testing, two assertions about the actual level of radioactivity associated with the M&E are formulated. The two assertions are called the null hypothesis (H_0) and the alternative hypothesis (H_1). H_0 and H_1 together describe all possible radionuclide concentrations or levels of radioactivity under consideration. The survey data are evaluated to choose which hypothesis to reject or not reject, and by implication which to accept.[2] In any given situation, one and only one of the hypotheses must be true. The null hypothesis is assumed to be true within the established tolerance for making decision errors (Section 7.2.5). Thus, the choice of the null hypothesis also determines the burden of proof for the test.

If the action level (AL) is not zero, the planning team generally assumes the radionuclide concentration or level of radioactivity (X) exceeds the action level unless the survey results provide evidence to the contrary. In other words, surveys are designed to provide sufficient evidence to disprove H_0. In this case, the null hypothesis is that the radionuclide concentration or level of radioactivity is greater than or equal to the action level (i.e., H_0: $X \geq$ AL). The alternative hypothesis is the radionuclide concentration or level of radioactivity is less than the action level (i.e., H_1: $X <$ AL). MARSSIM and NUREG-1505 (NRC 1998a) describe this as Scenario A, and the burden of proof falls on the owner of the M&E. Scenario A is sometimes referred to as "presumed not to comply" or "presumed not clean."

On the other hand, the planning team may choose to assume the action level has not been exceeded unless the survey results provide evidence to the contrary. The null hypothesis becomes H_0: $X \leq$ AL, and the alternative hypothesis is H_1: $X >$ AL. MARSSIM and NUREG-1505 (NRC 1998a) describe this as Scenario B, and the burden of proof falls on the regulator. Scenario B is sometimes referred to as "indistinguishable from background" or "presumed

---

[2] In hypothesis testing, to "accept" the null hypothesis only means not to reject it. For this reason many statisticians avoid the word "accept." A decision not to reject the null hypothesis does not imply the null hypothesis has been shown to be true.

clean." This is the only practical approach when the action level is equal to zero (above background); because it is technically impossible to obtain statistical evidence that the radionuclide concentration or level of radioactivity is exactly zero. However, Scenario B can be applied to situations other than "indistinguishable from background." The example in Section 8.4 uses Scenario B to support an interdiction decision.

7.2.2   Discrimination Limit

Action levels were defined in Section 3.3 based on the selected disposition option and applicable regulatory requirements. The planning team also chooses another radionuclide concentration or level of radioactivity that can be reliably distinguished from the action level by performing measurements (i.e., direct measurements, scans, in situ measurements, samples and laboratory analyses). This radionuclide concentration or level of radioactivity is called the discrimination limit (DL). An example where the discrimination limit is defined is provided in Section 8.4.5. The gray region is defined as the interval between the action level and the discrimination limit (Figures 7.1, 7.2, 7.3, and 7.4 provide visual descriptions of the gray region). The width of the gray region is called the shift and denoted as Δ. The objective of the disposition survey is to decide whether the concentration of radioactivity is more characteristic of the DL or of the AL, i.e., whether action should be taken, or if action is not necessary. Figures 7.1 and 7.2 show examples that would fall under Scenario A (discussed in Section 7.2.3). In Figure 7.1 (top) the difference in concentration between the AL and the DL (i.e., Δ) is large; but the variability in the measured concentration (i.e., σ) is also large. In Figure 7.2 (bottom) the difference in concentration between the AL and the DL (i.e., Δ) is relatively small. However, the variability in the measured concentration (i.e., σ) is also smaller. Figures 7.1 and 7.2 illustrate that determining the level of survey effort depends not just on the width of the gray region, but also in the ratio of that width to the expected variability of the data. This ratio, Δ/σ, is called the relative shift in MARSSIM. In situations where Δ/σ is small, i.e., less than 1, it may be impracticable to achieve the required accuracy of measurements or the number of samples to meet the Type I error rate in the DQOs. Section 4.4.4 presents options for relaxing project constraints to optimize the survey design in such cases. In Figure 7.1, Δ/σ is greater than 4; while in Figure 7.2, Δ/σ is approximately 1.

As discussed in MARSSIM, generally, the larger Δ/σ, the easier the survey effort. When Δ/σ is greater than three, the survey effort will be minimal, and any effort to increase it by either widening the gray region or reducing the measurement variability usually would not be worthwhile.

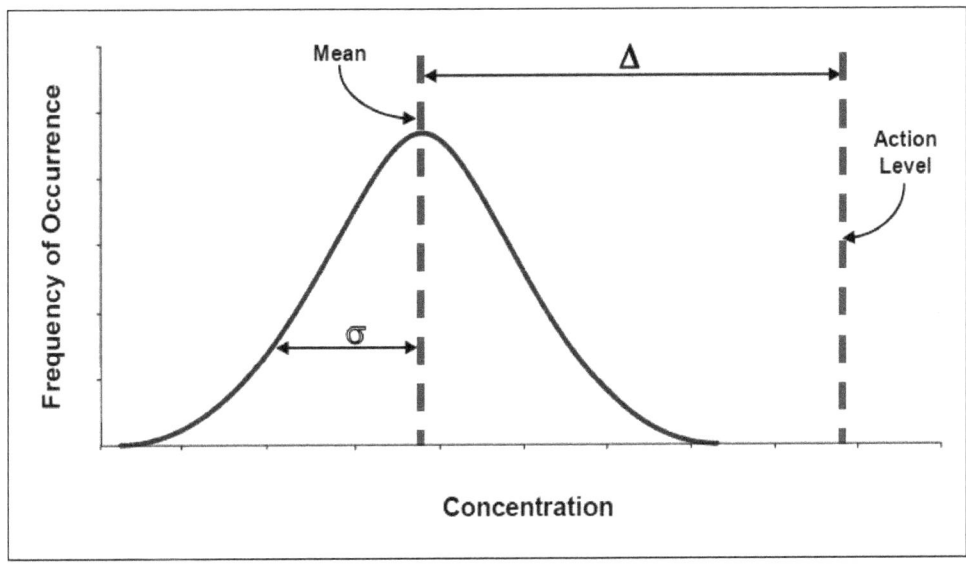

**Figure 7.1 Relative Shift, Δ/σ, Comparison for Scenario A:
σ is Large, but the Large Δ Results in a Large Δ/σ and Fewer Samples**

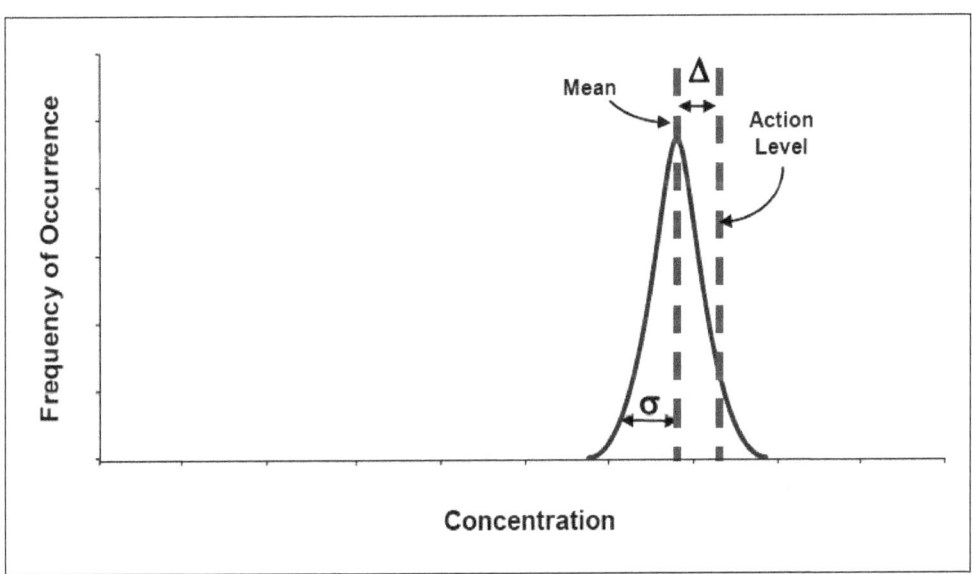

**Figure 7.2 Relative Shift, Δ/σ, Comparison for Scenario A:
σ is Small, but the Small Δ Results in a Small Δ/σ and More Samples**

On the other hand, when Δ/σ is less than one, the survey effort will become substantial, and any effort to increase it by either widening the gray region or reducing the measurement variability will be worthwhile. The measurement variability is thus just as important as the width of the gray region when designing disposition surveys. In MARSSIM surveys, the total variability had two components: sampling and analytical. For some MARSAME surveys this will also be the case. However, in many MARSAME surveys the sampling variability will be of less importance, either because 100% of the survey unit is being measured, or because disposition decisions are being made on the basis of single measurements on single items or single locations. In such cases, the required measurement method uncertainty discussed in Section 3.8.1 will be of paramount importance in the survey planning. The details for determining the required measurement method uncertainty and how to determine if it is being met are discussed in detail in Section 7.7.

Depending on the survey design, the combination of action levels, expected radionuclide concentrations or levels of radioactivity, instrument sensitivity, and local radiation background contribute to defining the width of the gray region. Reducing the radionuclide concentrations or levels of radioactivity known or assumed to be associated with the M&E can affect the selection of a discrimination limit, so remediation costs may need to be considered. Increasing the sensitivity of a measurement method to reduce the measurement method uncertainty generally involves increased instrument costs or increased counting times.

The lower bound of the gray region is denoted by LBGR and the upper bound of the gray region is denoted by UBGR. The association of either the UBGR or the LBGR with the DL or AL will depend on the scenario selected (see Sections 7.2.3 and 7.2.4). The width of the gray region (UBGR – LBGR) is denoted by "Δ" and is called the "shift" or the "required minimum detectable difference" in activity or concentration (MARSSIM Section 5.5.2 and Section D.6, MARLAP Section C.2, NRC 1998a, and EPA 2006a,).

7.2.3   Scenario A

The null hypothesis for Scenario A specifies that the radionuclide concentration or level of radioactivity associated with the M&E is equal to or exceeds the action level. For Scenario A (H_0: $X \geq$ AL), the UBGR is equal to the AL and the LBGR is equal to the DL. As a general rule for applying Scenario A, the DL should be set no higher than the expected radionuclide concentration associated with the M&E. The DL and the AL should be reported in the same units. Figure 7.3 illustrates Scenario A. Note that the Type I (α) and Type II (β) error rates need not be equal. This is discussed further in Section 7.2.5, and an example can be seen in Section 7.5.2.

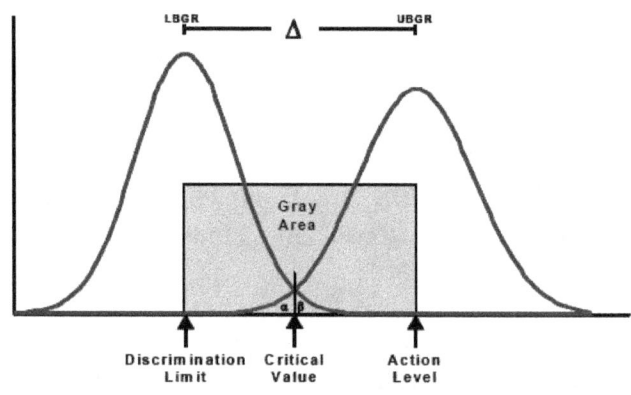

Figure 7.3 Illustration of Scenario A

7.2.4   Scenario B

The null hypothesis for Scenario B specifies the radionuclide concentration or level of radioactivity associated with the M&E is less than or equal to the action level. For Scenario B (H_0: $X \leq$ AL), the UBGR is equal to the DL and the LBGR is equal to the AL. For example, if the AL=0 (sometimes called indistinguishable from background), then the LBGR will be zero. The DL defines how hard the surveyor needs to look, and is determined through negotiations with the regulator.[3] In some cases, the DL will be set equal to a regulatory limit (e.g., 10 CFR 36.57 and DOE 1993). The DL and the AL should be reported in the same units. Figure 7.4 illustrates Scenario B. As above, note that the Type I (α) and Type II (β) error rates need not be equal. This is discussed further in Section 7.2.5, and an example can be seen in Section 7.5.2.

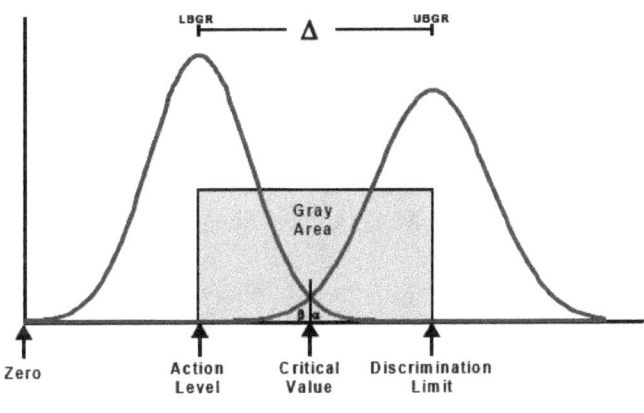

Figure 7.4 Illustration of Scenario B

This description of Scenario B is based on information in MARLAP and is fundamentally different from the description of Scenario B in NUREG-1505 (NRC 1998a).

In NUREG-1505 (NRC 1998a) the gray region is defined as being below the AL in both Scenario A and Scenario B. In MARSAME and MARLAP the gray region is defined as being above the AL in Scenario B. The difference lies in how the action level is defined.

7.2.5   Specify Limits on Decision Errors

There are two possible types of decision errors:

- Type I error: rejecting the null hypothesis when it is true.
- Type II error: failing to reject the null hypothesis when it is false.

---

[3] In some cases setting the discrimination limit may include negotiations with stakeholders.

Because there is always uncertainty associated with the survey results, the possibility of decision errors cannot be eliminated. So instead, the planning team specifies the maximum Type I decision error rate (α) that is allowable when the radionuclide concentration or level of radioactivity is at or above the action level. This maximum usually occurs when the true radionuclide concentration or level of radioactivity is exactly equal to the action level. The planning team also specifies the maximum Type II decision error rate (β) that is allowable when the radionuclide concentration or level of radioactivity equals the discrimination limit. Equivalently, the planning team can set the "power" ($1-\beta$) when the radionuclide concentration or level of radioactivity equals the discrimination limit. See MARSSIM Appendix D, Section D.6, for a more detailed description of error rates and statistical power.

The definition of decision errors depends on the selection of the null hypothesis. For Scenario A the null hypothesis is the radionuclide concentration or level of radioactivity exceeds the action level. A Type I error for Scenario A occurs when the decision maker decides the radionuclide concentration or level of radioactivity is below the action level when it is actually above the action level (i.e., mistakenly decides the M&E are clean when they are actually not clean). A Type II error for Scenario A occurs when the decision maker decides the radionuclide concentration or level of radioactivity is above the action level when it is actually below the action level (i.e., mistakenly decides the M&E are not clean when they are actually clean).

For Scenario B, the null hypothesis is that the radionuclide concentration or level of radioactivity is less than or equal to the action level. A Type I error for Scenario B occurs when the decision maker decides the radionuclide concentration or level of radioactivity is above the action level when it is actually below the action level (i.e., mistakenly decides the M&E are not clean when they are actually clean). A Type II error for Scenario B occurs when the decision maker decides the radionuclide concentration or level of radioactivity is below the action level when it is actually above the action level (i.e., mistakenly decides the M&E are clean when they are actually not clean). It is important to clearly define the scenario (i.e., A or B) and the decision errors for the survey being designed.

Once the decision errors have been defined, the planning team should determine the consequences of making each type of decision error. This process should be revisited as more information is obtained. For example, incorrectly deciding the activity is less than the action level may result in increased health and ecological risks. Incorrectly deciding the activity is above the action level when it is actually below may result in increased economic and social risks. The consequences of making decision errors are specific to the actual situation at a particular site and could vary significantly from one site to another, reflecting the major concerns of the various stakeholders.

Once the consequences of making both types of decision errors have been identified, acceptable decision error rates can be assigned for both Type I and Type II decision errors. Historically, a decision error rate of 0.05, or 5%, often has been acceptable for decision errors. However, assigning the same tolerable decision error rate to all projects does not account for the differences in consequences of making decision errors. This becomes evident with M&E where there are wide ranges of disposition options generating a wide range of consequences. For example, a Type I decision error for Scenario A could have different consequences for a clearance decision compared to a low-level radioactive waste disposal decision. Not all consequences of decision errors are the same, and it is unlikely that applying a fixed value to all decision error rates will result in reasonable survey designs resulting in comparable decisions. Project-specific decision error rates should be selected based on the project-specific consequences of making decision errors.

7.2.6   Develop an Operational Decision Rule

The theoretical decision rule developed in Section 3.7 was based on the assumption that the true radioactivity concentrations or radiation levels associated with the M&E were known. Since the disposition decision will be made based on measurement results and not the true but unknown concentration level, an operational decision rule needs to be developed to replace the theoretical decision rule. The operational decision rule is a statement of the statistical hypothesis test, which is based on comparing some function of the measurement results to some critical value. The theoretical decision rule is developed during Step 5 of the DQO Process (Chapter 3), while the operational decision rule is developed as part of Step 6 and Step 7 of the DQO Process. For example, a theoretical decision rule might be "if the results of any measurement identify surface radioactivity in excess of background, the front loader will be refused access to the site; if no surface radioactivity in excess of background is detected, the front loader will be granted access to the site." The related operational decision rule might be "any result that exceeds the critical value associated with the MDC, set at the discrimination limit, will result in rejection of the null hypothesis and the front loader will not be allowed on the site" (see more examples in Chapter 8).

Chapter 6 provides guidance on using statistical tests to evaluate data collected during the disposition survey to support a disposition decision. The planning team should evaluate the statistical tests and possible operational decision rules and select one that best matches the intent of the theoretical decision rule with the statistical assumptions. Each operational decision rule will have a different formula for determining the number of measurements or fraction of M&E to be measured to meet the DQOs.

Developing an operational decision rule incorporates all relevant information available concerning the M&E (Section 2.4.3), selected instrumentation and measurement technique (Section 5.9), selected statistical tests (Section 6.2.3), and any constraints on collecting data identified by the planning team. The operational decision rule will need to specify a measurement technique (e.g., scan-only, in situ, sample collection and analysis) and a statistical test. Examples of statistical tests include comparison to the UBGR (Section 6.3), comparison to an upper confidence interval (Section 6.4), the Sign test (Section 6.5), the Wilcoxon Rank Sum test (Section 6.6), and the Quantile test (Section 6.7). At this point in the survey design process it is not necessary to select a specific instrument to perform the measurements. However, selection of a measurement technique will assist the planning team in identifying the appropriate statistical test. For example, if a scan-only measurement method is selected it is not appropriate to select the Wilcoxon Rank Sum test to determine the number of measurements. However, if no scan-only or in situ measurement methods are available that meet the measurement quality objectives (MQOs), a MARSSIM-type survey (which combines scan and static measurements, see Section 4.4.3) should be developed.

The planning team uses the combination of the selected instrumentation and measurement technique (see Section 5.9) with a data evaluation method (see Section 6.2.5) to establish an operational decision rule. Then, from the operational decision rule, the planning team can determine the number of measurements or the fraction of the M&E that needs to be measured during the disposition survey. There is no formal structure for stating an operational decision rule. The structure of the operational decision rule is generally defined in terms that meet the needs of a particular project. An operational decision rule can be simple or complex. A simple example could be "If 100% of the surfaces of hand tools are surveyed using a scan-only technique that meets the DQOs, and none of the results exceed the action level for release, then the tools can be released." The statistical test for this simple example is a comparison of the mean to the action level; however, since all of the values are below the action level, the mean value must also be below the action level. Therefore, it is not necessary to perform the actual statistical test. This represents a conservative approach to data interpretation that may not always be appropriate. More complex operational decision rules can–

- Account for different types of measurements and multiple radionuclides of concern,
- Specify critical values and test statistics for the statistical tests, and
- Incorporate multiple decisions (e.g., average and maximum values, fixed and removable radioactivity) depending on the project.

7.3   Set Measurement Quality Objectives

Section 4.2 briefly discussed the DQO process for developing statistical hypothesis tests for the implementation of disposition decision rules using measurement data. This included formulating the null and alternative hypotheses, defining the gray region using the action level and discrimination limit, and setting the desired limits on potential Type I and Type II decision error probabilities that a decision maker is willing to accept for project results. Decision errors are possible, at least in part, because measurement results have uncertainties. The effect of these uncertainties is expressed in the size of the relative shift, Δ/σ, introduced in Section 7.2.2. The overall uncertainty, σ, has components that may be due to sampling variability in radioactivity concentration, σ_S, but also because of uncertainty in the measurement method, σ_M. Because DQOs apply to both sampling and measurement activities, what are needed from a measurement perspective are method performance characteristics specifically for the measurement process of a particular project. These method performance characteristics (see Section 3.8) are the measurement quality objectives (MQOs).

DQOs define the performance criteria that limit the probabilities of making decision errors by–

- Considering the purpose of collecting the data,
- Defining the appropriate type of data needed, and
- Specifying tolerable probabilities of making decision errors.

DQOs apply to both sampling and measurement activities.

MQOs can be viewed as the measurement portion of the overall project DQOs (see Section 3.8). MQOs are:

- The part of the project DQOs that apply to the measured result and its associated uncertainty.
- Statements of measurement performance objectives or requirements for a particular measurement method performance characteristic, for example, measurement method uncertainty and detection capability.
- Used initially for the selection and evaluation of measurement methods.
- Subsequently used for the ongoing and final evaluation of the measurement data.

A number of MQOs were introduced in Section 3.8, but for survey planning the single most important MQO is the required measurement method uncertainty, u_{MR}. Other MQOs, such as range, ruggedness, and specificity, if not controlled, will lead to increased measurement uncertainty. In this sense, the required measurement method uncertainty encompasses many of the effects of other MQO parameters that could impact decision making. MDCs and MQCs are closely related to the measurement uncertainty, have a long history of use for comparing the appropriateness of competing measurement techniques, and can contribute much to survey planning. These concepts are developed further in the later sections of this chapter (Sections 7.5 and 7.6). However, essentially the same information can be conveyed by specifying the required measurement method uncertainty, which is a more general concept. Thus, in this section and the next, it is this MQO that will be emphasized.

Measurement method uncertainty refers to the *predicted* uncertainty of a measured value that would be calculated if the method were applied to a hypothetical sample with a specified radioactivity concentration or radiation level. Measurement method uncertainty is a characteristic of the measurement method and the measurement process. Measurement uncertainty, as opposed to sampling uncertainty, is a characteristic of an individual measurement.

The true measurement method standard deviation, σ_M, is a theoretical quantity and is never known exactly, but it may be estimated using the methods described in Section 7.4. The estimate of σ_M will be denoted here by u_M and called the "measurement method uncertainty." The measurement method uncertainty, when estimated by uncertainty propagation, is the predicted value of the combined standard uncertainty (CSU, or "one-sigma" uncertainty) of the measurement for material with concentration equal to the UBGR. Note that the term "measurement method uncertainty" and the symbol u_M actually apply not just to the measurement method but also to the entire measurement process, that is, it should include uncertainties in how the measurement method is actually implemented. This definition of measurement method uncertainty is independent of the null hypothesis and applies to both Scenario A and Scenario B.

The true standard deviation of the measurement method, σ_M, is unknown, but the required measurement method uncertainty, σ_{MR}, is intended to be an upper bound for σ_M. In practice, σ_{MR} is actually used as an upper bound for the method uncertainty, u_M, which is an estimate of σ_M. Therefore, the value of σ_{MR} will be called the "required measurement method uncertainty" and denoted by u_{MR}.

The principal MQOs in any project will be defined by the required measurement method uncertainty, u_{MR}, at and below the UBGR and the relative required measurement method uncertainty, φ_{MR}, at and above the UBGR, $\varphi_{MR} = u_{MR}/\text{UBGR}$. See Section 7.3.2 for further discussion.

When making decisions about individual measurement results u_{MR} should ideally be 0.3Δ, and when making decisions about the mean of several measurement results u_{MR} should ideally be 0.1Δ, where Δ is the width of the gray region, $\Delta = \text{UBGR} - \text{LBGR}$.

7.3.1   Determine the Required Measurement Method Uncertainty at the UBGR

This section provides the rationale and guidance for establishing project-specific MQOs for controlling σ_M. This control is achieved by establishing a desired maximum measurement method uncertainty, u_{MR}, at the upper boundary of the gray region. This control also will assist in both the measurement method selection process and in the evaluation of measurement data. Approaches applicable to several situations are detailed below.

Four basic survey designs were described in Chapter 4: scan-only, in situ, MARSSIM-type, and method-based. The relative shift, Δ/σ, is important in determining the level of survey effort required in the first three survey designs. For a given width of the gray region, Δ, the relative shift, Δ/σ, can only be controlled by controlling σ. The overall standard deviation of the measurement results, σ, may have both a measurement component, σ_M, and a sampling component, σ_S. Segregation and classification may help in controlling σ_S (Sections 4.3 and 5.4).

7.3.1.1   Scan-Only Survey Designs

For 100% scan-only surveys, the decision uncertainty associated with σ_S is essentially eliminated because the entire survey unit is measured. In class 2 survey units, the scan coverage can vary from 10% to nearly 100% depending on the value of Δ/σ. This is a reflection of the fact that for a fixed measurement variability, σ_M, smaller values of Δ/σ imply larger sampling variability. Larger sampling variability demands higher scan coverage to reduce the decision uncertainty. That is, more of the survey unit must be measured to lower the standard deviation of the mean. In such cases, it will be desirable to reduce σ_M until it is negligible in comparison to σ_S. σ_M can be considered negligible if it is no greater than $\sigma_S/3$. Therefore, MARSAME recommends the requirement $u_{MR} \le \sigma_S/3$.

7.3.1.2   In Situ Survey Designs

For in situ survey designs, either the entire survey unit, or a large portion of it (e.g., greater than 10%), is covered with a single measurement. Thus, sampling variability will tend to be averaged out. When decisions are to be made by comparing such single measurements to an action level, the total variance of the data equals the measurement variance, σ_M^2, and the data distribution in most instances should be approximately normal. In these cases the DQOs will be met if

$$u_{MR} \le \frac{\text{UBGR-LBGR}}{z_{1-\alpha} + z_{1-\beta}} = \frac{\Delta}{z_{1-\alpha} + z_{1-\beta}} \qquad (7\text{-}8)$$

where $z_{1-\alpha}$, is the $(1-\alpha)$-quantile of the standard normal distribution and $z_{1-\beta}$, is the $(1-\beta)$-quantile of the standard normal distribution.

If $\alpha = \beta = 0.05$, then

$$u_{Mr} \leq \frac{\Delta}{z_{0.95} + z_{0.95}} = \frac{\Delta}{1.645 + 1.645} = \frac{\Delta}{3.29} \sim 0.3\,\Delta \qquad (7\text{-}9)$$

Therefore, MARSAME recommends the requirement $u_{MR} \leq 0.3\Delta$. The details are discussed in Section 7.7.2.

For the special case where the LBGR = 0, then Δ = UBGR and $\sigma_{MR} = \Delta / (z_{1-\alpha} + z_{1-\beta})$ implies

$$u_{MR} \leq \frac{UBGR}{z_{0.95} + z_{0.95}} = \frac{UBGR}{1.645 + 1.645} = \frac{UBGR}{3.29} \sim 0.3\,UBGR \qquad (7\text{-}10)$$

This is equivalent to requiring that the MDC (see Section 7.9.2) be less than the action level. The MDC is defined as the concentration at which the probability of detection is $1 - \beta$ and the probability of false detection in a sample with zero concentration is at most α.

Example 1: Suppose the action level is 10,000 Bq/m² and the lower bound of the gray region is 5,000 Bq/m², α = 0.05, and β = 0.10. If decisions are to be made about individual items, then the required measurement method uncertainty at 10,000 Bq/m² is

$$u_{MR} = \frac{\Delta}{z_{1-\alpha} + z_{1-\beta}} = \frac{10{,}000 \text{ Bq/m}^2 \text{-} 5{,}000 \text{ Bq/m}^2}{z_{0.95} + z_{0.90}} = \frac{5{,}000 \text{ Bq/m}^2}{1.645 + 1.282} = 1{,}700 \text{ Bq/m}^2$$

7.3.1.3   MARSSIM-Type Survey Designs

When a decision is to be made about the mean of a sampled population, generally the average of a set of measurements on a survey unit is compared to the disposition criterion. For MARSSIM-type designs, the ratio Δ/σ, called the "relative shift," determines the number of measurements required to achieve the desired decision error rates α and β. The target range for this ratio should be between 1 and 3, as explained in MARSSIM (MARSSIM 2002) and NUREG-1505 (NRC 1998a). Ideally, to keep the required number of measurements low, the DQOs are aimed at establishing $\Delta/\sigma \approx 3$. The cost in number of measurements rises rapidly as the ratio Δ/σ falls below 1, but there is little benefit from increasing the ratio much above 3. One of the main objectives in optimizing survey design is to achieve a relative shift, Δ/σ, of at least one and ideally three. Values of Δ/σ greater than three, while desirable, should not be pursued at additional cost. If Δ/σ is 3 and σ_M is negligible in comparison to σ_S, then σ_M will be $\Delta/10$. The details are discussed in Section 7.7.1.

Therefore, MARSAME recommends the requirement $u_{MR} \leq \Delta / 10$ by default when decisions are being made about the mean of a sampled population. If the LBGR is zero, this is equivalent to requiring that the MQC be less than the UBGR (Section 7.7.1).

Example 2: Suppose the action level is 10,000 Bq/m^2 and the lower bound of the gray region is 2,000 Bq/m^2. If decisions are to be made about survey units based on measurements at several locations, then the required measurement method uncertainty (u_{MR}) at 10,000 Bq/m^2 is

$$\mu_{MR} = \frac{\Delta}{10} = \frac{10,000 - 2,000}{10} = 800 \text{ Bq/m}^2$$

Example 3: Suppose the action level is 10,000 Bq/m^2, but this time assume the lower bound of the gray region is 0 Bq/m^2. In this case the required method measurement uncertainty, u_{MR}, at 10,000 Bq/m^2 is

$$\mu_{MR} = \frac{\Delta}{10} = \frac{10,000 - 0}{10} = 1,000 \text{ } Bq/m^2$$

The recommended values of u_{MR} are based on the assumption that any known bias in the measurement process has been corrected and that any remaining bias is well less than 10% of the shift, Δ, when a concentration near the gray region is measured.

Achieving a required measurement method uncertainty u_{MR} less than the recommended limits may be difficult in some situations. When the recommended requirement for u_{MR} is too difficult to meet, project planners may allow u_{MR} to be larger. In this case, project planners may choose u_{MR} to be as large as $\Delta/3$ or any calculated value that allows the data quality objectives to be met at an acceptable effort. Two situations that may make this possible are if σ_S is believed to be less than $\Delta/10$ or if it is not difficult to make the additional measurements required by the larger overall data variance ($\sigma_M^2 + \sigma_S^2$).

Example 4: Suppose the uncertainty in Example 2 of $u_{MR} = 800$ Bq/m^2 cannot be achieved because of the variability in instrument efficiency with surface roughness. A required measurement method uncertainty, u_{MR}, as large as $\Delta/3 \approx 2,700$ Bq/m^2 may be possible if σ_S is small or if more measurements are taken per survey unit.

7.3.2 Determine the Required Measurement Method Uncertainty at Concentrations Other Than the UBGR

The most important MQO for data evaluation is the one for measurement method uncertainty at a specified concentration. This MQO is expressed as the required measurement method uncertainty (u_{MR}) at the UBGR. However, to properly evaluate the data usability of measurement results at concentrations other than the UBGR, the implications of this requirement must be extended both above and below the UBGR.

When the concentration is less than or equal to the UBGR, the combined standard uncertainty (CSU), u_c, of a measured result should not exceed the required measurement method uncertainty, u_{MR}, specified at the UBGR. When the concentration is greater than the UBGR, the relative combined standard uncertainty (RCSU), φ_{MR}, of a measured result should not exceed the required relative measurement method uncertainty at the UBGR.

$$\varphi_{MR} = u_{MR}/\text{UBGR} \tag{7.11}$$

This is illustrated in Example 5 and Figure 7.5.

Example 5: Suppose the action level is 10,000 Bq/m^2 and the discrimination limit is 3,000. Scenario A is used, so the UBGR = AL = 10,000 Bq/m^2 and the LBGR = DL = 3,000 Bq/m^2. Thus the width of the gray region, Δ= 10,000 – 3,000 = 7,000. If decisions are to be made about individual items, α = 0.05, and β = 0.05, then the required measurement uncertainty at 10,000 Bq/m^2 is

$$u_{MR} = \frac{\Delta}{z_{1-\alpha} + z_{1-\beta}} = \frac{10{,}000 \text{ Bq/m}^2 \text{-}3{,}000 \text{ Bq/m}^2}{z_{0.95} + z_{0.95}} = \frac{7{,}000 \text{ Bq/m}^2}{1.645 + 1.645} \approx 2{,}000 \text{ Bq/m}^2$$

The required measurement method uncertainty, u_{MR}, is 2,000 Bq/m^2 at 10,000 Bq/m^2. Thus, for any measured result less than 10,000 Bq/m^2, the reported CSU, u_c, should be less than or equal to 2,000 Bq/m^2. For example, a reported result of 4,500 Bq/m^2 with a CSU of 1,900 Bq/m^2 would meet the requirement. A reported result of 7,700 Bq/m^2 with a CSU 2,500 Bq/m^2 would not meet the requirement.

The required relative measurement method uncertainty (φ_{MR}) is 2,000 Bq/m^2 / 10,000 Bq/m^2 = 20% at 10,000 Bq/m^2. Thus, for any measured result greater than 10,000 Bq/m^2, the reported RCSU should be less than or equal to 20%. For example, a reported result of 14,500 Bq/m^2 with a CSU of 2,900 Bq/m^2 would meet the requirement because 2,900/14,500 = 20%. A reported result of 18,000 Bq/m^2 with a CSU 4,500 Bq/cm^2 would not meet the requirement because 4,500/18,000 = 25%.

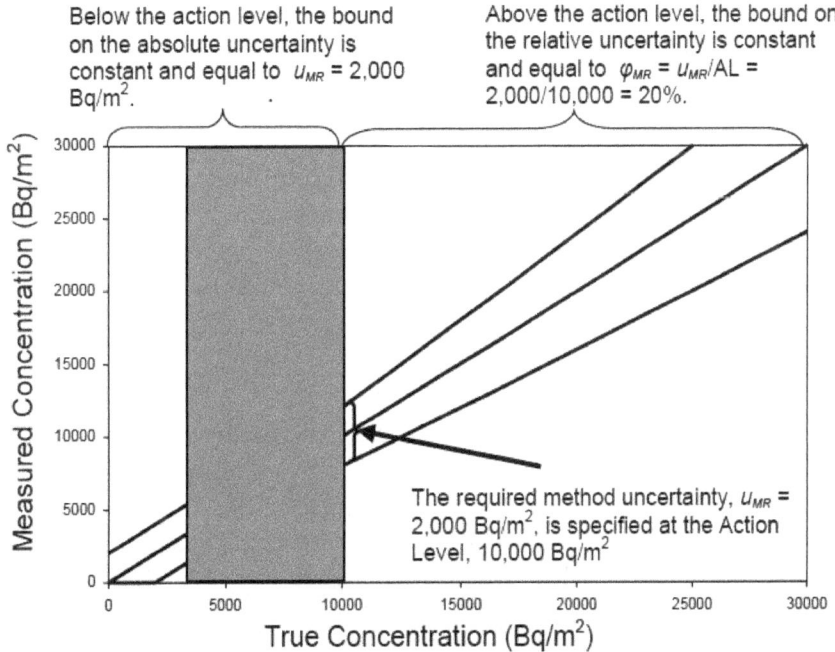

Figure 7.5 Example of the Required Measurement Uncertainty at Concentrations other than the UBGR. In this Example the UBGR Equals the Action Level

7.4   Determine Measurement Uncertainty

Checking the measurement quality against the required measurement method uncertainty relies on having realistic estimates of the measurement uncertainty. Often reported measurement uncertainties are underestimated, particularly if they are confined to the estimated Poisson counting uncertainty (Section 7.8). Tables of results are sometimes presented with a column listing "±" without indicating how these numbers were obtained. Often, the "±" represents the square root of the number of counts obtained during the measurement. The method for evaluation calculation and reporting of measurement uncertainty, approved by both the International Organization for Standardization (ISO) and the National Institute of Standards and Technology (NIST) is discussed in this section. Further details of the method are given in Section 7.8.

Measurements always involve uncertainty, which must be considered when measurement results are used as part of a basis for making decisions. Every measured and reported result should be accompanied by an explicit uncertainty estimate. One purpose of this section is to give users of data an understanding of the causes of measurement uncertainty and of the meaning of uncertainty statements; another is to describe procedures that can be used to estimate uncertainties. Much of this material is derived from MARLAP Chapter 19.

In 1980, the Environmental Protection Agency published a report entitled *Upgrading Environmental Radiation Data*, which was produced by an ad hoc committee of the Health Physics Society (EPA 1980). Two of the recommendations of this report were that:

1. Every reported measurement result (x) should include an estimate of its overall uncertainty (u_x) that is based on as nearly a complete an assessment as possible.
2. The uncertainty assessment should include every significant source of inaccuracy in the result.

The concept of traceability is also defined in terms of uncertainty. Traceability is defined as the "property of the result of a measurement or the value of a standard whereby it can be related to stated references, usually national or international standards, through an unbroken chain of comparisons all having stated uncertainties" (ISO 1996). Thus, to realistically make the claim that a measurement result is "traceable" to a standard, there must be a chain of comparisons (each measurement having its own associated uncertainty) connecting the result of the measurement to that standard.

This section considers only measurement variability, σ_M. Reducing sampling variability, σ_S, by segregating M&E was discussed in Section 5.4. Sampling variability due to field sampling uncertainties is often larger than measurement uncertainties. Although this statement may be true in some cases, this is not an argument for failing to perform a full evaluation of the measurement uncertainty. A realistic estimate of the measurement uncertainty is one of the most useful data quality indicators for a result (Section 3.8).

Although the need for reporting uncertainty has sometimes been recognized, often it consists of only the estimated component due to Poisson counting statistics. The component of uncertainty resulting from the random nature of radioactive decay is only one component of measurement method uncertainty. If only this component of uncertainty is accounted for, rather than performing a full uncertainty analysis, the result will be misleading because it is at best only a lower bound of the uncertainty and may lead to incorrect decisions based on overconfidence in the measurement. Software is available to perform the mathematical operations for uncertainty evaluation and propagation, eliminating much of the difficulty in implementing the mathematics of uncertainty calculations. There are several examples of such software (McCroan 2006, GUM Workbench 2006, Kragten 1994, and Vetter 2006).

7.4.1 Use Standard Terminology

The methods, terms, and symbols recommended by MARSAME for evaluating and expressing measurement uncertainty are described in the GUM (ISO 1995). The ISO methodology is summarized in the NIST Technical Note TN-1297 (NIST 1994).

The result of a measurement is generally used to estimate some particular quantity called the measurand. The difference between the measured result and the actual value of the measurand is the error of the measurement. Both the measured result and the error may vary with each repetition of the measurement, while the value of the measurand (the true value) remains fixed. The error of a measurement is unknowable, because one cannot know the error without knowing the true value of the quantity being measured (the measurand). For this reason, the error is primarily a theoretical concept. However, the uncertainty of a measurement is a concept with practical uses. According to the GUM and NIST Technical Note 1297, the term "uncertainty of measurement" denotes the values that could reasonably be attributed to the measurand. In practice, there is seldom a need to refer to the error of a measurement, but an uncertainty should be stated for every measured result.

The first step in defining a measurement process is to define the measurand clearly. The specification of the measurand is always ambiguous to some extent, but it should be as clear as necessary for the intended purpose of the data. For example, when measuring the activity of a radionuclide on a surface, it is generally necessary to specify the activity, the date and time, what area of the surface was measured, and where.

Often the measurand is not measured directly but instead an estimate is calculated from the measured values of other input quantities, which have a known mathematical relationship to the measurand. For example, input quantities in a measurement of radioactivity may include the gross count, blank or background count, counting efficiency, and area measured. The mathematical model measurement process specifies the relationship between the output quantity, Y, and measurable input quantities, $X_1, X_2, ... X_N$, on which its value depends: $Y = f(X_1, X_2, ... X_N)$.

The mathematical model for a radioactivity measurement may have the simple form:

$$\text{Measurement} = \frac{(\text{Gross Instrument Signal}) - (\text{Blank Signal})}{\text{Efficiency}} \qquad (7\text{-}12)$$

Each of the quantities shown here may actually be a more complicated expression. For example, the efficiency may be the product of factors such as surveyor efficiency, surface roughness efficiency correction, and the instrument counting efficiency. Interferences may be due to ambient background or other radionuclides that have interactions with the detector in a manner that contributes spuriously to the gross instrument signal.

When a measurement is performed, a specific value x_i is estimated for each input quantity, X_i, and an estimated value, y, of the measurand is calculated using the relationship $y = f(x_1, x_2, \ldots, x_N)$. Since there is an uncertainty in each input estimate, x_i, there is also an uncertainty in the output estimate, y. Determining the uncertainty of the output estimate y requires that the uncertainties of all the input estimates x_i be determined and expressed in comparable forms. The uncertainty of x_i is expressed in the form of an estimated standard deviation, called the standard uncertainty and denoted by $u(x_i)$. The ratio $u(x_i) / |X_i|$ is called the relative standard uncertainty of x_i, where $|X_i|$ is the absolute value of x_i.

The partial derivatives, $\partial f / \partial x_i$, are called sensitivity coefficients, and are usually denoted by c_i. The c_i measure how much f changes when x_i changes. The standard uncertainties are combined with sensitivity coefficients to obtain the component of the uncertainty in y due to x_i, $c_i u(x_i)$.

The square of the CSU, denoted by $u_c^2(y)$, is called the combined variance. It is obtained using the formula for the propagation of uncertainty:[4]

$$u_c^2(y) = \sum_{i=1}^{N} \left(\frac{\partial f}{\partial x_i} \right)^2 u^2(x_i) = \sum_{i=1}^{N} c_i^2 u^2(x_i) \qquad (7\text{-}13)$$

The square root of the combined variance is the CSU of y, denoted by $u_c(y)$. Further details of this process are given in Section 7.8.1.

7.4.2   Consider Sources of Uncertainty

The following sources of uncertainty should be considered:

- The random nature of radioactive decay (e.g., counting statistics),
- Instrument calibration (e.g., counting efficiency),
- Variable instrument backgrounds,
- Variable counting efficiency (e.g., due to the instrument or to source geometry and placement), and
- Interferences, such as crosstalk and spillover.

Other sources of uncertainty could include:

- Temperature and pressure.
- Volume and mass measurements,
- Determination of counting time and correction for dead time,

---

[4] If the input estimates are potentially correlated, covariance estimates $u(x_i, x_j)$ must also be determined. The covariance $u(x_i, x_j)$ is often recorded and presented in the form of an estimated correlation coefficient, $r(x_i, x_j)$, which is defined as the quotient $u(x_i, x_j) / u(x_i)u(x_j)$. See Section 7.8.

- Time measurements used in decay and ingrowth calculations,
- Approximation errors in simplified mathematical models, and
- Published values for half-lives and radiation emission probabilities.
-

There are a number of sources of measurement uncertainty in gamma-ray spectroscopy, including:

- Poisson counting uncertainty,
- Compton baseline determination,
- Background peak subtraction,
- Multiplets and interference corrections,
- Peak-fitting model errors,
- Efficiency calibration model error,
- Summing,
- Density-correction factors, and
- Dead time.

Additional discussion of some major sources of uncertainty may be found in Section 7.8.2.2.

The following example may appear complex, but all but the most casual users will use software to perform these calculations. Some possibilities are listed after the example. A complete example is worked out to here to illustrate the underlying principles.

Example 6: Consider a simple measurement of a sample. The activity will be calculated from

$$y = \frac{(N_S/t_S) - (N_B/t_B)}{\varepsilon}$$

Where:

y = sample activity (Bq)

ε = counting efficiency 0.4176 (s^{-1}/Bq)

N_S = gross count observed during the measurement of the source, (11578)

t_S = source count time (300 s)

N_B = observed background count (87)

t_B = background count time (6,000 s)

The CSU of ε is given by $u_C(\varepsilon) = 0.005802$. This is shown in Example 2 in Section 7.8.2.2. Assume the radionuclide is long-lived; so, no decay corrections are needed. The uncertainties of the count times are also assumed to be negligible. The standard uncertainties in N_S and N_B will be estimated as $\sqrt{N_S}$ and $\sqrt{N_B}$ using the Poisson assumption.

Then, $y = \dfrac{(N_S/t_S) - (N_B/t_B)}{\varepsilon} = \dfrac{(11578/300) - (87/6000)}{0.4179} = 92.316$

$$u_c^2(y) = \sum_{i=1}^{N} \left(\frac{\partial f}{\partial x_i}\right)^2 u^2(x_i) = \sum_{i=1}^{N} c_i^2 u^2(x_i)$$

$$= \left(\frac{\partial \dfrac{(N_S / t_S) - (N_B / t_B)}{\varepsilon}}{\partial N_S} \right)^2 u^2(N_S) + \left(\frac{\partial \dfrac{(N_S / t_S) - (N_B / t_B)}{\varepsilon}}{\partial N_B} \right)^2 u^2(N_B)$$

$$+ \left(\frac{\partial \dfrac{(N_S / t_S) - (N_B / t_B)}{\varepsilon}}{\partial \varepsilon} \right)^2 u^2(\varepsilon)$$

$$= \left(\frac{1/t_S}{\varepsilon} \right)^2 u^2(N_S) + \left(\frac{-1/t_B}{\varepsilon} \right)^2 u^2(N_B) + \left(\frac{-((N_S / t_S) - (N_B / t_B))}{\varepsilon^2} \right)^2 u^2(\varepsilon)$$

$$= \left(\frac{1/300}{0.4176} \right)^2 \sqrt{11578}^2 + \left(\frac{-1/6000}{0.4176} \right)^2 \sqrt{87}^2 + \left(\frac{-(11578/300) + (87/6000)}{0.4176^2} \right)^2 0.005802^2$$

$$= 0.73768 + 0.00001 + 1.64745 = 2.38515.$$

Note that these calculations show which input quantities are contributing the most to the combined variance. N_S contributes $0.73768/2.38515 \sim 31\%$. N_B contributes virtually nothing. The uncertainty in the efficiency contributes $1.64745/2.38515 \sim 69\%$. An analysis such as this is called an uncertainty budget, and quickly points out where improvements in the measurement may be made.

Taking the square root of the combined variance we find $u_c(y) = 1.54439$. Usually the CSU is rounded to two significant figures and the result is rounded to match the same number of decimal places. So the result would be reported as 92.3 Bq with a CSU of 1.5 Bq.

Note that if the uncertainty in the efficiency had been neglected, the CSU would have been underestimated as 0.86 Bq, and would have been attributed entirely to the uncertainty in the sample counts. This illustrates the importance of including all significant sources of uncertainty in the calculations. Many of these calculations can be done using computer software programs mentioned earlier.

A much more detailed and involved example is given in Section 7.8.3.

Again, it should be noted that software (e.g., McCroan 2006, GUM Workbench 2006, Kragten 1994, Vetter 2006) is available to perform the partial derivatives, insert the proper mean and standard uncertainty for each input, and perform the algebra for uncertainty evaluation and propagation. This eliminates much of the tedium in implementing the uncertainty calculations, and frees the analyst to carefully examine the model equation to be sure that significant sources of uncertainty are not omitted.

7.4.3   Recommendations for Uncertainty Calculation and Reporting

- Use the terminology and methods of the GUM (ISO 1995) for evaluating and reporting measurement uncertainty.

- Follow QC procedures that ensure the measurement process remains in a state of statistical control, which is a prerequisite for uncertainty evaluation.
- Account for possible blunders or other spurious errors. Spurious errors indicate a loss of statistical control of the process and are not part of the uncertainty analysis described above.
- Report each measured value with either its CSU (or its expanded uncertainty, see Section 7.8.1.7).
- Reported measurement uncertainties should be clearly explained. (In particular, when an expanded uncertainty is reported, the coverage factor should be stated and the basis for the coverage probability should also be given, see Section 7.8.1.7).
- Consider all possible sources of measurement uncertainty and evaluate and propagate the uncertainties from all sources believed to be potentially significant in the final result.
- Each uncertainty should be rounded to either one or two significant figures and the measured value should be rounded to the same number of decimal places as its uncertainty.
- Results should be reported as obtained together with their uncertainties (whether positive, negative, or zero).

7.5   Determine Measurement Detectability

This section summarizes issues related to measurement detection capabilities. Much of this material is derived from MARLAP Chapter 20. More detail may be found in see Section 7.9.

Environmental radioactivity measurements may involve material with very small amounts of the radionuclide of interest. Measurement uncertainty often makes it difficult to distinguish such small amounts from zero. Therefore, an important MQO of a measurement process is its detection capability, which is usually expressed as the smallest concentration of radioactivity that can be reliably distinguished from zero. Effective project planning requires knowledge of the detection capabilities of the measurement method that will be or could be used. This section explains an MQO called the minimum detectable concentration (MDC) and describes radioactivity detection capabilities, as well as methods for calculating it.

The method most often used to make a detection decision about radiation or radioactivity involves the principles of statistical hypothesis testing. It is a specific example of a Scenario B hypothesis testing procedure described in Section 7.2.4. To "detect" the radiation or radioactivity requires a decision on the basis of the measurement data that the radioactivity is present. The detection decision involves a choice between the null hypothesis (H_0): There is no radiation or radioactivity present (above background), and the alternative hypothesis (H_1): There is radiation or radioactivity present (above background). In this context, a Type I error is to conclude that radiation or radioactivity is present when it actually is not, and a Type II error is to conclude that radiation or radioactivity is not present when it actually is.[5] Making the choice between these hypotheses requires the calculation of a critical value. If the measurement result exceeds this critical value, the null hypothesis is rejected and the decision is that radiation or radioactivity is present.

---

[5] Note that in any given situation only one of the two types of decision error is possible. If the sample *does not* contain radioactivity, a Type I error is possible. If the sample *does* contain radioactivity, a Type II error is possible.

7.5.1   Calculate the Critical Value

The critical value defines the lowest value of the net instrument signal[6] (count) that is too large to be compatible with the premise that there is no radioactivity present. It has become standard practice to make the detection decision by comparing the net instrument count to its critical value, S_C. The net count is calculated from the gross count by subtracting the estimated background and any interferences.[7]

The mean value of the net instrument count typically is positive when there is radioactivity present (i.e., above background). The gross count must be corrected by subtracting an estimate of the count produced under background conditions. See Section 7.8.2 for more information on instrument background.

Table 7.5 lists some formulas that are commonly used to calculate the critical value, S_C, together with the major assumptions made in deriving them. Note that the Stapleton formulas given in rows 3 through 5 especially are appropriate when the total background is less than 100 counts. These formulas depend on N_B (background count), t_B (background count time), t_S (sample count time), and $z_{1-\alpha}$ (the $(1 - \alpha)$-quantile of the standard normal distribution). The value of α determines the sensitivity of the test. It is the probability that a detection decision is made when no radioactivity above background is actually present.

More detail on the calculation of critical values is given in Section 7.9.3. Software (Strom 1999) is available for calculating S_C using the equations recommended here, among others.

Table 7.5 Recommended Approaches for Calculating the Critical Value of the Net Instrument Signal (Count), S_C[8]

	Critical Value Equation	Assumptions	Background Count
1	$$S_C = z_{1-\alpha}\sqrt{N_B \frac{t_S}{t_B}\left(1 + \frac{t_S}{t_B}\right)}$$	Poisson	> 100
2	$$S_C = 2.33\sqrt{N_B}$$	Poisson $\alpha = 0.05$ $t_B = t_S$	> 100

---

[6] "Net instrument signal," is used here as a general term, because many radiation-detection instruments may have output other than "counts" (e.g., current for ionization chambers). In cases where the instrument output *is* in counts, the term "net counts" can be substituted for the term "net instrument signal."

[7] "Interference" is the presence of other radiation or radioactivity or electronic signals that hinder the ability to analyze for the radiation or radioactivity of interest.

[8] These particular expressions for the critical value of the net instrument signal (in this case the net count) depend for their validity on the assumption of Poisson counting statistics. If the variance of the blank signal is affected by interferences, or background instability, then the Equation 20.7 of MARLAP may be more appropriate.

	Critical Value Equation	Assumptions	Background Count
3	$S_C = d \times \left(\dfrac{t_S}{t_B} - 1 \right) + \dfrac{z_{1-\alpha}^2}{4} \times \left(1 + \dfrac{t_S}{t_B} \right) + z_{1-\alpha}\sqrt{(N_B + d)\dfrac{t_S}{t_B}\left(1 + \dfrac{t_S}{t_B} \right)}$	Stapleton $t_B \neq t_S$	< 100
4	$S_C = 0.4 \times \left(\dfrac{t_S}{t_B} - 1 \right) + \dfrac{1.645^2}{4} \times \left(1 + \dfrac{t_S}{t_B} \right) + 1.645\sqrt{(N_B + 0.4)\dfrac{t_S}{t_B}\left(1 + \dfrac{t_S}{t_B} \right)}$	Stapleton $t_B \neq t_S$ $\alpha = 0.05$ $d = 0.4$	< 100
5	$S_C = 1.35 + 2.33\sqrt{N_B + 0.4}$	Stapleton $t_B = t_S$ $\alpha = 0.05$ $d = 0.4$	< 100

d = the critical value of the net instrument signal parameter in the Stapleton Equation

---

Example 7: A 600-second background measurement is performed on a proportional counter and 108 beta counts are observed. A sample is to be counted for 300 s. Estimate the critical value of the net instrument signal (i.e., net count) when $\alpha = 0.05$.

$$S_C = z_{1-\alpha}\sqrt{N_B \frac{t_S}{t_B}\left(1 + \frac{t_S}{t_B} \right)}$$

$$S_C = 1.645\sqrt{108 \times \left(\frac{300\ \text{s}}{600\ \text{s}} \right)\left(1 + \frac{300\ \text{s}}{600\ \text{s}} \right)} = 14.8\ \text{net counts}$$

Therefore, if 15 or more net counts are observed, the decision will be made that the sample contains radioactivity above background. Values of S_C should be rounded up when necessary to make sure that the specified Type I error probability, α, is not exceeded.

7.5.2   Calculate the Minimum Detectable Value of the Net Instrument Signal or Count

Table 7.6 lists some formulas that are commonly used to calculate the minimum detectable net count, S_D, together with the major assumptions made in deriving them. S_D, is defined as the mean value of the net instrument signal or count that gives a specified probability, $1 - \beta$, of yielding an observed net instrument signal or count greater than its critical value S_C. Therefore, S_C must be calculated before S_D. Note specifically that the Stapleton formulas given in rows 4 and 5 are especially appropriate when the total background is less than 100 counts. Generally, the Stapleton methods may be used for both high and low total background counts as they agree well with the more traditional methods when the background counts are over 100. The simpler, more familiar formulas have been included for completeness.

It is important that the assumptions used to calculate S_D are consistent with those that were used to calculate S_C. The equations for S_D depend on the same variables as S_C, namely N_B, t_B, and t_S. Notice that neither α nor $z_{1-\alpha}$ appears explicitly, rather they enter the calculation through S_C. However, β now enters the calculation of S_D through $Z_{1-\beta}$. The value of β, like α, is usually chosen to be 0.05 or is assumed to be 0.05 by default if no value is specified.

Table 7.6 Recommended Approaches for Calculating the Minimum Detectable Net Instrument Signal or Count[9]

	Minimum Detectable Net Signal Equation	Assumptions	Background Count
1	$S_D = S_C + \dfrac{z_{1-\beta}^2}{2} + z_{1-\beta}\sqrt{\dfrac{z_{1-\beta}^2}{4} + S_C + N_B \dfrac{t_S}{t_B}\left(1 + \dfrac{t_S}{t_B}\right)}$	Poisson $t_B \neq t_S$	> 100
2	$S_D = z_{1-\beta}^2 + 2S_C$	Poisson $t_B \neq t_S$ $\alpha = \beta$	> 100
3	$S_D = 2.71 + 2S_C = 2.71 + 2(2.33\sqrt{N_B}) = 2.71 + 4.66\sqrt{N_B}$	Poisson $\alpha = \beta = 0.05$ $t_B = t_S$	> 100
4	$S_D = \dfrac{(z_{1-\alpha} + z_{1-\beta})^2}{4}\left(1 + \dfrac{t_S}{t_B}\right) + (z_{1-\alpha} + z_{1-\beta})\sqrt{N_B \dfrac{t_S}{t_B}\left(1 + \dfrac{t_S}{t_B}\right)}$	Stapleton	< 100
5	$S_D = 5.41 + 4.65\sqrt{N_B}$	Stapleton $\alpha = \beta = 0.05$ $t_B = t_S$	< 100

Example 8 A 600-second background measurement on a proportional counter produces 108 beta counts and a source is to be counted for 300 s. Assume the background measurement gives the available estimate of the true mean background count rate and use the value 0.05 for Type I and Type II error probabilities. From section 7.5.1, Example 7, the critical net count, S_C, equals 14.8, so $S_D = z_{1-\beta}^2 + 2S_C = 1.645^2 + 2(14.8) = 32.3$ net counts. Values of S_D should be rounded up when necessary to make sure that the specified Type II error probability, β, is not exceeded.

The relationship between the critical value of the net instrument signal (or count), S_C, and the minimum detectable net instrument signal (or count), S_D, is shown in Figure 7.6. Figure 7.6 illustrates a case where alpha is greater than beta. The net instrument signal (or count) obtained for a blank sample will usually be distributed around zero as shown. Occasionally, a net count rate above S_C may be obtained by chance. The probability that this happens is controlled by the value of α, shown as the lightly shaded area in Figure 7.6. Smaller values of α result in larger values of S_C and vice versa. The minimum detectable value of the net instrument signal (or count) S_D is that value of the mean net instrument signal (or count) that results in a detection decision with probability $1-\beta$. That is, there is only a probability equal to β, shown as the more

---

[9] These expressions for the critical value of the net count depend for their validity on the assumption of Poisson counting statistics. If the variance of the blank signal is affected by interferences, or background instability, then Equation 20.7 of MARLAP may be more appropriate. "Interference" is the presence of other radiation or radioactivity or electronic signals that hinder the ability to analyze for the radiation or radioactivity of interest.

darkly shaded area in Figure 7.6, of yielding an observed count less than S_C. Smaller values of β result in larger values of S_D and vice versa.

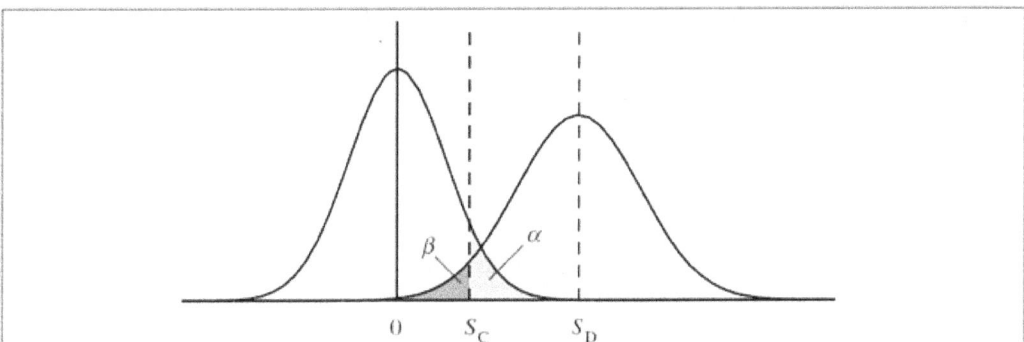

Figure 7.6 The Critical Value of the Net Instrument Signal (S_C) and the Minimum Detectable Net Signal (S_D)

More information detail on the calculation of the minimum detectable value of the net instrument signal, S_D, is given in Section 7.9.

7.5.3   Calculate the Minimum Detectable Concentration

The MDC is usually obtained from the minimum detectable value of the net instrument signal (or count), S_D. The MDC is by definition an estimate of the true concentration of the radiation or radioactivity required to give a specified high probability that the measured response will be greater than the critical value. The common practice of comparing a measured concentration to the MDC, instead of to the S_C, to make a detection decision is incorrect. To calculate the MDC, the minimum detectable value of the net signal (or count), S_D, must first be converted to the detectable value of the net instrument signal per unit time (or count rate), $S_D / t_S(s^{-1})$. This in turn must be divided by the counting efficiency, ε (s^{-1})/Bq to get the minimum detectable activity, y_D. Finally, the minimum detectable activity can be divided by the sample volume or mass to obtain the MDC. At each stage in this process, additional uncertainty may be introduced by the uncertainties in time, efficiency, volume, mass, etc. Thus, prudently conservative values of these factors should be used so that the desired detection power, $1-\beta$, at the MDC is maintained. Another approach would be to recognize that y_D itself has an uncertainty which can be calculated using the methods of Section 7.4. Thus any input quantity that is used to convert from S_D to y_D that has significant uncertainty can be incorporated to assess the overall uncertainty in the MDC. Additional discussion of the calculation of the MDCs is given in Section 7.9.5.

Example 9: Continuing Example 8, $S_D = 32.3$ net counts.

Assuming negligible uncertainty in the count time, the net count rate is
$S_D / t_S = 32.3/300 = 0.1077 \ (s^{-1})$.

The mean efficiency from Example 6 in Section 7.4.2 was 0.4176 (s^{-1})/(Bq) with a CSU of u_C $(\varepsilon) = 0.005802$.

In Example 8, the value 0.05 was specified for both Type I and Type II error probabilities. So the specified power was $1-\beta = 1 - 0.05 = 0.95$.

Assume a normal distribution for ε, to obtain a 95% probability of detection for the MDC. To account for the variability in the efficiency, the value used for ε should be the 5th percentile, i.e., $0.4176 - 1.645(0.005802) = 0.4081$.

Thus, the minimum detectable activity, $y_D = \dfrac{S_D / t_s}{\varepsilon} = 0.1077 / 0.4081 = 0.2639$ Bq.

Using the mean value of the efficiency would potentially underestimate the minimum detectable activity as $y_D = \dfrac{S_D / t_s}{\varepsilon} = 0.1077 / 0.4176 = 0.2578$ Bq.

These values for y_D would then be divided by the mass or volume of the sample to yield the MDC.

7.5.4   Summary of Measurement Detectability

The concepts surrounding the MDC and the critical value are illustrated in Figure 7.7, using familiar formulae for S_C and S_D discussed above, assuming a background count of $N_B = 100$ with $\alpha = \beta = 0.5$. In this case, the equation in row 2 of Table 7.5 was used to obtain $S_C = 23.3$, and the corresponding equation in row 3 of Table 7.6 to obtain $S_D = 49.3$. The use of these equations implies $\alpha = \beta = 0.05$ and $t_B = t_S$. It is important to note that traditionally the values $\alpha = \beta = 0.05$ are used for MDC calculations, so that the MDCs for different methods are comparable. However, when developing a standard operating procedure for a survey, other values for α and β may be more appropriate. A case where this typically occurs is in the calculation of scan MDCs (Section 7.11.6) where α may be much greater than β, because the consequences associated with misidentifying a background area as elevated are much lower than the consequences associated with missing a true elevated area.

Note, the upper abscissa scale is in concentration and the lower abscissa scale is in net count. These are related by the efficiency at the point where the MDC corresponds to the minimum detectable net instrument signal (or count), S_D. Each of the curves illustrates the distribution of mean net counts (or concentration) that may exist for a measurement. The width of these curves represents the variation due to counting statistics. The variability due to other factors is associated with uncertainty in ε. Changes in the relationship between the lower and the upper scales result from changes in ε. This illustrates the importance of choosing realistic, or even conservative, values of ε. Note that the probability of making a detection decision (which is proportional to the area of each curve to the right of S_C) depends on the concentration, increasing from 5% at background to 95% at the MDC, passing through 50% at S_C. This is perhaps more clearly shown in Figure 7.8, which plots the probability of making a detection decision as a function of net instrument signal, count, or concentration.

Figure 7.8 shows that for concentrations corresponding to net counts between 0 and S_C the probability of a non-detect is greater than 50%. For concentrations corresponding to net counts between S_C and S_D the probability of detection is greater than 50%, but less than 95%.

Concentrations above the MDC (with net counts greater than S_D) are highly likely to be detected, but will have relative standard uncertainties that are somewhat large.

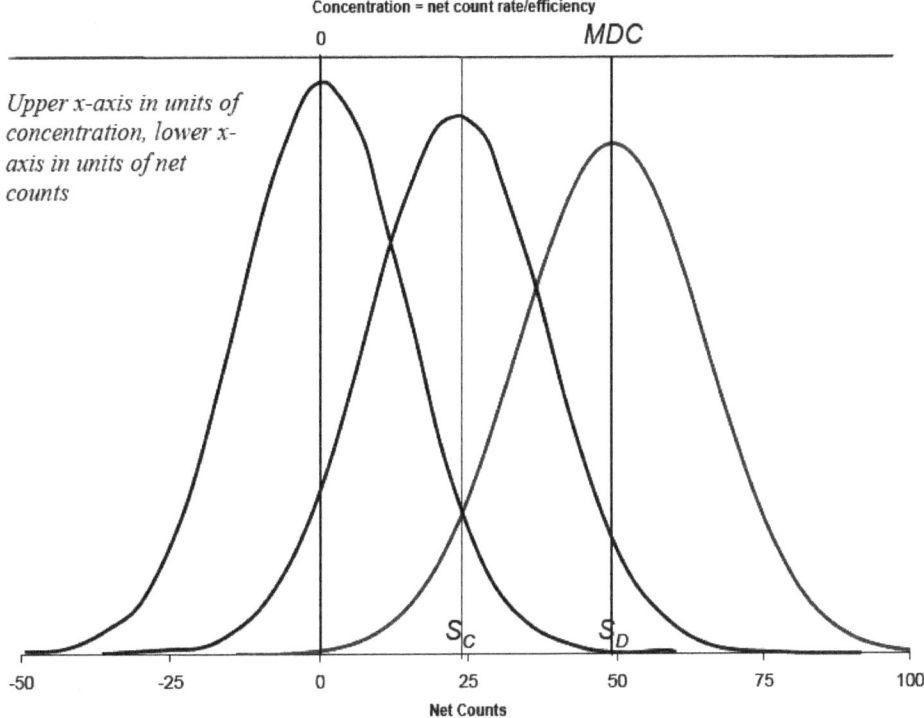

Figure 7.7 Relationship Between the Critical Value of the Net Count, the Minimum Detectable Net Counts and the MDC

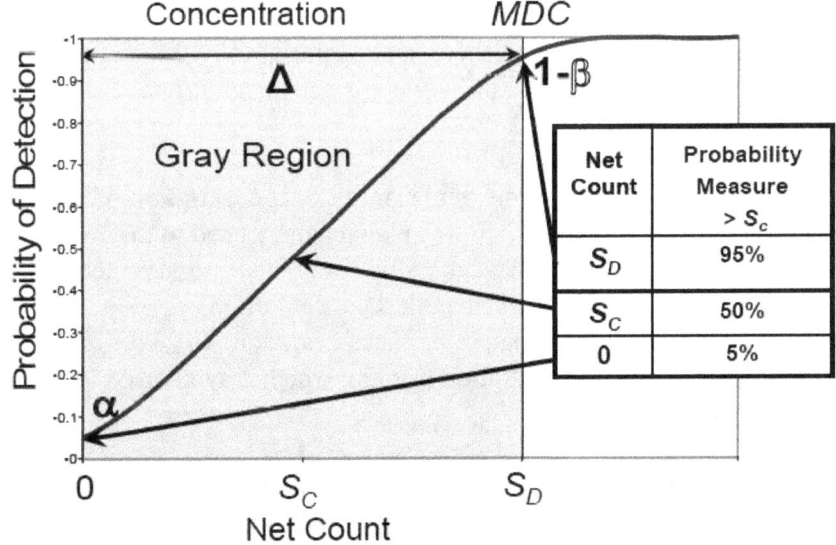

Figure 7.8 Probability of Detection as a Function of Net Count (Lower X-Axis) and Concentration (Upper X-Axis)

7.5.5   Measurement Detectability Recommendations

- When a detection decision is required, generally it should be made by comparing the net instrument signal (or count) to its corresponding critical value.
- Expressions from Tables 7.5 and 7.6 for S_C and S_D should be chosen to match the assumptions and background for the measurement method.
- An appropriate background should be used to predict the instrument signal produced when there is no radioactivity present in the sample.
- The Minimum Detectable Concentration (MDC) should be used only as a MQO for the measurement method. To make a detection decision, a measurement result should be compared the critical value and never to the MDC.
- The validity of the Poisson approximation for the measurement process should be confirmed using the methods described in MARLAP Chapter 20 before using an expression for the critical value that is based on Poisson statistics. When the Poisson approximation is inappropriate for determining the critical value, estimating σ by the sample standard deviation of replicated background measurements is preferable to using the square root of the number of counts.
- Consider all significant sources of variance in the instrument signal (or other response variable) when calculating the critical value, S_C, and minimum detectable value, S_D.
- Report each measurement result and its uncertainty as obtained even if the result is less than zero. Never report a result as "less than MDC" or "less than S_C."
- The MDC should not be used for projects where the issue is a quantitative comparison of measurements to a limit rather than just a detection decision made for a single measurement. For these projects, the minimum quantifiable concentration is a more relevant MQO for the measurement process (see Section 7.6).

7.6   Determine Measurement Quantifiability

This section discusses issues related to measurement quantifiability. Much of this material is derived from the MARLAP Chapter 20. Further details and an additional example are given in Section 7.10.

Action levels are frequently stated in terms of a quantity or concentration of radioactivity, rather than in terms of detection. In these cases, project planners may need to know the quantification capability of a measurement method, or its capability for precise measurement. The quantification capability is expressed as the smallest concentration of radiation or radioactivity that can be measured with a specified relative standard deviation. This section explains an MQO called the minimum quantifiable concentration (MQC), which may be used to describe quantification capabilities.

The MQC of the concentration, y_Q, is defined as the concentration at which the measurement process gives results with a specified relative standard deviation $1/k_Q$ where k_Q is usually chosen to be 10 for comparability.

Historically much attention has been given to the detection capabilities of radiation and radioactivity measurement processes, but less attention has been given to quantification capabilities. For some projects, quantification capability may be a more relevant issue. For example, suppose the purpose of a project is to determine whether the ^{226}Ra concentration on material at a site is below an action level. Since ^{226}Ra can be found in almost any type of naturally occurring material, it may be assumed to be present in every sample, making detection decisions unnecessary. The MDC of the measurement process obviously should be less than the action level, but a more important question is whether the MQC is less than the action level.

A common practice in the past has been to select a measurement method based on the minimum detectable concentration (MDC), which is defined in Section 7.5. For example, MARSSIM (2002) says:

> During survey design, it is generally considered good practice to select a measurement
> system with an MDC between 10-50% of the DCGL [action level].

Such guidance implicitly recognizes that for cases when the decision to be made concerns the mean of a population that is represented by multiple measurements, criteria based on the MDC may not be sufficient and a somewhat more stringent requirement is needed. The requirement that the MDC (approximately 3-5 times σ_M) be 10% to 50% of the action level is tantamount to requiring that σ_M be 0.02 to 0.17 times the action level – in other words, the relative standard deviation should be approximately 10% at the action level. However, the concentration at which the relative standard deviation is 10% is the MQC when k_Q assumes its conventional value of 10. Thus, a requirement that is often stated in terms of the MDC may be more naturally expressed in terms of the MQC, e.g., by saying that the MQC should not exceed the action level.

7.6.1   Calculate the MQC

The minimum quantifiable concentration, when there are no interferences, can be calculated from:

$$y_Q = \frac{k_Q^2}{2t_S\varepsilon(1-k_Q^2\phi_{\hat{\varepsilon}}^2)}\left(1+\sqrt{1+\frac{4(1-k_Q^2\phi_{\hat{\varepsilon}}^2)}{k_Q^2}\left(N_B\frac{t_S}{t_B}\left(1+\frac{t_S}{t_B}\right)\right)}\right) \qquad (7\text{-}14)$$

Where:

t_S = count time for the source, s
t_B = count time for the background, s
N_B = background count
$\phi_{\hat{\varepsilon}}^2$ = relative variance of the measured efficiency, $\hat{\varepsilon}$ (see Section 7.8.2.2)
k_Q = relative percent standard deviation at the MQC, usually assumes a conventional value of 10 for purposes of comparison among methods

If $k_Q^2\phi_{\hat{\varepsilon}}^2 \geq 1$, this equation has no solution.

Example 10: Continuing Example 9, $t_S = 300$, $t_B = 600$, $N_B = 108$, $\phi_{\varepsilon}^2 = (0.005802/0.4176)^2 = 0.0001932$, and $k_Q = 10$. So,

$$y_Q = \frac{k_Q^2}{2t_S\varepsilon(1-k_Q^2\phi_{\varepsilon}^2)}\left(1+\sqrt{1+\frac{4(1-k_Q^2\phi_{\varepsilon}^2)}{k_Q^2}\left(N_B\frac{t_S}{t_B}\left(1+\frac{t_S}{t_B}\right)\right)}\right)$$

$$= \frac{100}{2(300)(0.4176)(1-100(0.0001932))}\left(1+\sqrt{1+\frac{4(1-100(0.0001932))}{100}\left(108\frac{300}{600}\left(1+\frac{300}{600}\right)\right)}\right)$$

$= 1.239$ Bq. This value for y_Q would then be divided by the mass or volume of the sample to yield the MQC.

The next example is given to verify that the equation for y_Q does indeed produce a value with a relative uncertainty of 10%. It also provides an opportunity to give another illustration of the methodology for the calculation of measurement uncertainty developed in Sections 7.4 and 7.8. Additional information on the calculation of MQCs is given in Section 7.10.

Example 11: The calculations of Example 10 can be verified by calculating the uncertainty of a measurement made at the MQC. The expected number of counts for a sample at the MQC counted for 300 s:

$$N_S = y_Q t_S \varepsilon + N_B(t_S/t_B) = (1.239 \text{ Bq})(300 \text{ s})(0.4176) + (108 \text{ s}^{-1})(300/600) = 209,$$

rounded to the nearest whole number.

The model equation is the same as was used in Example 6, Section 7.4.2:

$y = \dfrac{(N_S/t_S)-(N_B/t_B)}{\varepsilon}$, so the equation for the CSU is the same:

$$u_c^2(y) = \left(\frac{1/t_S}{\varepsilon}\right)^2 u^2(N_S) + \left(\frac{-1/t_B}{\varepsilon}\right)^2 u^2(N_B) + \left(\frac{-((N_S/t_S)-(N_B/t_B))}{\varepsilon^2}\right)^2 u^2(\varepsilon)$$

$$= \left(\frac{1/300}{0.4176}\right)^2 (209) + \left(\frac{-1/600}{0.4176}\right)^2 (108) + \left(\frac{-(209/300)+(108/600)}{0.4176^2}\right)^2 (0.005802)^2$$

$$= 1.332\times10^{-2} + 1.72\times10^{-3} + 2.95\times10^{-4} = 1.534\times10^{-2}$$

$u_c(y) = \sqrt{1.534\times10^{-2}} = 0.124$. Thus, the relative uncertainty at the MQC is $0.124/1.239 = 0.09995$. This means, apart from some small difference due to rounding, the relative measurement uncertainty at y_Q is 10%, as should be the case for the MQC.

7.6.2   Summary of Measurement Quantifiability

Figure 7.9 is a modification of Figure 7.8, illustrating the relationships between the critical value, the MDC, the MQC and the probability of exceeding the critical value. As can be seen, the issue of detection is almost moot at the MQC. The probability of detection is near 100%. However, the

MQC specifies a concentration with a defined relative standard uncertainty, making comparisons between measurements or comparisons between measurements and regulatory criteria meaningful.

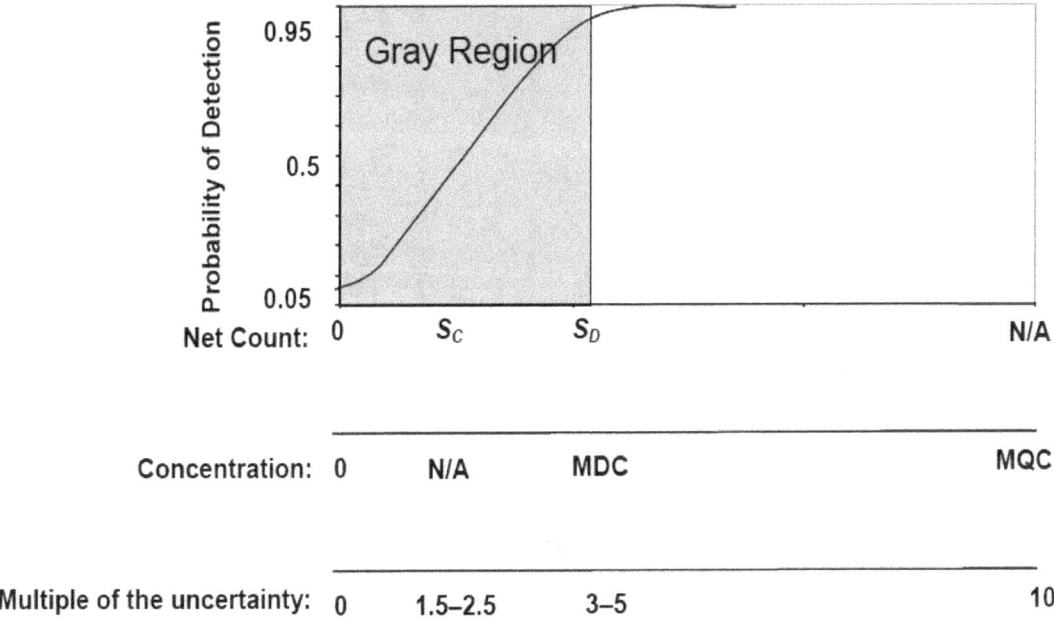

Figure 7.9 Relationships Among the Critical Value, the MDC, the MQC, and the Probability of Exceeding the Critical Value

Three x-axis scales are shown in Figure 7.9 for net count, concentration, and multiple of measurement uncertainty. This figure emphasizes, for example, that the minimum detectable net count, S_D, corresponds to the MDC, but has different units. It also shows that the MQC is by convention 10 times the measurement uncertainty at that concentration. The critical value of the net count, S_C, has no corresponding common term in concentration units. This is because detection decisions are usually made on the basis of the net counts (instrument reading). These are inherently qualitative "yes or no" decisions. The relationship between S_C and S_D and the multiple of the uncertainty varies according to which set of assumptions are used and which equations in Table 7.5 and Table 7.6 are appropriate to those assumptions. Therefore, an approximate range is shown for these quantities on the multiple of uncertainty axis.

7.7   Establish a Required Measurement Method Uncertainty

This section provides the rationale and guidance for establishing project-specific MQOs for controlling σ_M and expands on the material in Section 7.3. Control of σ_M is achieved by establishing a desired maximum measurement method uncertainty at the upper boundary of the gray region. This control also will assist in both the measurement method selection process and in the evaluation of measurement data. Approaches applicable to several situations are detailed below.

7.7.1   Developing a Requirement for Measurement Method Uncertainty for MARSSIM-Type Surveys

When, as in MARSSIM-type surveys, a decision is to be made about the mean of a sampled population, generally the average of a set of measurements on a survey unit is compared to the disposition criterion.

The total variance of the data, σ^2, is the sum of two components

$$\sigma^2 = \sigma_M^2 + \sigma_S^2$$

(7-15)

Where:

σ_M^2 = measurement method variance (M for measurement)

σ_S^2 = variance of the radionuclide concentration or activity concentration in the sampled population (S for sampling)

The spatial and temporal distribution of the concentration (i.e., the variation of the true but unknown concentrations from place to place and from time to time), the extent of the survey unit, the physical sizes of the measured material, and the choice of measurement locations may affect the sampling standard deviation, σ_S. The measurement standard deviation, σ_M, is affected by the measurement methods. The value of σ_M is estimated in MARSAME by the CSU of a measured value for a measurement of material whose concentration equals the hypothesized population mean concentration. The calculation of measurement uncertainties is covered in Sections 7.4 and 7.8.

Four cases are considered below where target values for σ_M can be suggested depending on what is known about σ_S. Cases 1 and 2 treat the desired overall objective of keeping $\Delta/\sigma \approx 3$ or higher. When this is not possible, Cases 3 and 4 treat the less desirable alternative of attempting to prevent Δ/σ from going lower than 1. If $\Delta/\sigma < 1$ then a large number of measurements will be required to meet the Type I and II decision error rates specified in the DQO process. If $\sigma \gg \Delta$, it may be necessary to re-evaluate the error rates specified in the DQO process.

Case 1: σ_S is known relative to $\Delta / 3$
Generally, it is easier to control σ_M than σ_S. If σ_S is known (approximately), a target value for σ_M can be determined.

> Case 1a: $\sigma_S \leq \Delta/3$
> If $\sigma_S \leq \Delta/3$, then a value of σ_M no greater than $\sqrt{(\Delta^2/9) - \sigma_S^2}$ ensures that $\sigma \leq \Delta / 3$, because we have $\sigma^2 = \sigma_M^2 + \sigma_S^2 \leq (\Delta^2/9 - \sigma_S^2) + \sigma_S^2 = \Delta^2/9$, as desired.

> Case 1b: $\sigma_S > \Delta/3$
> If $\sigma_S > \Delta/3$, the requirement that the total σ be less than $\Delta/3$ cannot be met regardless of σ_M. In this case, it is sufficient to make σ_M negligible in comparison to σ_S. Generally, σ_M can be considered negligible in comparison to σ_S if $\sigma_M < \sigma_S/3$.

Case 2: σ_S is not known relative to $\Delta/3$

Often one needs a method for choosing σ_M in the absence of specific information about σ_S. Since it is desirable to have $\sigma \leq \Delta/3$, this condition is adopted as a primary requirement. Assume for the moment that σ_S is large. Then σ_M should be made negligible by comparison. As mentioned above, σ_M can be considered negligible if it is no greater than $\sigma_S/3$. When this condition is met, further reduction of σ_M has little effect on σ and therefore is usually not cost-effective. So, $\sigma_M \leq \sigma_S/3$ is adopted as a secondary requirement.

Starting with the definition $\sigma^2 = \sigma_M^2 + \sigma_S^2$ and substituting the secondary requirement $\sigma_M \leq \sigma_S/3$ we get $\sigma^2 \geq \sigma_M^2 + 9\sigma_M^2 = 10\sigma_M^2$, thus

$$\sigma_M \leq \frac{\sigma}{\sqrt{10}} \qquad (7\text{-}16)$$

Substituting the primary requirement that $\Delta/\sigma \geq 3$ (i.e., $\sigma \leq \Delta/3$) we get $\sigma_M \leq \dfrac{\sigma}{\sqrt{10}} \leq \dfrac{\Delta/3}{\sqrt{10}}$, thus

$$\sigma_M \leq \frac{\Delta}{3\sqrt{10}} \qquad (7\text{-}17)$$

Or approximately

$$\sigma_M \leq \frac{\Delta}{10} \qquad (7\text{-}18)$$

The required upper bound for the standard deviation σ_M will be denoted by σ_{MR}. MARSAME recommends the equation

$$\sigma_{MR} = \frac{\Delta}{10} \qquad (7\text{-}19)$$

by default as a requirement when σ_S is unknown and a decision is to be made about the mean of a sampled population.

This upper bound was derived from the assumption that σ_S was large, but it also ensures that the primary requirement $\sigma \leq \Delta/3$ (i.e., $\Delta/\sigma \geq 3$) will be met if σ_S is small. When the measurement standard deviation σ_M is less than σ_{MR}, the primary requirement will be met unless the sampling variance, σ_S^2, is so large that σ_M^2 is negligible by comparison, in which case little benefit can be obtained from further reduction of σ_M.

It may be that the primary requirement that Δ/σ be at least 3 is not achievable. Suppose that the primary requirement is relaxed to achieving Δ/σ at least 1 (i.e., $\sigma \leq \Delta$). This leads to consideration of–

Case 3: σ_S is known relative to Δ

As in Case 1, it is generally easier to control σ_M than σ_S. If σ_S is known (approximately), a target value for σ_M can be determined.

> Case 3a: $\sigma_S \leq \Delta$
>
> If $\sigma_S \leq \Delta$, then a value of σ_M no greater than $\sqrt{\Delta^2 - \sigma_S^2}$ ensures that $\sigma \leq \Delta$, because we have $\sigma^2 = \sigma_M^2 + \sigma_S^2 \leq (\Delta^2 - \sigma_S^2) + \sigma_S^2 = \Delta^2$ as desired.
>
> Case 3b: $\sigma_S > \Delta$
>
> If $\sigma_S > \Delta$, the requirement that the total σ be less than Δ cannot be met regardless of σ_M. In this case, it is sufficient to make σ_M negligible in comparison to σ_S. Generally, σ_M can be considered negligible if it $\sigma_M < \sigma_S/3$.

Case 4: σ_S is not known relative to Δ

Suppose $\sigma \leq \Delta$ is adopted as the primary requirement. As in Case 2, if σ_S is large, then σ_M should be made negligible by comparison. As mentioned above, σ_M can be considered negligible if it is no greater than $\sigma_S/3$. When this condition is met, further reduction of σ_M has little effect on σ and therefore is usually not cost-effective. So, $\sigma_M \leq \sigma_S/3$ is adopted as a secondary requirement.

Starting with the definition $\sigma^2 = \sigma_M^2 + \sigma_S^2$ and substituting the secondary requirement $\sigma_M \leq \sigma_S/3$ we get $\sigma^2 \geq \sigma_M^2 + 9\sigma_M^2 = 10\sigma_M^2$, thus–

$$\sigma_M \leq \frac{\sigma}{\sqrt{10}} \tag{7-20}$$

Substituting the primary requirement that $\Delta/\sigma \geq 1$ (i.e., $\sigma \leq \Delta$) we get $\sigma_M \leq \frac{\sigma}{\sqrt{10}} \leq \frac{\Delta}{\sqrt{10}}$, thus–

$$\sigma_M \leq \frac{\Delta}{\sqrt{10}} \approx \frac{\Delta}{3} \tag{7-21}$$

7.7.2 Developing a Requirement for Measurement Method Uncertainty When Decisions are to be Made About Individual Items

When decisions are to be made about individual items, the total variance of the data equals the measurement variance, σ_M^2, and the data distribution in most instances should be approximately normal. The decision in this case may be made by comparing the measured concentration, x, plus or minus a multiple of its CSU, to the action level. The CSU, $u_c(x)$, is assumed to be an estimate of the true standard deviation of the measurement process as applied to the item being measured;

so, the multiplier of $u_c(x)$ equals $z_{1-\alpha}$, the $(1-\alpha)$-quantile of the standard normal distribution (see MARLAP appendix C).

Alternatively, if AL = 0, so that any detectable amount of radioactivity is of concern, the decision may involve comparing the net instrument signal (e.g., count rate) to the critical value of the net instrument signal, S_C, as defined in Section 7.5.1.

Two cases are considered below where target values for σ_M can be suggested depending on what is known about the width of the gray region and the desired Type I and Type II decision error rates. Case 5 is for Scenario A, and Case 6 is for Scenario B.

Case 5: Suppose the null hypothesis is $X \geq AL$ (see Scenario A in Chapter 4), so that the action level is the upper bound of the gray region. Given the measurement variance σ_M^2, only a measured result that is less than (UBGR $- z_{1-\alpha}\sigma_M$) will be judged to be clearly less than the action level. Then the desired power of the test $1-\beta$ is achieved at the lower bound of the gray region only if the LBGR $\leq$ UBGR $- z_{1-\alpha}\sigma_M - z_{1-\beta}\sigma_M$. Algebraic manipulation transforms this requirement to

$$\sigma_M \leq \frac{\text{UBGR - LBGR}}{z_{1-\alpha} + z_{1-\beta}} = \frac{\Delta}{z_{1-\alpha} + z_{1-\beta}} \tag{7-22}$$

Case 6: Suppose the null hypothesis is $X \leq AL$ (see Scenario B in Chapter 4), so that the action level is the lower bound of the gray region. In this case, only a measured result that is greater than LBGR $+ z_{1-\alpha}\sigma_M$ will be judged to be clearly greater than the action level. The desired power of the test $1 - \beta$ is achieved at the upper bound of the gray region only if the UBGR $\geq$ LBGR $+ z_{1-\alpha}\sigma_M + z_{1-\beta}\sigma_M$. Algebraic manipulation transforms this requirement to:

$$\sigma_M \leq \frac{\text{UBGR - LBGR}}{z_{1-\alpha} + z_{1-\beta}} = \frac{\Delta}{z_{1-\alpha} + z_{1-\beta}} \tag{7-23}$$

So, in either Scenario A or Scenario B, the requirement remains that:

$$\sigma_M \leq \frac{\Delta}{z_{1-\alpha} + z_{1-\beta}} \tag{7-24}$$

Therefore, MARSAME uses the equation:

$$u_{MR} = \sigma_{MR} = \frac{\Delta}{z_{1-\alpha} + z_{1-\beta}} \tag{7-25}$$

as an MQO for method uncertainty when decisions are to be made about individual items or locations and not about population parameters.

If $\alpha = \beta = 0.05$, one may use the value $u_{MR} = 0.3\Delta$. Other combinations of α and β may lead to a similar result, but the relationship is nonlinear (depending on the standard normal distribution function) so one cannot simply apply a proportionality factor. Equation 7-25 must be used,

The recommended value of u_{MR} is based on the assumption that any known bias in the measurement process has been corrected and that any remaining bias is well less than a third of the method uncertainty.

7.8   Calculate the Combined Standard Uncertainty of a Measurement

This section expands upon the material in Section 7.4. Calculations of combined standard uncertainties (CSUs) can be complex, and typically would be carried out using a software package. For the purpose of illustration and clarity, fully worked out examples are included in this section.

7.8.1   Procedures for Evaluating Uncertainty

The usual eight steps for evaluating and reporting the uncertainty of a measurement are summarized in the following subsections (adapted from Chapter 8 of GUM):

7.8.1.1   Identify the Measurand, Y, and all the Input Quantities, X_i, for the Mathematical Model

Include all quantities whose variability or uncertainty could have a potentially significant effect on the result. Express the mathematical relationship, $Y = f(X_1, X_2,\ldots,X_N)$, between the measurand and the input quantities.

The procedure for assessing the uncertainty of a measurement begins with listing all significant sources of uncertainty in the measurement process. A good place to begin is with the input quantities' mathematical model $Y = f(X_1, X_2,\ldots,X_N)$. When an effect in the measurement process that is not explicitly represented by an input quantity has been identified and quantified, an additional quantity should be included in the mathematical measurement model to correct for it. The quantity, called a correction (additive with a nominal value of zero) or correction factor (multiplicative with a nominal value of one), will have an uncertainty that should also be evaluated and propagated. Each uncertainty that is potentially significant should be evaluated quantitatively.

7.8.1.2   Determine an Estimate, x_i, of the Value of Each Input Quantity, X_i

This involves simply determining for the particular measurement at hand, the specific value, x_i, that should be substituted for the input quantity X_i in the mathematical relationship, $Y = f(X_1, X_2,\ldots,X_N)$.

7.8.1.3   Evaluate the Standard Uncertainty, $u(x_i)$, for Each Input Estimate, x_i, Using a Type A Method, a Type B Method, or a Combination of Both

Methods for evaluating standard uncertainties are classified as either "Type A" or "Type B" (NIST 1994). Both types of uncertainty need to be taken into consideration. A Type A evaluation of an uncertainty uses a series of measurements to estimate the standard deviation empirically. Any other method of evaluating an uncertainty is a Type B method. A Type B evaluation of standard uncertainty is usually based on scientific judgment using all the relevant information available, which may include:

- Previous measurement data,
- Experience with, or general knowledge of, the behavior and property of relevant materials and instruments,
- Manufacturer's specifications,
- Data provided in calibration and other reports, and
- Uncertainties assigned to reference data taken from handbooks.

The Type A standard uncertainty of the input estimate x_i is defined to be the experimental standard deviation of the mean:

$$u(x_i) = \sqrt{\frac{1}{n(n-1)} \sum_{k=1}^{n} (x_{i,k} - \overline{x_i})^2} = s(x_i)/\sqrt{n} \qquad (7\text{-}26)$$

Example 12: Type A uncertainty calculation using Equation 7-26:

Ten independent one-minute measurements of the counts from a check source X_i were made with a digital survey meter, yielding the values:

12,148, 12,067, 12,207, 12,232, 12,284, 12,129, 11,862, 11,955, 12,044, and 12,150.

The estimated value x_i is the arithmetic mean of the values $X_{i,k}$.

$$x_i = X_i \frac{1}{n} \sum_{k=1}^{n} x_{i,k} = \frac{121078}{10} = 12107.8$$

The standard uncertainty of x_i is

$$u(x_i) = \sqrt{\frac{1}{n(n-1)} \sum_{k=1}^{n} (x_{i,k} - \overline{x_i})^2} = \sqrt{\frac{1}{10(10-1)} \sum_{k=1}^{10} (x_{i,k} - 12107.8)^2}$$

$$= \sqrt{16628.84} = 128.95$$

There are other Type A methods, but all are based on repeated measurements.

Any evaluation of standard uncertainty that is not a Type A evaluation is a Type B evaluation. Sometimes a Type B evaluation of uncertainty involves making a best guess based on all available information and professional judgment. Despite the reluctance to make this kind of evaluation, it is almost always better to make an informed guess about an uncertainty component than to ignore it completely.

There are many ways to perform Type B evaluations of standard uncertainty. One example of a Type B method is the estimation of counting uncertainty using the square root of the observed counts. If the observed count is N, when the Poisson approximation is used, the standard uncertainty of N may be evaluated as $u(N) = \sqrt{N}$.

Example 13: The standard uncertainty of the first value in Example 12, (12,148 counts), could be estimated as $\sqrt{12148} = 110.218$ counts. When N may be very small or even zero, the equation $u(N) = \sqrt{N+1}$ may be preferable.

Another Type B evaluation of an uncertainty $u(x)$ consists of estimating an upper bound, $a,$ for the magnitude of the error of x based on professional judgment and the best available information. If nothing else is known about the distribution of the measured result, then after a is estimated, the standard uncertainty may be calculated using the equation

$$u(x) = \frac{a}{\sqrt{3}} \tag{7-27}$$

which is the standard deviation of a random variable uniformly distributed over the interval $(x - a, x + a)$. The variable a is called the half-width of the interval.

Example 14: Suppose in Example 12, all that was given was the observed range of the data from an analog survey meter dial (i.e., from 11,862 to 12,284), a difference of 422. If it was assumed that the data came from a uniform distribution across this range, then the average is

$$(11{,}862 + 12{,}284)/2 = 12{,}073$$

the half-width is 211, and an estimate of the standard uncertainty would be

$$u(x) = \frac{211}{\sqrt{3}} = 121.821$$

Given the same information on the range, if values near the middle of the range were considered more likely than those near the endpoint, a triangular distribution may be more appropriate. The standard uncertainty for a triangular distribution is calculated using the equation

$$u(x) = \frac{a}{\sqrt{6}} \tag{7-28}$$

which represents the standard deviation of a random variable with a triangular distribution over the interval $(x - a, x + a)$. Given the same information on the range, if values near the middle of the range were considered more likely than those near the endpoints, a triangular distribution may be more appropriate. The mean would be the same as above, 12,073. However the standard uncertainty then be calculated using the equation

$$u(x) = \frac{a}{\sqrt{6}} = \frac{211}{\sqrt{6}} = 86.14 \tag{7-29}$$

Example 15: As in Example 14, all that is given was the observed range of the data from an analog survey meter dial, i.e., from 11,862 to 12,284, a difference of 422. If it was assumed that the data came from a triangular distribution across this range, then the average is $(11{,}862 + 12{,}284)/2 = 12{,}073$, the half-width is 211, and an estimate of the standard uncertainty would be

$$u(x) = \frac{a}{\sqrt{6}} = \frac{211}{\sqrt{6}} = 86.14$$

When the estimate of an input quantity is taken from an external source, such as a book or a calibration certificate, the stated standard uncertainty can be used.

7.8.1.4 Evaluate the Covariances, $u(x_i, x_j)$, for all Pairs of Input Estimates with Potentially Significant Correlations

A Type A evaluation of the covariance of the input estimates $x_i =$ and $x_j =$ is

$$u(x_i, x_j) = \frac{1}{n(n-1)} \sum_{k=1}^{n} (x_{i,k} - \overline{x_i})(x_{j,k} - \overline{x_j}) \qquad (7\text{-}30)$$

An evaluation of variances and covariances of quantities determined by the method of least squares may also be a Type A evaluation. Evaluation of the covariance of two input estimates, x_i and x_j, whose uncertainties are evaluated by Type B methods may require expert judgment. In such cases it may be simpler to estimate the correlation coefficient,

$$r(x_i, x_j) = [u(x_i, x_j) / u(x_i) \cdot u(x_j)] \qquad (7\text{-}31)$$

first and then multiply it by the standard uncertainties, $u(x_i)$ and $u(x_j)$ to obtain the covariance, $u(x_i, x_j)$.

A covariance calculation is demonstrated in Example 16 in Section 7.8.2.2.

7.8.1.5 Calculate the Estimate, y, of the Measurand from the Relationship $y = f(x_1, x_2, \ldots, x_N)$

This involves simply substituting, for the particular measurement at hand, the specific values of x_i for the input quantity X_i into the mathematical relationship, $Y = f(X_1, X_2, \ldots, X_N)$, and calculating the result $y = f(x_1, x_2, \ldots, x_N)$.

7.8.1.6 Determine the Combined Standard Uncertainty, $u_c(y)$, of the Estimate, y

The CSU of y is obtained using the following formula:

$$u_c^2(y) = \sum_{i=1}^{N} \left(\frac{\partial f}{\partial x_i} \right)^2 u^2(x_i) + 2 \sum_{i=1}^{N-1} \sum_{j=i+1}^{N} \frac{\partial f}{\partial x_i} \frac{\partial f}{\partial x_j} u(x_i, x_j) \qquad (7\text{-}32)$$

Here, $u^2(x_i)$ denotes the estimated variance of x_i, or the square of its standard uncertainty; $u(x_i, x_j)$ denotes the estimated covariance of x_i and x_j; $\partial f / \partial x_i$ (or $\partial y / \partial x_i$) denotes the partial derivative of f with respect to x_i evaluated at the measured values $x_1, x_2, \ldots, x_N$; and $u_c^2(y)$ denotes the combined variance of y, whose positive square root, $u_c(y)$, is the CSU of y. The partial derivatives, $\partial f / \partial x_i$, are called sensitivity coefficients, usually denoted c_i. The sensitivity coefficient measures how much f changes when x_i changes. Equation 7-32 is called the "law of propagation of uncertainty" in GUM (ISO 1995).

If the input estimates $x_1, x_2, \ldots, x_N$ are uncorrelated, the uncertainty propagation formula reduces to

$$u_c^2(y) = \sum_{i=1}^{N} \left(\frac{\partial f}{\partial x_i} \right)^2 u^2(x_i) \tag{7-33}$$

Suppose the values $x_1, x_2, \ldots, x_N$ are composed of two groups $w_1, w_2, \ldots, w_n$ and $z_1, z_2, \ldots, z_m$ with $N = n + m$. If all the variables, w and z, are uncorrelated and nonzero, the CSU of $y = \dfrac{w_1 w_2 \ldots w_n}{z_1 z_2 \ldots z_m}$ may be calculated from the formula:

$$u_c^2(y) = y^2 \left(\frac{u^2(w_1)}{w_1^2} + \frac{u^2(w_2)}{w_2^2} + \ldots + \frac{u^2(w_n)}{w_n^2} + \frac{u^2(z_1)}{z_1^2} + \frac{u^2(z_2)}{z_2^2} + \ldots + \frac{u^2(z_m)}{z_m^2} \right) \tag{7-34}$$

The symbols $z_1, z_2, \ldots, z_m$ have been introduced simply to differentiate those values appearing in the denominator of the model equation from the $w_1, w_2, \ldots, w_n$ appearing in the numerator.

If $y = \dfrac{f(w_1, w_2, \ldots, w_n)}{z_1 z_2 \ldots z_m}$, where f is some specified function of $w_1, w_2, \ldots, w_n$, all the z_i are nonzero,

and all the input estimates are uncorrelated. Then:

$$u_c^2(y) = \frac{u_c^2(f(w_1, w_2, \ldots, w_n))}{z_1 z_2 \ldots z_m} + y^2 \left(\frac{u^2(z_1)}{z_1^2} + \frac{u^2(z_2)}{z_2^2} + \ldots + \frac{u^2(z_m)}{z_m^2} \right) \tag{7-35}$$

An alternative to uncertainty propagation is the use of computerized Monte Carlo methods to propagate not the uncertainties of input estimates but their distributions. Given assumed distributions for the input estimates, the method provides an approximate distribution for the output estimate, from which the CSU or an uncertainty interval may be derived.

7.8.1.7   Optionally Multiply $u_c(y)$ by a Coverage Factor k to Obtain the Expanded Uncertainty U

The interval $[y - U, y + U]$, constructed using the expanded uncertainty $U = k \cdot u_c(y)$, can be expected to contain the value of the measurand with a specified probability, p. The specified probability, p, is called the "level of confidence" or the "coverage probability" and generally is only an approximation of the true probability of coverage. When the distribution of the measured result is approximately normal, the coverage factor often is chosen to be $k = 2$ for a coverage probability of approximately 95%. An expanded uncertainty calculated with $k = 2$ or 3 is sometimes informally called a "two-sigma" or "three-sigma" uncertainty, respectively. The GUM recommends the use of coverage factors in the range of 2 to 3 when the CSU represents a good estimate of the true standard deviation. Attachment 19D of MARLAP describes a more general procedure for calculating the coverage factor that gives a desired coverage probability p when there is substantial uncertainty in the value of $u_c(y)$.

7.8.1.8   Report the Result as $y \pm U$ with the Unit of Measurement

At a minimum, state the coverage factor used to compute U and the estimated coverage probability. Alternatively, report the result, y, and its CSU, $u_c(y)$, with the unit of measurement.

The number of significant figures that should be reported for the result of a measurement depends on the uncertainty of the result. A common convention, recommended by MARLAP, is to round the uncertainty (standard uncertainty or expanded uncertainty) to two significant figures and to report both the measured value and the uncertainty to the same number of decimal places. Only final results should be rounded in this manner. Intermediate results in a series of calculation steps should be carried through all steps with additional figures to prevent unnecessary round-off errors. Additional figures are also recommended when the data are stored electronically. Rounding should be performed only when the result is reported. Many of the values in the examples given in MARSAME carry more significant digits so that the calculations can be reasonably reproduced by the reader. All results, whether positive, negative, or zero, should be reported as obtained, together with their uncertainties.

A measured value y of a quantity Y that is known to be positive may be so far below zero that it indicates a possible blunder, procedural failure, or other quality control problem. Usually, if $y + 3u_c(y) < 0$, the result may be invalid. For example, if $y = -10$ and $u_c(y) = 1$, this would imply that Y is negative with high probability, which is known to be impossible. However, if $y = -1$ and $u_c(y) = 1$, the expanded uncertainty covers positive values with reasonable probability. The accuracy of the uncertainty estimate $u_c(y)$ must be considered in evaluating such results, especially in cases where only few counts are observed during the measurement and counting uncertainty is the dominant component of $u_c(y)$. (See MARLAP Chapter 18 and Attachment 19D).

7.8.2   Examples of Some Parameters that Contribute to Uncertainty

The sources of uncertainty described in the following sections, drawn from MARLAP Section 19.5, should be considered.

7.8.2.1   Instrument Background

Single-channel background measurements are usually assumed to follow the Poisson model, in which the uncertainty in the number of counts obtained, N, is given by $\sqrt{N}$. There may be effects that increase the variance beyond what the model predicts. For example, cosmic radiation and other natural sources of instrument background may vary between measurements, the instrument may become contaminated, or the instrument may simply be unstable. Generally, the variance of the observed background is somewhat greater than the Poisson counting variance, although for certain types of instruments, the Poisson model may overestimate the background variance (Currie et al., 1998). If the background does not closely follow the Poisson model, its variance should be estimated by repeated measurements.

The "instrument background," or "instrument blank," is usually measured under the same conditions that will be encountered in the field. Ambient background sources should be minimized, and kept constant during the measurements of M&E. Periodic checks should be made to ensure that the instrument has not picked up additional radioactivity from the M&E during the measurements. If the background drifts or varies non-randomly over time (i.e., is non-stationary), it is important to minimize the consequences of the drift by performing frequent background measurements.

If repeated measurements demonstrate that the background level is stable, then the average, $\overline{x}$, the results of n similar measurements performed over a period of time may give the best estimate of the background. In this case, if all measurements have the same duration, the experimental standard deviation of the mean, $s(\overline{x})$, is also a good estimate of the measurement uncertainty.

Given the Poisson assumption, the best estimate of the uncertainty is still the Poisson estimate, which equals the square root of the summed counts, divided by the number of measurements, $\sqrt{n\overline{x}}\Big/n = \sqrt{\overline{x}\Big/n}$, but the experimental standard deviation may be used when the Poisson assumption is invalid. It is always wise to compare the value of $s(\overline{x})$ to the value of the Poisson uncertainty when possible to identify any discrepancies.

7.8.2.2   Counting Efficiency

The counting efficiency for a measurement of radioactivity (usually defined as the detection probability for a particle or photon of interest emitted by the source) may depend on many factors, including source geometry, placement, composition, density, activity, radiation type and energy and other instrument-specific factors. The estimated efficiency is sometimes calculated explicitly as a function of such variables (in gamma-ray spectroscopy, for example). In other cases a single measured value is used (e.g., alpha-particle spectrometry). If an efficiency function is used, the uncertainties of the input estimates, including those for both calibration parameters and sample-specific quantities, must be propagated to obtain the CSU of the estimated efficiency. Calibration parameters tend to be correlated; so, estimated covariances must also be included. If a single value is used instead of a function, the standard uncertainty of the value is determined when the value is measured. An example of the calculation of the uncertainty in counting efficiency is given in Example 16.

Example 16: A radiation counter is calibrated, taking steps to ensure that the geometry of the source position, orientation of the source, pressure, temperature, relative humidity, and other factors that could contribute to uncertainty are controlled, as described below:

The standard source is counted 15 times on the instrument for 300 s.
The radionuclide is long-lived; so, no decay corrections are needed. The uncertainties of the count times are assumed to be negligible.

Within the range of linearity of the instrument, the mathematical model for the calibration is:

$$\varepsilon = \frac{1}{n}\sum_{i=1}^{n} \frac{(N_{S,i}/t_S)-(N_B/t_B)}{a_s} \tag{7-36}$$

Where:

ε	=	counting efficiency
n	=	number times the source is counted (15)
$N_{S,i}$	=	gross count observed during the ith measurement of the source

---

t_S = source count time (300 s)
N_B = observed background count (87)
t_B = background count time (6,000 s)
a_S = activity of the standard source (150.0 Bq). (The standard uncertainty of the source, 2.0 Bq, was given in the certificate for the source.)

The CSU of ε can be evaluated using Equation 7-36. For the purpose of uncertainty evaluation, it is convenient to rewrite the model as:

$$\varepsilon = \frac{\overline{R}}{a_s}$$

Where:

$$\overline{R} = \frac{1}{n}\sum_{i=1}^{n} R_i \quad \text{and} \quad R_i = (N_{S,i}/t_S) - (N_B/t_B), \quad i = 1,2,\ldots,n$$

The values R_i and their average, $\overline{R}$, are estimates of the count rate produced by the standard, while $\overline{R}/a_S$ is an estimate of the count rate produced by 1 Bq of activity. The standard uncertainty of $\overline{R}$ can be evaluated experimentally from the 15 repeated measurements:

$u^2(\overline{R}) = s^2(\overline{R}) = \dfrac{1}{n(n-1)}\sum_{i=1}^{n}(R_i - \overline{R})^2$. Since only one background measurement was made, the

input estimates R_i are correlated with each other. The uncertainty of N_B, $u(N_B) = \sqrt{87}$, using a Type B evaluation based on an assumption of a Poisson distribution for the number of background counts.

The covariance between R_i and R_j, for $i \neq j$, may be estimated as

$$u(R_i, R_j) = \frac{\partial R_i}{\partial N_B}\frac{\partial R_j}{\partial N_B}u^2(N_B) = \frac{-1}{t_B}\frac{-1}{t_B}u^2(N_B) = \frac{u^2(N_B)}{t_B^2} = \frac{\sqrt{87}^2}{6000^2} \cong 2\times10^{-6}$$

However, the correlation is negligible here because the uncertainty of the background count, N_B, is much smaller than the uncertainty of each source count, $N_{S,i}$. So, the correlation of the input estimates R_i will be approximated as zero (i.e., treated as if they were uncorrelated), and the correlation terms dropped from Equation 7-32. This means the evaluation used to calculate the CSU of ε can proceed using equation 7-33.

$u_c^2(y) = \sum_{i=1}^{N}\left(\dfrac{\partial f}{\partial x_i}\right)^2 u^2(x_i)$, so since $\varepsilon = \dfrac{\overline{R}}{a_s}$,

$$u_c^2(\varepsilon) = \left(\frac{\partial(\frac{\overline{R}}{a_s})}{\partial\overline{R}}\right)^2 u^2(\overline{R}) + \left(\frac{\partial(\frac{\overline{R}}{a_s})}{\partial a_s}\right)^2 u^2(a_s) = \left(\frac{1}{a_s}\right)^2 u^2(\overline{R}) + \left(\frac{-\overline{R}}{a_s^2}\right)^2 u^2(a_s)$$

$= \left(\dfrac{u^2(\overline{R})}{a_s^2}\right) + \varepsilon^2\left(\dfrac{u^2(a_s)}{a_s^2}\right)$. Therefore, $u_c(\varepsilon) = \sqrt{\dfrac{u^2(\overline{R})}{a_S^2} + \varepsilon^2\dfrac{u^2(a_S)}{a_S^2}}$

Assume the following data were obtained for the 15 separate counts of the calibration source.		
Count Number, i	**Gross count, $N_{S,i}$**	**R_i (s^{-1})**
1	18,375	61.236
2	18,644	61.236
3	18,954	61.236
4	19,249	64.149
5	19,011	63.356
6	18,936	63.106
7	18,537	61.776
8	18,733	62.429
9	18,812	62.692
10	18,546	61.806
11	18,810	62.686
12	19,273	64.229
13	18,893	62.962
14	18,803	62.662
15	18,280	60.919
Average, $\overline{R}$ (s^{-1})		62.6202
Experimental standard deviation, $s(R_i)$ (s^{-1})		0.9483
Experimental standard deviation of the mean, $s(\overline{R})$ (s^{-1})		0.2449

Then the estimated counting efficiency is:

$$\varepsilon = \frac{\overline{R}}{a_s} = \frac{62.6202 \text{ s}^{-1}}{150.0 \text{ Bq}} = 0.4176$$

And the CSU of ε is given by

$$u_c(\varepsilon) = \sqrt{\frac{(0.2449 \text{ s}^{-1})^2}{(150.0 \text{ Bq})^2} + 0.4176^2 \times \frac{(2.0 \text{ Bq})^2}{(150.0 \text{ Bq})^2}} = 0.005802$$

Which may be rounded to 0.0058.

The true counting efficiency may vary because of variations in geometry, position and other influence quantities not explicitly included in the model. These sources of uncertainty may not be controlled as they were in the above example. If this is the case, the standard uncertainty of ε should include not only the standard uncertainty of the estimated mean, as calculated in the example, but also another component of uncertainty due to variations of the true efficiency during subsequent measurements. The additional component may be written as $\varepsilon\phi$, where ϕ is the coefficient of variation (i.e., the standard deviation divided by the mean) of the true efficiency. Then the total uncertainty of ε is obtained by squaring the original uncertainty estimate, adding $\varepsilon^2\phi^2$, and taking the square root of the sum.

$$u_c(\varepsilon) = \sqrt{\frac{u^2(\overline{R})}{a_S^2} + \varepsilon^2\left(\frac{u^2(a_S)}{a_S^2} + \phi^2\right)} \qquad (7\text{-}37)$$

In the example above, the experimental variance of the count rates, R_i, may be used to estimate ϕ. Section 18B.2 of Attachment 18B of MARLAP describes an approach for estimating such "excess" variance in a series of measurements.

Variations in counting efficiency due to source placement should be reduced as much as possible through the use of positioning devices that ensure a source with a given geometry is always placed in the same location relative to the detector. If such devices are not used, variations in source position may significantly increase the measurement uncertainty.

Calibrating an instrument under conditions different from the conditions under which M&E sources are counted may lead to large uncertainties in the activity measurements. Source geometry in particular tends to be an important factor for many types of radiation counters. If correction factors are used, their uncertainties should be evaluated and propagated, as mentioned in Section 7.8.1.1.

7.8.2.3   Digital Displays and Rounding

If a measuring device has a digital display with readability[10] δ, the standard uncertainty of a measured value is at least $\delta/2\sqrt{3}$, which is the variance of a random variable uniformly distributed over the interval $(x - \delta/2, x + \delta/2)$. Note that this is the same result as given by equation 7-24 with $a = \delta/2$. This uncertainty component exists even if the instrument is completely stable.

A similar Type B method may be used to evaluate the standard uncertainty due to computer round-off error. When a value x is rounded to the nearest multiple of 10^n (where n is an integer), the component of uncertainty generated by round-off error is $10^n/(2\sqrt{3})$. This component of uncertainty should be kept small in comparison to the total uncertainty of x by performing rounding properly and printing with an adequate number of figures. In a long calculation involving mixed operations, carry as many digits as possible through the entire set of calculations and then round the final result appropriately as described in MARLAP Section 19.3.7 (MARLAP 2004).

Example 17: The readability of a digital survey dose rate meter is 1 nGy/h. Therefore, the minimum standard uncertainty of a measured absorbed dose rate is $1/2\sqrt{3} = 0.29$ nGy/h.

---

[10] **Readability is the** smallest difference that can still be read on a display. For instruments with an analog indicating device, the readability is equal to the smallest fraction of a scale interval that can still be estimated with reasonable reliability or which can be determined by an auxiliary device. For instruments with a numeric indicator (digital display), the readability is equal to one digital step.

Example 18: Suppose the results for R_i in Example 16 had been rounded to the nearest whole number before the analysis. Then the average would be computed as 62.6 instead of 62.6202 and the standard deviation would be computed as 0.9103 instead of 0.9483. This demonstrates the effect that rounding intermediate results can have on subsequent calculations. If this rounding to the nearest positive integer had already occurred prior to receiving the data, and the original data were no longer available, a correction for it could be made when estimating the CSU of R_i. The component of uncertainty generated by round-off error is $1/(2\sqrt{3})$:

$$u(R_i) = \sqrt{0.9103^2 + \left(\frac{1}{2\sqrt{3}}\right)^2} = 0.9549.$$

7.8.3   Example Uncertainty Calculation

To illustrate how the uncertainty calculations are performed in practice, the following example is given based on that of Lewis et al. (2005). The calculation will be that of the CSU in the calibration of a surface contamination monitor.

7.8.3.1   Model Equation and Sensitivity Coefficients

Surface contamination monitors are calibrated in terms of their response to known rates of radioactive emissions. In practice this is achieved by using large-area, planar sources that have a defined area and whose emission rates have been determined in a traceable manner. The calibration is usually determined in terms of response per emission rate per unit area. In this example, the source is positioned with its active face parallel to and at a distance of 3 mm from the face of the detector. The monitor detector area (50 cm^2) is smaller than the area of the calibration source, which is a 10 cm × 10 cm layer of ^{14}C on a thick aluminum substrate. The monitor has an analog display and has a means to set the detector voltage.

The efficiency, ε, is defined by:

$$\varepsilon = \frac{(M - B) \times f_V \times f_d \times f_u \times f_{bs}}{\left(E/A\right)} \tag{7-38}$$

Where:

M	=	observed monitor reading, s^{-1}
B	=	background reading, s^{-1}
E	=	emission rate of the calibration source, s^{-1}
A	=	area of the active portion of the calibration source, cm^2
f_V	=	plateau voltage factor
f_d	=	source-detector separation factor
f_u	=	source uniformity factor
f_{bs}	=	backscatter factor

The sensitivity coefficients of Equation 7-38 are given by:

$$\frac{\partial \varepsilon}{\partial M} = (A/E) \times f_V \times f_d \times f_u \times f_{bs} = \frac{\varepsilon}{(M - B)} \tag{7-39}$$

$$\frac{\partial \varepsilon}{\partial B} = -(A/E) \times f_V \times f_d \times f_u \times f_{bs} = \frac{-\varepsilon}{(M-B)} \tag{7-40}$$

$$\frac{\partial \varepsilon}{\partial E} = -(M-B)(A/E^2) \times f_V \times f_d \times f_u \times f_{bs} = \frac{-\varepsilon}{E} \tag{7-41}$$

$$\frac{\partial \varepsilon}{\partial A} = (M-B)(1/E) \times f_V \times f_d \times f_u \times f_{bs} = \frac{\varepsilon}{A} \tag{7-42}$$

$$\frac{\partial \varepsilon}{\partial f_V} = (M-B)(A/E) \times f_d \times f_u \times f_{bs} = \frac{\varepsilon}{f_V} \tag{7-43}$$

$$\frac{\partial \varepsilon}{\partial f_d} = (M-B)(A/E) \times f_V \times f_u \times f_{bs} = \frac{\varepsilon}{f_d} \tag{7-44}$$

$$\frac{\partial \varepsilon}{\partial f_u} = (M-B)(A/E) \times f_V \times f_d \times f_{bs} = \frac{\varepsilon}{f_u} \tag{7-45}$$

$$\frac{\partial \varepsilon}{\partial f_{bs}} = (M-B)(A/E) \times f_V \times f_d \times f_u = \frac{\varepsilon}{f_{bs}} \tag{7-46}$$

Under normal conditions, the factors f_V, f_d, f_u and f_{bs} are each assumed to have a value of one. If the uncertainties are to be calculated in relative terms, the uncertainty equation becomes (see Equation 7-34):

$$\left(\frac{\sigma_C}{\varepsilon}\right)^2 = \left(\frac{M}{M-B}\right)^2 \left(\frac{\sigma_M}{M}\right)^2 + \left(\frac{B}{M-B}\right)^2 \left(\frac{\sigma_B}{B}\right)^2 + \left(\frac{\sigma_E}{E}\right)^2 + \left(\frac{\sigma_A}{A}\right)^2 + \left(\frac{\sigma_{f_V}}{f_V}\right)^2 + \left(\frac{\sigma_{f_d}}{f_d}\right)^2 + \left(\frac{\sigma_{f_u}}{f_u}\right)^2 + \left(\frac{\sigma_{f_{bs}}}{f_{bs}}\right)^2 \tag{7-47}$$

If the relative uncertainties are all expressed as percentages, $\left(\frac{\sigma_{x_i}}{x_i}\right)$, where x_i is an input quantity, then the CSU will be a percentage. The relative sensitivity coefficients, c_i, are the terms multiplying each relative uncertainty term $\left(\frac{\sigma_{x_i}}{x_i}\right)$ in Equation 7-47. This approach produces relative sensitivity coefficients of unity for the last 6 terms.

7.8.3.2   Uncertainty Components

<u>Monitor Reading of Source, M (Type A)</u>

Several techniques can be used to determine the mean observed monitor reading, M, and its uncertainty. Assume a snap-shot technique is used whereby six successive, but randomly timed, readings are recorded, giving 350, 400, 400, 325, 350, 350 s^{-1}. The mean and standard deviation of the mean becomes $362.5 \pm 12.5 \text{ s}^{-1}$. This equates to a percentage uncertainty in M of 3.45% and the relative sensitivity coefficient from Equation 7-47, $\dfrac{M}{(M-B)}$, is 362.5/(362.5 – 32.5), which is equal to 1.10. The distribution is assumed to be normal.

Monitor Reading of Background, B (Type B)

In this case, an eye-averaging technique was used whereby the highest and lowest count rates were recorded over a given period of time. These count rates were 40 and 25 s^{-1} respectively, giving a mean value of 32.5 s^{-1}. This value is assumed to have a rectangular distribution with a half-width of 7.5 s^{-1}, and an uncertainty of $7.5/\sqrt{3} = 4.330$, equating to a percentage uncertainty of $4.330/32.5 = 0.1332$ or 13%. The relative sensitivity coefficient from Equation 7-47, $\dfrac{B}{(M-B)}$, is 32.5/(362.5 - 32.5), which gives a value of 0.098.

Emission Rate of Calibration Source, E (Type B)

The emission rate of the source and its uncertainty were provided on the calibration certificate by the laboratory that calibrated the source using a windowless proportional counter. The statement on the certificate was:

"The measured value of the emission rate is E = 2,732±13 s^{-1}"

The reported uncertainty is based on a standard uncertainty multiplied by a coverage factor of $k = 2$, which provides a level of confidence of approximately 95%. The standard uncertainty on E is therefore 13/2 = 6.5 s^{-1} or 0.24%. Unless the certificate provides information to the contrary, it is assumed that the uncertainty has a normal distribution.

Source Area, A (Type B)

In the absence of an uncertainty statement by the manufacturer, the only information available is the product drawing that shows the active area dimensions to be 10 cm × 10 cm. On the assumption that the outer bounds of the length, L, and the width, W, are 9.9 and 10.1 cm, the uncertainty of the linear dimensions may be taken to be a rectangular distribution with a half-width of 0.1 cm.

$L = 10$ and $u(L) = 0.1/\sqrt{3} = 0.0577$. $W = 10$ and $u(W) = 0.1/\sqrt{3} = 0.0577$. Because $A = LW$, we get $u^2(A) = u^2(LW) = L^2 u^2(W) + W^2 u^2(L) = 2(10)^2(0.0577)^2 = 0.665858$, therefore $u(A) = 0.816 \text{ cm}^2$ or 0.816%.

Plateau Voltage Factor, f_V (Type B)

This applies only to those instruments where voltage adjustments are possible. If the setting is not checked and/or adjusted between calibrations, then this has no effect. Changing the plateau voltage without performing a recalibration is not recommended. If, however, the user is allowed to do this, the setting may not be returned to exactly that used during the calibration. In this particular example, the slope of the response curve in this region is taken to be 10% / 50 v. It is assumed that an operator is more likely to set the voltage nearer to the optimum than the extremes and that ± 50 v represents the range at the 100% confidence level. Accordingly, a triangular distribution is assumed with a half-width of 50 v, equating to an uncertainty for the voltage of $50/\sqrt{6} = 20.4124$ and an uncertainty for the voltage factor of $20.4124(10\%)/50 = 4.0825\%$.

Source-Detector Separation Factor, f_d (Type B)

This effect arises from the uncertainty in mounting the calibration source exactly 3 mm from the detector face. Experimental evidence has shown that, for the particular ^{14}C source at 3 mm source-detector separation, the change in response was 2.6% per mm. It is assumed that the deviation from the nominal 3-mm separation is no greater than 1 mm but that all values are equally probable between 2 and 4 mm, a rectangular distribution. The uncertainty in the separation is thus $1/\sqrt{3} = 0.5774$. The uncertainty of the separation factor is thus 0.5774 mm × 2.6% / mm, equal to 1.5011%.

Non-Uniformity of Calibration Source, f_u (Type B)

Large area sources may have a non-uniform activity distribution across their surfaces. For the ^{14}C source, the uniformity is assumed to be better than ± 10%. This is based on comparing 10 cm^2 sections of the source. For a typical monitor with a detector area of 50 cm^2 and a calibration source area of 100 cm^2, a worst-case condition could be that the area under the detector has an activity per unit area that is 10% greater than the mean value for the whole source. (The outer area correspondingly will be 10% less than mean value.) Assuming a rectangular distribution, this represents an uncertainty of $10/\sqrt{3} = 5.774\%$ for the source non-uniformity factor.

Backscatter Factor, f_{bs} (Type B)

Variations in backscatter effects arise from factors such as the nature of the surface on which the calibration source is resting and the proximity to scattering surfaces such as walls. This effect can be quite marked for photon emitters, but for ^{14}C on aluminum substrates the effect is negligible.

7.8.3.3   Uncertainty Budget

An important part of the uncertainty analysis is to determine which factors are contributing the most to the overall uncertainty.

To do this, each component of uncertainty $u_i(y)=c_i\,u_i(x_i)$ is squared to give its component of variance $(u_i(y))^2$. These are totaled to obtain the total variance, 69.07. Finally, the ratio of each component of variance to the total is computed. The relative sensitivity coefficients, c_i, are the terms multiplying each relative uncertainty term $\left(\dfrac{\sigma_{x_i}}{x_i}\right)$ in Equation 7-47.

The last column of the uncertainty budget (Table 7.7) shows that the major source of uncertainty is due to source non-uniformity (48%) followed by the voltage factor (24%) and the reading of the source (21%). Thus, to decrease the overall uncertainty, attention should be paid to those factors first.

Table 7.7 Uncertainty Budget for the Efficiency Example

Source of Uncertainty	Type	Probability Distribution	Relative Sensitivity Coeffient, c_i	$u_i(x_i)$ (%)	$u_i(y) = c_i\, u_i(x_i)$ (%)	$(u_i(y))^2$	$(u_i(y))^2$/Total
Standard deviation of mean of M	A	Normal	1.10	3.45	3.80	14.44	0.21
Standard deviation of mean of B	B	Rectangular	0.098	13.32	1.31	1.72	0.02
Standard uncertainty of calibration source emission rate, E	B	Normal	1.0	0.24	0.24	0.06	0.00
Half-width of source length, L and width W on the area A	B	Product of 2 independent rectangular	1.0	0.816	0.816	0.666	0.01
Half-width of voltage factor, f_V	B	Triangular	1.0	4.08	4.08	16.65	0.24
Half-width of source-detector separation factor, f_d	B	Rectangular	1.0	1.50	1.50	2.25	0.03
Half-width of calibration source non-uniformity factor, f_u	B	Rectangular	1.0	5.77	5.77	33.29	0.48
Uncertainty of backscatter factor, f_{bs}	B	n.a.	1.0	0.0	0.0	0.00	0.00
Combined standard uncertainty		Normal	—	—	$8.31 = \sqrt{69.07}$	Total= 69.07	0.99
Expanded uncertainty (k=2)		Normal	—	—	$2 \cdot 8.31 = 16.6$	—	—

7.8.3.4   Reported Result

Using the formula above, the calibration factor in terms of emission rate becomes:

$$\varepsilon = \frac{(M - B) \times f_V \times f_d \times f_u \times f_{bs}}{\left(E/A\right)} = \frac{(362.5 - 32.5) \times 1 \times 1 \times 1 \times 1}{\left(2732/100\right)} = 12.1 \ (\text{counts} \times \text{s}^{-1})/(\text{s}^{-1} \times \text{cm}^{-2}) \quad (7\text{-}48)$$

The CSU is $(12.1)\times(0.0831) = 1.0056$. The reported expanded uncertainty will be 2.0, based on a standard uncertainty of 1.0 multiplied by a coverage factor of $k = 2$, which provides a level of confidence of approximately 95%.

7.9   Calculate the Minimum Detectable Concentration

This section is intended to expand on the material in Section 7.5. It contains more statistical detail and more complex examples. This advanced material may be deferred on a first reading of MARSAME.

7.9.1   Critical Value

In the terminology of ISO 11843-1 (1997), the measured concentration is the state variable, denoted by Y, which represents the state of the material being analyzed. The state variable usually cannot be observed directly, but it is related to an observable response variable, denoted by X, through a calibration function F, the mathematical relationship being written as $X = F(Y)$. The response variable X is most often an instrument signal, such as the number of counts observed. The inverse, $Y = F^{-1}(X)$ of the calibration function is sometimes called the evaluation function. The evaluation function, which gives the value of the net concentration in terms of the response variable, is closely related to the mathematical model $Y = f(X_1, X_2, \ldots, X_N)$ described in Section 7.8.1.1.

Either the null or the alternative hypothesis is chosen on the basis of the observed value of the response variable, X. The value of X must exceed a certain threshold value to justify rejection of the null hypothesis and acceptance of the alternative hypothesis. This threshold is called the critical value of the response variable and is denoted by x_C.

The calculation of x_C requires the choice of a significance level for the test. The significance level is a specified upper bound for the probability, α, of a Type I error. The significance level is usually chosen to be 0.05. This means that when there is no radiation or radioactivity present (above background), there should be at most a 5% probability of incorrectly deciding that it is present.

The critical value of the concentration, y_C, is defined as the value obtained by applying the evaluation function, F^{-1}, to the critical value of the response variable, x_C. Thus, $y_C = F^{-1}(x_C)$. When x is the gross instrument signal, this formula typically involves subtraction of the background signal and division by the counting efficiency, and possibly other factors. A detection decision can be made by comparing the observed gross instrument signal to its critical value, x_C, as indicated above. However, it has become standard practice to make the decision by comparing the net instrument signal to its critical value, S_C. The net signal is calculated from the gross signal by subtracting the estimated blank value and any interference. The critical value of the net instrument signal, S_C, is calculated from the critical gross signal, x_C, by subtracting the same correction terms; so, in principle, either approach should lead to the same detection decision.

Because the term "critical value" alone is ambiguous, one should specify the variable to which the term refers. For example, one may discuss the critical (value of the) radionuclide concentration, the critical (value of the) net signal, or the critical (value of the) gross signal. In this document, the signal is usually a count, and the critical value generally refers to the net count.

The response variable is typically an instrument signal, whose mean value generally is positive even when there is radioactivity present (i.e., above background). The gross signal must be corrected by subtracting an estimate of the signal produced under those conditions. See Section 7.8.2.1 (Instrument Background).

7.9.2   Minimum Detectable Concentration

The minimum detectable concentration (MDC) is the minimum concentration of radiation or radioactivity that must be present in a sample to give a specified power, $1 - \beta$. It may also be defined as:

- The minimum radiation or radioactivity concentration that must be present to give a specified probability, $1 - \beta$, of detecting the radiation or radioactivity; or
- The minimum radiation or radioactivity concentration that must be present to give a specified probability, $1 - \beta$, of measuring a response greater than the critical value, leading one to conclude correctly that there is radiation or radioactivity present.

The *power* of any hypothesis test is defined as the probability that the test will reject the null hypothesis when it is false, i.e., the correct decision. Therefore, if the probability of a Type II error is denoted by β, the power is $1 - \beta$. In the context of radiation or radioactivity detection, the power of the test is the probability of correctly detecting the radiation or radioactivity (concluding that the radiation or radioactivity is present), which happens whenever the response variable exceeds its critical value. The power depends on the concentration of the radiation or radioactivity and other conditions of measurement; so, one often speaks of the "power function" or "power curve." Note that the power of a test for radiation or radioactivity detection generally is an increasing function of the radiation or radioactivity concentration (i.e., the greater the radiation or radioactivity concentration, the higher the probability of detecting it).

In the context of MDC calculations, the value of β that appears in the definition, like α, is usually chosen to be 0.05 or is assumed to be 0.05 by default if no value is specified. The minimum detectable concentration is denoted in mathematical expressions by y_D. The MDC is usually obtained from the minimum detectable value of the net instrument signal, S_D. S_D is defined as the mean value of the net signal that gives a specified probability, $1 - \beta$, of yielding an observed signal greater than its critical value S_C. The relationship between the critical value of the net instrument signal, S_C, and the minimum detectable net signal, S_D, is shown in Figure 7.6 in Section 7.5.2.

The term MDC must be carefully and precisely defined to prevent confusion. The MDC is by definition an estimate of the true concentration of the radiation or radioactivity required to give a specified high probability that the measured response will be greater than the critical value.

The common practice of comparing a measured concentration to the MDC, instead of to the S_C, to make a detection decision is incorrect. If this procedure were used, then there would be only a a 50% chance of deciding that radioactivity was present when the concentration was actually at the MDC. This is in direct contradiction to the definition of MDC. See MARLAP Appendix B, Attachment B1 for a further discussion of this issue.

Since the MDC is calculated from measured values of input quantities such as the counting efficiency and background level, the MDC estimate has a CSU, which in principle can be obtained by uncertainty propagation. To avoid confusion, it may be useful to remember that a detection decision is usually made by comparing the instrument response to the critical value, and that the critical value generally does not even have the units of radiation or radioactivity concentration.

7.9.3   Calculation of the Critical Value

If the net signal is a count, then in many circumstances the uncertainty in the count can be estimated by a Type B evaluation using the fact that for a Poisson distribution with mean N_B, the variance is also N_B. Thus, the uncertainty in the background count is estimated as $\sqrt{N_B}$ and the critical value is often an expression involving $\sqrt{N_B}$.

The most commonly used approach for calculating the critical value of the net instrument signal, S_C, is given by the following equation.[11]

$$S_C = z_{1-\alpha}\sqrt{N_B \frac{t_S}{t_B}\left(1+\frac{t_S}{t_B}\right)}$$

(7-49)

Where:

N_B = background count
t_S = count time for the sample
t_B = count time for the background
$z_{1-\alpha}$ = $(1-\alpha)$-quantile of the standard normal distribution

---

Example 19: A 6,000-second background measurement is performed on a proportional counter and 108 beta counts are observed. A sample is to be counted for 3,000 s. Estimate the critical value of the net count when $\alpha = 0.05$.

$$S_C = z_{1-\alpha}\sqrt{N_B \frac{t_S}{t_B}\left(1+\frac{t_S}{t_B}\right)}$$

---

[11] This expression for the critical net count depends for its validity on the assumption of Poisson counting statistics. If the variance of the blank signal is affected by interferences, or background instability, then Equation 20.7 of MARLAP may be more appropriate. Interference is the presence of other radiation or radioactivity or electronic signals that hinder the ability to analyze for the radiation or radioactivity of interest.

$$S_C = 1.645\sqrt{108 \times \left(\frac{3{,}000\text{ s}}{6{,}000\text{ s}}\right)\left(1 + \frac{3{,}000\text{ s}}{6{,}000\text{ s}}\right)} = 14.8\text{ net counts}$$

If $\alpha = 0.05$ and $t_B = t_S$, Equation 7-49 leads to the well-known expression $2.33\sqrt{N_B}$ for the critical net count (Currie, 1968).

When the background count is high (e.g., 100 or more), Equation 7-49 works well, but at lower background levels it can produce a high rate of Type I errors. Because this is a Scenario B hypothesis test, this means that too often a decision will be made that there is radiation or radioactivity present when it actually is not.

When the mean background counts are low and $t_B \neq t_S$, another approximation formula for S_C appears to out-perform all of the other approximations reviewed in MARLAP, namely the Stapleton approximation:

$$S_C = d \times \left(\frac{t_S}{t_B} - 1\right) + \frac{z_{1-\alpha}^2}{4} \times \left(1 + \frac{t_S}{t_B}\right) + z_{1-\alpha}\sqrt{(N_B + d)\frac{t_S}{t_B}\left(1 + \frac{t_S}{t_B}\right)} \qquad (7\text{-}50)$$

When $\alpha = 0.05$, setting the parameter $d = 0.4$ yields the best results. When, in addition, $t_B = t_S$, the Stapleton approximation gives the equation

$$S_C = 1.35 + 2.33\sqrt{N_B + 0.4} \qquad (7\text{-}51)$$

7.9.4   Calculation of the Minimum Detectable Value of the Net Instrument Signal

The traditional method for calculating the MDC involves three steps: first calculating critical value of the net instrument signal, then calculating the minimum detectable value of the net instrument signal and finally converting the result to a concentration using the mathematical measurement model.

The minimum detectable value of the net instrument signal, denoted by S_D, is defined as the mean value of the net signal that gives a specified probability, $1 - \beta$, of yielding an observed signal greater than its critical value, S_C.

Note: The MDC may be estimated by calculating the minimum detectable value of the net instrument signal, S_D, and converting the result to a concentration.

Counting data rarely, if ever, follow the Poisson model exactly, but the model can be used to calculate S_D if the variance of the background signal is approximately Poisson and a conservative value of the efficiency constant, ε, is used to convert S_D to y_D. The equation below shows how to calculate S_D using the Poisson model.

$$S_D = S_C + \frac{z_{1-\beta}^2}{2} + z_{1-\beta}\sqrt{\frac{z_{1-\beta}^2}{4} + S_C + R_B t_S \left(1 + \frac{t_S}{t_B}\right)} \qquad (7\text{-}52)$$

Where:

S_C =   critical value

R_B =   mean count rate of the blank, $R_B = \dfrac{N_B}{t_B}$

N_B =   background count

t_S =   count time for the test source

t_B =   count time for the background

$z_{1-\beta}$ = $(1 - \beta)$-quantile of the standard normal distribution

When Equation 7-49 is appropriate for the critical net count, and $\alpha = \beta$, this expression for S_D simplifies to $z_{1-\beta}^2 + 2S_C$. If in addition, $\alpha = \beta = 0.05$ and $t_B = t_S$ then

$$S_D = 2.71 + 2S_C = 2.71 + 2(2.33\sqrt{N_B}) = 2.71 + 4.66\sqrt{N_B} \qquad (7\text{-}53)$$

Example 20: A 6,000-s background measurement on a proportional counter produces 108 beta counts and a source is to be counted for 3,000 s. Assume the background measurement gives the available estimate of the true mean background count rate, R_B and use the value 0.05 for Type I and Type II error probabilities. From Section 7.9.3, Example 19, the critical net count, S_C, equals 14.8, so $S_D = z_{1-\beta}^2 + 2S_C = 1.645^2 + 2(14.8) = 32.3$ net counts.

When the Stapleton approximation (Equation 7-51) is used for S_C, the minimum detectable net count S_D may be calculated using the Equation 7-53, but when the Poisson model is assumed, a better estimate is given by the equation:

$$S_D = \frac{(z_{1-\alpha} + z_{1-\beta})^2}{4}\left(1 + \frac{t_S}{t_B}\right) + (z_{1-\alpha} + z_{1-\beta})\sqrt{R_B t_S\left(1 + \frac{t_S}{t_B}\right)} \qquad (7\text{-}54)$$

This equation is the same as that recommended by ISO 11929-1 (ISO 2000) in a slightly different form.

When $\alpha = \beta = 0.05$ and $t_B = t_S$, the preceding equation becomes:

$$S_D = 5.41 + 4.65\sqrt{R_B t_S} \qquad (7\text{-}55)$$

Consult MARLAP Chapter 20 for a discussion of the calculation of S_D and y_D when both Poisson counting statistics and other sources of variance are considered.

7.9.5   Calculation of the Minimum Detectable Concentration

The MDC is often used to compare different measurement procedures against specified requirements. The calculation of the nominal MDC is complicated by the fact that some input quantities in the mathematical model, such as interferences, counting efficiency, and instrument background may vary significantly from measurement to measurement. Because of these variable quantities, determining the value of the radiation or radioactivity concentration that corresponds to the minimum detectable value of the net instrument signal, S_D, may be difficult in practice. One common approach to this problem is to make conservative choices for the values of the variable quantities, which tend to increase the value of the MDC.

The mean net signal, S, is usually directly proportional to Y, the true radiation or radioactivity concentration present. Hence, there is a efficiency constant, ε, such that $S = \varepsilon Y$. The constant ε is typically the mean value of the product of factors such as the source count time, decay-correction factor, and counting efficiency. Therefore, the value of the minimum detectable concentration, y_D, is

$$y_D = \frac{S_D}{\varepsilon} \tag{7-56}$$

The preceding equation is only true if all sources of variability are accounted for when determining the distribution of the net signal, $\hat{S}$. Note that ensuring the MDC is not underestimated also requires that the value of ε not be overestimated.

Using any of the equations in Section 7.5.2 to calculate S_D is only appropriate if a conservative value of the efficiency constant, ε, is used when converting S_D to the MDC.

Example 21: Consider a scenario where $t_B = 6,000$ s, $t_S = 3,000$ s, and $R_B \approx 0.018$ s^{-1}. Let the measurement model be $Y = \dfrac{N_S - (N_B t_S / t_B)}{t_S \varepsilon}$

Where:

Y = activity of the radionuclide in the sample and

ε = counting efficiency (counts per second)/(Bq/cm^2)

Assume the source count time, t_S, has negligible variability, the counting efficiency has mean 0.42 and a 10% relative CSU, and from Example 20, $S_D = 32.3$ net counts.

The mean minimum detectable concentration is $y_D = \dfrac{S_D}{t_S \varepsilon} = \dfrac{32.3}{(3000)(0.42)} = 0.0256$ Bq/cm^2.

Adjusting for the 10% variability in the counting efficiency, the uncertainty is $(0.10) \times (0.42) = 0.042$. Assuming that the efficiency is normally distributed, the lower 5th percentile for ε is $(0.42) - (1.645)(0.042) = 0.35$, where -1.645 is the 5th percentile of a standard normal distribution. Therefore, a conservative estimate of the efficiency constant is $\varepsilon = 0.35$ and a conservative estimate of the minimum detectable concentration is:

$$y_D = \frac{S_D}{t_S \varepsilon} = \frac{32.3}{(3000)(0.35)} = 0.0308 \text{ Bq/cm}^2.$$

An alternative procedure could be to recognize that because of the uncertainties in the input estimates entered into the measurement model to convert from S_D to Y, that the MDC is actually a random variable. Then the methods for propagation of uncertainty given in Section 7.8 can be applied. Using the same assumptions as above, we would find that $y_D = 0.0256 \pm 0.0051$ with 95% confidence based on a coverage factor of 2. Therefore the 95% upper confidence level for y_D would be 0.0307 Bq.

More conservative (higher) estimates of the MDC may be obtained by following NRC recommendations (NRC 1984), in which formulas for the MDC include estimated bounds for relative systematic error in the background determination (Δ_B) and the sensitivity (Δ_A). The critical net count S_C is increased by $\Delta_B N_B (t_S/t_B)$, and the minimum detectable net count S_D is increased by $2 \Delta_B N_B (t_S/t_B)$. Next, the MDC is calculated by dividing S_D by the efficiency and multiplying the result by $1 + \Delta_A$. The conservative approach presented in NRC 1984 treats random errors and systematic errors differently to ensure that the MDC for a measurement process is unlikely to be consistently underestimated, which is an important consideration if it is required by regulation or contract to achieve a specified MDC.

7.10   Calculate the Minimum Quantifiable Concentration

This section is intended to expand on the material in Section 7.6. It contains more statistical detail and more complex examples. This advanced material may be deferred on a first reading of MARSAME.

Calculation of the MQC requires that one be able to estimate the standard deviation for the result of a hypothetical measurement performed on a sample with a specified radionuclide concentration. The MQC is defined symbolically as the value y_Q that satisfies the relation:

$$y_Q = k_Q \sqrt{\sigma^2(y \mid Y = y_Q)} \qquad (7\text{-}57)$$

Where the specified relative standard deviation of y_Q is $1/k_Q$ (usually chosen to be 10% so that $k_Q = 10$). $\sigma^2(y \mid Y = y_Q)$ is the variance of the estimator y given the true concentration Y equals y_Q. If the function $\sigma^2(y \mid Y = y_Q)$ has a simple form, it may be possible to solve the above equation for y_Q using only algebraic manipulation. Otherwise, fixed-point iteration, or other more general approaches, may be used, as discussed in MARLAP Section 20.4.3.

When Poisson counting statistics are assumed, and the mathematical model for the radionuclide concentration is

$$Y = S / \varepsilon \qquad (7.58)$$

Where:

S	=	net count
t_S	=	count time for the source
S / t_S	=	net count rate
ε	=	efficiency of the measurement

Then Equation 7-57 may be solved for y_Q to obtain:

$$y_Q = \frac{k_Q^2}{2t_S\varepsilon(1-k_Q^2\phi_{\hat\varepsilon}^2)}\left(1+\sqrt{1+\frac{4(1-k_Q^2\phi_{\hat\varepsilon}^2)}{k_Q^2}\left(R_B t_S\left(1+\frac{t_S}{t_B}\right)+R_I t_S+\sigma^2(\hat{R}_I)t_S^2\right)}\right) \qquad (7\text{-}59)$$

Where:

t_S	=	count time for the source, s
t_B	=	count time for the background, s
R_B	=	mean background count rate, s^{-1}
R_I	=	mean interference count rate, s^{-1}
$\sigma(\hat{R}_I)$	=	standard deviation of the measured interference count rate, s^{-1}, and
$\phi_{\hat\varepsilon}^2$	=	relative variance of the measured efficiency, $\hat\varepsilon$

If the efficiency ε may vary, then a conservative value, such as the 0.05-quantile $\varepsilon_{0.05}$, should be substituted for ε in the formula. Note that $\phi_{\hat\varepsilon}^2$ denotes only the relative variance of $\hat\varepsilon$ due to subsampling and measurement error; it does not include any variance of the efficiency ε itself (see discussion in Section 7.8).

Note that Equation 7-59 defines the MQC only if $1-k_Q^2\phi_{\hat\varepsilon}^2 > 0$. If $1-k_Q^2\phi_{\hat\varepsilon}^2 \le 0$, the MQC is infinite, because there is no concentration at which the relative standard deviation of y fails to exceed $1/k_Q$. In particular, if the relative standard deviation of the measured efficiency $\hat\varepsilon$ exceeds $1/k_Q$, then $1-k_Q^2\phi_{\hat\varepsilon}^2 < 0$ and the MQC is infinite.

If there are no interferences, Equation 7-59 simplifies to:

$$y_Q = \frac{k_Q^2}{2t_S\varepsilon(1-k_Q^2\phi_{\hat\varepsilon}^2)}\left(1+\sqrt{1+\frac{4(1-k_Q^2\phi_{\hat\varepsilon}^2)}{k_Q^2}\left(R_B t_S\left(1+\frac{t_S}{t_B}\right)\right)}\right) \qquad (7\text{-}60)$$

Example 22: Consider the scenario of Example 21, where $t_B = 6{,}000$ s, $t_S = 3{,}000$ s, and $R_B \approx 0.018$ s^{-1}. Suppose the measurement model is $Y = \dfrac{N_S - (N_B t_S / t_B)}{t_S\varepsilon}$

Where:

Y	=	specific activity of the radionuclide in the sample
ε	=	counting efficiency (cps/Bq)/(Bq/cm^2)

Assume the source count time, t_S, has negligible variability, the counting efficiency has a mean of 0.42 and a 5% relative CSU, and $S_D = 32.3$ net counts.

$S_D/t_S = 32.3/3000$ is the net count rate and the counting efficiency, ε, is 0.42.

The mean minimum detectable concentration is $y_D = \dfrac{S_D}{t_S\varepsilon} = \dfrac{32.3}{(3000)(0.42)} = 0.0256$ Bq/cm^2.

Also assume:

k_Q	$= 10$

$$\phi_{\hat{\varepsilon}} \quad = 0.05$$

$$\phi_{\hat{\varepsilon}}^2 \quad = 0.05^2$$

$$1 - k_Q^2 \phi_{\hat{\varepsilon}}^2 = 1 - 100 \times (0.05^2) = 0.75$$

There are no interferences so that Equation 7-60 can be used.

Note that if the counting efficiency had a mean of 0.42 and a 10% relative standard uncertainty as in Example 11, then $1 - k_Q^2 \phi_{\hat{\varepsilon}}^2 = 1 - 100 \times (0.10^2) = 0$ and the MQC would be infinite. Therefore it was necessary to change the procedure for evaluating the efficiency in this example so that the relative CSU could be reduced. In this example it is assumed to be 5%.
The MQC can be calculated as:

$$y_Q = \frac{k_Q^2}{2 t_S \varepsilon (1 - k_Q^2 \phi_{\hat{\varepsilon}}^2)} \left(1 + \sqrt{1 + \frac{4(1 - k_Q^2 \phi_{\hat{\varepsilon}}^2)}{k_Q^2} \left(R_B t_S \left(1 + \frac{t_S}{t_B}\right) + 0 \right)} \right)$$

$$y_Q = \frac{100}{2\,(3000)(0.42)(0.75)} \left(1 + \sqrt{1 + \frac{4\,(0.75)}{100} \left((0.018 \text{ s}^{-1})(3000 \text{ s}) \left(1 + \frac{(3000 \text{ s})}{(6000 \text{ s})}\right) + 0 \right)} \right)$$

$$= 0.151 \text{ Bq/cm}^2$$

As a check, y_Q can be calculated in a different way. If y_Q is the MQC and $k_Q = 10$, then the relative CSU of a measurement of concentration y_Q is 10%. The procedure described in Section 7.4 can be used to predict the CSU of a measurement made on a hypothetical sample whose concentration is exactly $y_Q = 0.151 \text{ Bq/cm}^2$.

The measurement model is $Y = \dfrac{N_S - (N_B t_S / t_B)}{t_S \varepsilon}$.

Recall from Section 7.8.1.6 that if $y = \dfrac{f(x_1, x_2, \ldots, x_n)}{z_1 z_2 \ldots z_m}$, where f is some specified function of $x_1, x_2, \ldots, x_n$, all the z_i are nonzero, and all the input estimates are uncorrelated that the CSU may be calculated using Equation 7-35:

$$u_c^2(y) = \frac{u_c^2(f(x_1, x_2, \ldots, x_n))}{z_1 z_2 \ldots z_m} + y^2 \left(\frac{u^2(z_1)}{z_1^2} + \frac{u^2(z_2)}{z_2^2} + \ldots + \frac{u^2(z_m)}{z_m^2} \right)$$

Substituting
$y = Y$,
$f(x_1, x_2, \ldots, x_n) = f(N_S, N_B, t_S, t_B) = N_S - (N_B t_S / t_B)/t_S$,
$z_1 = \varepsilon$, and

$$u_c^2(N_S - (N_B t_S / t_B)/t_S) = u_c^2(N_S / t_S) + u_c^2((N_B t_S / t_B)/t_S) = \frac{u_c^2(N_S) + (t_S / t_B)^2 u_c^2(N_B)}{t_S^2} =$$

$$\frac{\sqrt{N_S}^2 + \sqrt{N_B}^2 (t_S^2 / t_B^2)}{t_S^2} = \frac{N_S + N_B(t_S^2 / t_B^2)}{t_S^2}$$

Results in:

$$u_c^2(Y) = \frac{N_S + (N_B t_S^2 / t_B^2)}{t_S^2 \varepsilon^2} + Y^2 \left(\frac{u^2(\varepsilon)}{\varepsilon^2} \right) \text{ or}$$

$$u_c(Y) = \sqrt{\frac{N_S + (N_B t_S^2 / t_B^2)}{t_S^2 \varepsilon^2} + Y^2 \left(\frac{u^2(\varepsilon)}{\varepsilon^2} \right)}$$

Inserting the values

$Y = y_Q = 0.151 \text{ Bq/cm}^2$

$t_B = 6{,}000 \text{ s}$

$t_S = 3{,}000 \text{ s}$

$\varepsilon = 0.42$ (counts per second)/(Bq/cm^2)

$N_B = R_B t_B = (0.018 \text{ s}^{-1})(3{,}000 \text{ s}) = 108$ and

$N_S = x_Q t_S \varepsilon + R_B t_B = (0.151 \text{ Bq})(3000 \text{ s})(0.42) + (0.018 \text{ s}^{-1})(3{,}000 \text{ s}) = 244.26$

yields

$$u_c(Y) = \sqrt{\frac{244.26 + (108)(3{,}000)^2 / (6{,}000)^2}{(3000)^2 (0.42)^2} + (0.151)^2 (0.05^2)} = 0.0151 \text{ Bq/cm}^2$$

Thus, the uncertainty at $y_Q = 0.151$ is 0.0151 and the relative uncertainty is 0.1, so y_Q is verified to be the MQC.

As above in this example, we adjust for the (now) 5% relative CSU in the counting efficiency. The uncertainty is $(0.05) \times (0.42) = 0.02142$. Assuming that the efficiency is normally distributed, the lower 5th percentile is $(0.42) - (1.645)(0.021) = 0.385$. Therefore a conservative estimate of the efficiency is $\varepsilon = 0.385$ and a conservative estimate of the minimum detectable concentration is: $y_Q = \dfrac{(0.151)(0.42)}{0.385} = 0.165 \text{ Bq/cm}^2$.

7.11   Calculate Scan MDCs

The methodology used to determine the scan MDC is based on NUREG-1507 (NRC 1998b). This procedure is quite complex as it requires, among other skills, a familiarity with radiation transport calculations for its implementation. The information developed here will be used in the example in Section 8.2, "Mineral Processing Facility Concrete Rubble." However, the details given in this section are not essential to understanding the example.

The radionuclides of concern are the members of the natural uranium and thorium series. The instrument used is a "Field Instrument for the Detection of Low Energy Radiation" (FIDLER). The approach used would be similar for other instruments and radionuclides.

The approach to determine scan MDCs includes:

- Calculate the fluence rate relative to the exposure rate (FRER) for the range of energies of interest (Section 7.11.1).
- Calculate the probability of interaction (P) between the radiation of interest and the detector (Section 7.11.2).
- Calculate the relative detector response (RDR) for each of the energies of interest (Section 7.11.3).
- Determine the relationship between the detector's net count rate to net exposure rate in cpm/μR/h, Section 7.11.4).
- Determine the relationship between the detector response and the radionuclide concentration (Section 7.11.5).
- Obtain the minimum detectable count rate (MDCR) for the ideal observer, for a given level of performance, by postulating detector background and a scan rate or observation interval (Section 7.11.6).
- Relate the MDCR for the ideal observer to a radionuclide concentration (in Bq/kg) to calculate the scan MDC (Section 7.11.7).

7.11.1   Calculate the Relative Fluence Rate to Exposure Rate (FRER)

For particular gamma energies, the relationship of NaI scintillation detector count rate and exposure rate may be determined analytically (in cpm/μR/h). The approach is to determine the gamma fluence rate necessary to yield a fixed exposure rate (μR/h) as a function of gamma energy. The fluence rate, following NUREG-1507 (NRC 1998b), is directly proportional to the exposure rate and inversely proportional to the incident photon energy and mass energy absorption coefficient:

$$\text{Fluence Rate(FRER)} \propto \dot{X}\,\frac{1}{E_\gamma}\,\frac{1}{(\mu_{en}/\rho)_{air}} \qquad (7\text{-}61)$$

Where:

$\dot{X}$ = exposure rate (set equal to 1 μR/hr for these calculations)

E_γ = energy of the gamma photon of concern (keV)

$(\mu_{en}/\rho)_{air}$ = mass energy absorption coefficient in air at the gamma photon energy of concern (cm^2/g)

The mass energy absorption coefficients in air are presented in Table 7.8 (natural uranium) and Table 7.9 (natural thorium) along with the calculated fluence rates (up to a constant of proportionality, since only the ratios of these values are used in subsequent calculations). Note that while the mass energy absorption coefficients in air, $(\mu_{en}/\rho)_{air}$, are tabulated values (NIST 1996), the selected energies are determined by the calculation of the detector response based on radionuclide concentration (Section 7.11.5).

7.11.2   Calculate the Probability of Interaction

Assuming that the primary gamma interaction producing the detector response occurs through the end of the detector (i.e., through the beryllium window of the detector, as opposed to the sides), the probability of interaction (P) for a gamma may be calculated using Equation 7-52:

$$P = 1 - e^{-(\mu/\rho)_{NaI}(x)(\rho_{NaI})} = 1 - e^{-(0\ 117\ cm^2/g)(0\ 16\ cm)(3\ 67\ g/cm^3)} = 0.066 \text{ at } 400 \text{ keV} \qquad (7\text{-}62)$$

Where:

P	=	probability of interaction (unitless)
$(\mu/\rho)_{NaI}$	=	mass attenuation coefficient of FIDLER NaI crystal at the energy of interest (e.g., 0.117 cm^2/g at 400 keV)
x	=	thickness of the thin edge of the FIDLER NaI crystal (0.16 cm)
ρ	=	density of the NaI crystal (3.67 g/cm^3)

The mass attenuation coefficients for the NaI crystal and the calculated probabilities for each of the energies of interest are presented in Table 7.8 (natural uranium) and Table 7.9 (natural thorium). The mass attenuation coefficients for NaI were calculated using the XCOM program (NIST 1998).

Table 7.8 Calculation of Detector Response to Natural Uranium

Energy (keV)	$(\mu_{en}/\rho)_{air}$ (cm^2/g)	FRER (Section 7.11.1)	$(\mu/\rho)_{NaI}$ cm^2/g	P (Section 7.11.2)	RDR (Section 7.11.3)	cpm per μR/h (Section 7.11.4)
15	1.334	0.04998	47.4	1.000	0.04998	28,374
20	0.5389	0.09278	21.8	1.000	0.09278	52,678
30	0.1537	0.2169	7.36	0.9867	0.2140	121,498
40	0.06833	0.3659	18.8	1.000	0.3659	207,725
50	0.04098	0.4880	10.5	0.9979	0.4870	276,511
60	0.03041	0.5481	6.45	0.9773	0.5356	304,123
80	0.02407	0.5193	3.00	0.8282	0.4301	244,204
100	0.02325	0.4301	1.67	0.6249	0.2688	152,606
150	0.02496	0.2671	0.611	0.3015	0.08052	45,717
200	0.02672	0.1871	0.328	0.1752	0.03278	18,613
300	0.02872	0.1161	0.166	0.09288	0.01078	6,120
400	0.02949	0.08477	0.117	0.06640	0.005629	3,196
500	0.02966	0.06743	0.0950	0.05426	0.003659	2,077
600	0.02953	0.05644	0.0822	0.04712	0.002660	1,510
662	0.02931	0.05154	0.0766	0.04398	0.002267	1,287
800	0.02882	0.04337	0.0675	0.03886	0.001685	957
1,000	0.02789	0.03586	0.0588	0.03394	0.001217	691
1,500	0.02547	0.02617	0.0470	0.02722	0.0007125	405
2,000	0.02345	0.02132	0.0415	0.02407	0.0005133	291

Table 7.9 Calculation of Detector Response for Natural Thorium

Energy (keV)	$(\mu_{en}/\rho)_{air}$ (cm²/g)	FRER (Section 7.11.1)	$(\mu/\rho)_{NaI}$ cm²/g	P (Section 7.11.2)	RDR (Section 7.11.3)	cpm per µR/h (Section 7.11.4)
40	0.06833	0.3659	18.8	1.000	0.3659	207,725
60	0.03041	0.5481	6.45	0.9773	0.5356	304,123
80	0.02407	0.5193	3.00	0.8282	0.4301	244,204
100	0.02325	0.4301	1.67	0.6249	0.2688	152,606
150	0.02496	0.2671	0.611	0.3015	0.08052	45,717
200	0.02672	0.1871	0.328	0.1752	0.03278	18,613
300	0.02872	0.1161	0.166	0.09288	0.01078	6,120
400	0.02949	0.08477	0.117	0.06640	0.005629	3,196
500	0.02966	0.06743	0.0950	0.05426	0.003659	2,077
600	0.02953	0.05644	0.0822	0.04712	0.002660	1,510
662	0.02931	0.05154	0.0766	0.04398	0.002267	1,287
800	0.02882	0.04337	0.0675	0.03886	0.001685	957
1,000	0.02789	0.03586	0.0588	0.03394	0.001217	691
1,500	0.02547	0.02617	0.0470	0.02722	0.0007125	405
2,000	0.02343	0.02134	0.0415	0.02407	0.0005137	292
3,000	0.02057	0.01620	0.0368	0.02138	0.0003464	197

7.11.3   Calculate the Relative Detector Response

The relative detector response (RDR) for each of the energies of interest is determined by multiplying the FRER by P. The results are presented in Table 7.8 (natural uranium) and Table 7.9 (natural thorium).

7.11.4   Relationship Between Detector Response and Exposure Rate

Using the same methodology described in Sections 7.11.1 through 7.11.3, FRER, P, and RDR are calculated at the ^{137}Cs energy of 662 keV and are also presented in Table 7.8 and Table 7.9. The manufacturer of the FIDLER NaI detector provides an estimated response of the crystal in a known radiation field, which is 1,287 cpm per µR/h at the ^{137}Cs energy of 662 keV. The response at 662 keV can be used to determine the response at all other energies of interest using Equation 7-63:

$$\frac{cpm}{\mu R/h_{E_i}} = \left(\frac{1{,}287 \text{ cpm}}{\mu R/h} \right) \times \frac{RDR_{E_i}}{RDR_{^{137}Cs}} \qquad (7\text{-}63)$$

Where:

E_i =   energy of the photon of interest (keV)

$$\frac{\text{cpm}}{\mu R/h_{E_i}} = \text{response of the detector for energies of interest, Table 7.8 and Table 7.9}$$

$$RDR_{E_i} = \text{RDR at the energy of interest, Table 7.8 and Table 7.9}$$

$$RDR_{^{137}Cs} = \text{RDR for } ^{137}Cs, \text{ Table 7.8 and Table 7.9}$$

The responses in cpm per $\mu R/h$ for each of the decay energies of interest are presented in Tables 7.8 and 7.9.

7.11.5   Relationship Between Detector Response and Radionuclide Concentration

The minimum detectable exposure rate is used to determine the MDC by modeling a specific impacted area. The relationship between the detector response (in cpm) and the radionuclide concentration (in Bq/kg) uses a computer gamma dose modeling code to model the presence of a normalized 1 Bq/kg total activity source term for natural uranium and natural thorium. The following assumptions from NUREG-1507 (NRC 1998b) were used to generate the computer gamma dose modeling runs:

- Impacted media is concrete,
- Density of concrete is 2.3 g/cm^3,
- Activity is uniformly distributed into a layer of crushed concrete 15 cm thick,
- Measurement points are 10 cm above the concrete surface,
- Areas of elevated activity are circular with an area of 0.25 m^2 and a radius of 28 cm,
- 0.051 cm beryllium shield simulates the window of the FIDLER detector, and
- Normalized 1 Bq/kg source term decayed for 50 years to allow ingrowth of decay progeny.

The weighted cpm per $\mu R/h$ response (weighted instrument sensitivity [WS_i]) for each decay energy is calculated by multiplying the $\mu R/h$ at 1 Bq/kg (exposure rate with buildup, R_i) by the cpm per $\mu R/h$ and dividing by the total $\mu R/h$ (at 1 Bq/kg) for all decay energies of interest (Equation 7-64):

$$WS_i = \frac{R_i \times (\text{cpm per } \mu R/h)}{R_T} \tag{7-64}$$

Where:

WS_i = weighted instrument sensitivity (cpm per $\mu R/h$)
R_i = exposure rate with buildup ($\mu R/h$)
R_T = Total exposure rate with buildup ($\mu R/h$)

Calculate the percent of FIDLER response for each of the decay energies of interest by dividing WS_i by the total weighted cpm per $\mu R/h$ and multiplying by 100 percent (Equation 7-62):

$$\text{Percent of FIDLER response} = \frac{WS_i \times 100\%}{W_T} \tag{7-65}$$

Where:

W_T = Total WS_i weighted instrument sensitivity (cpm per $\mu R/h$)

The exposure rates for each of the decay energies of interest are presented in Table 7.10 (assuming natural uranium for the source term) and Table 7.11 (assuming natural thorium for the source term).

Table 7.10 Detector Response to Natural Uranium

Energy keV	R_i (µR/h) (Section 7.11.5)	cpm per µR/h (Section 7.11.4)	WS_i (cpm per µR/h) (Section 7.11.5)	Percent of FIDLER Response (Section 7.11.5)
15	4.473×10^{-10}	28,374	0	0.00%
20	3.597×10^{-12}	52,678	0	0.00%
30	2.623×10^{-07}	121,498	226	0.504%
40	1.299×10^{-10}	207,725	0	0.00%
50	1.052×10^{-07}	276,511	206	0.460%
60	5.065×10^{-06}	304,123	10903	24.3%
80	1.518×10^{-06}	244,204	2625	5.86%
100	2.309×10^{-05}	152,606	24938	55.7%
150	5.138×10^{-06}	45,717	1663	3.71%
200	2.881×10^{-05}	18,613	3796	8.48%
300	2.237×10^{-07}	6,120	10	0.0216%
400	2.434×10^{-07}	3,196	6	0.0123%
500	4.208×10^{-07}	2,077	6	0.0138%
600	2.048×10^{-06}	1,510	22	0.0489%
800	1.478×10^{-05}	957	100	0.224%
1,000	5.759×10^{-05}	691	282	0.629%
1,500	1.695×10^{-06}	405	5	0.0108%
2,000	2.841×10^{-07}	291	1	0.00131%
Total	1.413×10^{-04}		44,923	100%

Table 7.11 Detector Response to Natural Thorium

Energy keV	R_i (µR/h) (Section 7.11.5)	cpm per µR/h (Section 7.11.4)	WS_i (cpm per µR/h) (Section 7.11.5)	Percent of FIDLER Response (Section 7.11.5)
40	1.299×10^{-06}	207,725	10	0.266%
60	1.816×10^{-06}	304,123	21	0.544%
80	1.989×10^{-04}	244,204	1855	47.8%
100	5.027×10^{-05}	152,606	293	7.55%
150	5.862×10^{-05}	45,717	102	2.64%
200	1.135×10^{-03}	18,613	807	20.8%
300	8.922×10^{-04}	6,120	209	5.37%

Table 7.11 Detector Response to Natural Thorium (Continued)

Energy keV	R_i (µR/h) (Section 7.11.5)	cpm per µR/h (Section 7.11.4)	WS_i (cpm per µR/h) (Section 7.11.5)	Percent of FIDLER Response (Section 7.11.5)
400	1.105×10^{-04}	3,196	13	0.348%
500	8.146×10^{-04}	2,077	65	1.67%
600	2.218×10^{-03}	1,510	128	3.30%
800	2.892×10^{-03}	957	106	2.72%
1,000	6.443×10^{-03}	691	170	4.38%
1,500	2.062×10^{-03}	405	32	0.821%
2,000	5.822×10^{-05}	292	1	0.0167%
3,000	9.249×10^{-03}	197	69	1.79%
Total	2.619×10^{-02}		3881	100%

7.11.6 Calculation of Scan Minimum Detectable Count Rates

In the computer gamma dose modeling, an impacted area with a radius of 28 cm or approximately 0.25 m was assumed. Using a scan speed of 0.25 m/s provides an observation interval of one second.

A typical background exposure rate is 10 µR/h. Using a conversion factor based upon field measurements of 1,287 cpm per µR/h for ^{137}Cs (see 7.11.4) results in an estimated background count rate of 12,870 cpm. Converting this value from cpm to counts per second (cps) using Equation 7-66 results in a background of 214.5 cps.

$$b(\text{cpm}) \times \frac{1\,\text{min}}{60\,\text{sec}} \times i(\text{sec}) = \frac{1,287\,\text{cpm}}{1\,\mu R/h} \times 10\,\mu R/h \times \frac{1\,\text{min}}{60\,\text{sec}} \times 1\,\text{sec} = 214.5\,\text{cps} \qquad (7\text{-}66)$$

Where:
- b = background count rate (12,870 cpm)
- i = observation interval length (1 s)

The MDCR is calculated using the methodology in NUREG-1507 (NRC 1998b) shown in Equations 7-67 and 7-68:

$$s_i = d'\sqrt{b_i} = 1.38 \times \sqrt{214.5} = 20.21\,\text{counts} \qquad (7\text{-}67)$$

$$s_{i,\,surveyor} = \frac{d'\sqrt{b_i}}{\sqrt{p}} = \frac{1.38 \times \sqrt{214.5}}{\sqrt{0.5}} = 28.58\,\text{counts}$$

$$\text{MDCR} = s_i \times (60/i) = 20.21 \times (60/1) = 1,212\,\text{cpm} \qquad (7\text{-}68)$$

$$\text{MDCR}_{surveyor} = s_{i,\,surveyor} \times (60/i) = 28.58 \times (60/1) = 1,715 \text{ cpm}$$

Where:

b_i	=	average number of counts in the background interval (214.5 counts)
i	=	observation interval length (one second)
p	=	efficiency of a less than ideal surveyor, range of 0.5 to 0.75 from NUREG-1507 (NRC 1998b); a value 0.5 was chosen as a conservative value
d'	=	detectability index from Table 6.1 of NUREG-1507 (NRC 1998b); a value of 1.38 was selected, which represents a true positive detection rate of 95% and a false positive detection rate of 60%[12]
s_i	=	minimum detectable number of net source counts in the observation interval (counts)
$s_{i,surveyor}$	=	minimum detectable number of net source counts in the observation interval by a less than ideal surveyor
MDCR	=	minimum detectable count rate (cpm)
$\text{MDCR}_{surveyor}$	=	MDCR by a less than ideal surveyor (cpm)

7.11.7   Calculate the Scan Minimum Detectable Concentration

The scan minimum detectable concentration (MDC) can be calculated from the minimum detectable exposure rate (MDER). The MDER can be calculated using the previously calculated total weighted instrument sensitivities (WS_i), in cpm per μR/h, for natural uranium and natural thorium as shown in Equations 7-69 and 7-70:

$$\text{MDER} = \frac{\text{MDCR}_{surveyor}}{W_T} \tag{7-69}$$

$$\text{Scan MDC} = C \times \frac{\text{MDER}}{R_T} \tag{7-70}$$

Where:

MDER	=	MDER for the "i^{th}" source term, by a less than ideal surveyor, (μR/h)
$\text{MDCR}_{surveyor}$	=	MDCR rate by a less than ideal surveyor (cpm), from Section 7.11.6
W_T	=	Total weighted instrument sensitivity (cpm per μR/h, Table 7.10 and Table 7.11)
R_T	=	Total exposure rate with buildup (μR/h, Table 7.10 and Table 7.11)
C	=	concentration of source term (set at 1 Bq/kg in Section 7.11.5)

---

[12] A Type I error, misidentifying a background area as elevated will have the consequence that a longer reading will be needed to verify the initial decision. This will happen with probability α. A Type II error, missing a true elevated area, may lead to incorrectly exceeding the limit for the chosen disposition option. This will happen with probability β. Since in this instance the consequences of a Type I error are often considered much lower than the consequences associated with a Type II error. Thus, α may be set higher than β. Setting both very low could result in slow scanning speeds and operator fatigue.

Scan MDC   =   minimum detectable concentration (Bq/kg)

The Scan MDCs for the FIDLER were calculated using Equations 7-69 and 7-70 and the instrument response information from Table 7.10 (assuming natural uranium as the source term) and Table 7.11 (assuming natural thorium as the source term). The scan MDCs for natural uranium and natural thorium using a FIDLER are listed in Table 7.12.

Table 7.12 Scan MDCs for FIDLER

Source Term	$MDCR_{surveyor}$ (cpm) Section 7.11.6	W_T (cpm per µR/h) Section 7.11.5	MDER (µR/h) Section 7.11.7	R_T (µR/h) Section 7.11.5	C (Bq/kg) Section 7.11.5	Scan MDC (Bq/kg) Section 7.11.7
Natural Uranium	1,715	44,786	0.03829	1.413×10^{-04}	1	$271 \approx 300$
Natural Thorium	1,715	3,881	0.4419	2.619×10^{-02}	1	$16.9 \approx 20$

The scan MDCs of approximately 300 Bq/kg for uranium and 20 Bq/kg for thorium are both less than their respective action levels of 38,000 and 330 Bq/kg, respectively.

8   ILLUSTRATIVE EXAMPLES

8.1   Introduction

This chapter presents illustrative examples providing examples of applications of the information in the *Multi-Agency Radiation Survey and Assessment of Materials and Equipment* manual (MARSAME) supplement to the *Multi-Agency Radiation Survey and Site Investigation Manual* (MARSSIM). The purpose of these illustrative examples is to illustrate applications of the information in conditions that are frequently encountered and cover a broad range of situations. The general format for each illustrative example mirrors as closely as possible the information presented in MARSAME. References to information, tables, figures, and equations from Chapter 2 through Chapter 6 are provided throughout the illustrative examples.

MARSAME contains both procedural as well as informative sections. The illustrative examples provide a practical use of the MARSAME process and, as such, generally apply only the procedural sections. In addition, much of the information in MARSAME is designed to be applied iteratively. In some illustrative examples, the information is applied in a different sequence than it is presented in MARSAME because of this iterative nature.

Section 8.2 provides an example of a disposition survey for a large quantity of bulk material at a mineral processing facility. This example establishes gross activity action levels based on normalized effective dose equivalents. These action levels are applied with multiple decision rules using a MARSSIM-type survey design to collect scan survey data as well as systematic and judgmental samples for laboratory analysis.

Section 8.3 and Section 8.4 are based on the same mineral processing facility that serves as the basis for Section 8.2. Section 8.3 provides an example of an interdiction survey for rented heavy equipment that is designed to establish a "baseline" estimate of the residual radioactivity associated with a front loader before it is brought into a radiological control area (RCA) for the impacted bulk material. This baseline survey establishes zero net activity as the lower bound of the gray region (LBGR) and applies MARSAME processes to a Scenario B survey design.

Section 8.4 demonstrates the clearance of the same rented front loader that was brought on to the site in Section 8.3. Section 8.4 describes a Scenario A clearance survey based on the same surface activity action levels to clear the front loader. Sections that contain redundant information are presented in Section 8.3 only and are omitted from Section 8.4.

8.2   Mineral Processing Facility Concrete Rubble

This illustrative example is provided for information purposes only and presents a theoretical application of MARSAME guidance. This illustrative example discusses the process of designing and implementing a MARSSIM-type disposition survey design for a large quantity of bulk material at a mineral processing facility. This example includes discussions on most of the guidance provided in MARSAME, including establishing gross activity action levels based on normalized effective dose equivalents. Calculations of uncertainties associated with scanning measurements are included. The MARSSIM-type survey design includes scanning, systematic samples, and judgmental samples to support a disposition decision. The text is provided to illustrate the application of MARSAME guidance, and should not be considered an example survey plan. The amount of discussion provided in this example is based on the complexity of the problem and the relative difficulty expected from applying or interpreting specific portions of MARSAME guidance. The amount of discussion for this example is not related to, and should not be used as an estimate of, the level of effort associated with planning, implementing, or assessing an actual disposition survey.

8.2.1 Description

An abandoned mineral processing facility is being redeveloped for commercial/industrial use. The facility processed mineral ores for various metals for over 30 years and was abandoned more than 10 years ago. The processing equipment and existing stockpiles of ore were transferred to another facility when site renovations began. The receiving facility discovered radioactivity levels in excess of background on exterior portions of processing equipment using hand-held Geiger-Mueller (GM) "pancake" detectors.

Prior to discovery of the radioactivity on the processing equipment, the concrete floors had been removed from the processing buildings and stockpiled on-site. Note that if the buildings were still intact, they could be surveyed using a MARSSIM survey. An investigation is performed to trace the source of the radioactivity to the appropriate portion(s) of the mineral processing facility.

8.2.2 Objectives

The objective is to make an appropriate disposition decision regarding the concrete rubble from the impacted portions of the mineral processing facility. It is anticipated that leaks of potentially radioactive processing liquids could have occurred throughout the operating lifetime of the facility. Airborne radioactive concrete dust may have been released during demolition activities, which could have exposed construction personnel and contacted components of the demolition equipment.

8.2.3 Initial Assessment of the M&E

8.2.3.1 Categorize the M&E as Impacted or Non-Impacted

As part of the initial assessment (IA), it is necessary to determine whether the concrete rubble is impacted or not. A visual inspection of the concrete rubble was performed. Historical records from the facility concerning sources of ore, ore processing techniques, waste disposal practices, industrial accidents, as well as building and equipment repairs, modifications, and upgrades were reviewed. Interviews with key facility personnel were also performed. In addition, research into mineral processing techniques and radionuclide content of raw ores was performed to obtain additional process knowledge.

Process knowledge indicated the facility processed ilmenite ore (iron titanium oxide, $FeTiO_3$) and produced titanium dioxide. A sentinel measurement of a small amount of ilmenite ore remaining at the site was analyzed by alpha spectrometry and found to contain elevated levels of natural uranium and thorium. Additional measurements performed on the radioactive processing equipment reported concentrations of uranium and thorium greater than expected from background.

Site history indicates that the general layout of the process was unchanged over the lifetime of the facility, and it is likely that spills occurred repeatedly in discrete locations. Processing liquids and slurries were considered hazardous because of their low pH; radioactivity was not considered an issue. Limited information regarding site history and operations was obtained through interviews with former employees and review of historical documentation. Former employees stated that spills and leaks of process liquids and slurries occurred periodically in several areas of the processing plant; these represent the only potential source of radioactivity in the plant. Fluid spills were quickly corrected by neutralizing the acid to protect employees and equipment. Spills frequently resulted from seal failure within the various pumps in use at the processing operation.

Results from the visual inspection indicated there was a reasonable potential for radioactivity from plant activities to be associated with the concrete rubble. Several chunks of concrete rubble are obviously discolored from plant operations, indicating possible locations of spills. The facility floor consisted of reinforced concrete on a gravel base mat. Portions of the rubble contain possible evidence of staining. The rubble still contains rebar which, for operational reasons, must be segregated and treated as a separate waste stream.

The concrete rubble is considered to be impacted due to the discovery of residual radioactivity on exterior portions of the processing equipment, historical records that acidic process fluids may have spilled on the concrete floor, and process knowledge that the acidic process fluids were mixed with raw ore containing elevated levels of naturally occurring radioactive material (NORM) from the uranium and thorium radioactive decay series. The results of the sentinel measurement performed on the raw ore support the categorization as impacted.

8.2.3.2   Describe the M&E

Table 8.1 lists the physical attributes of the concrete rubble. No data gaps associated with the physical attributes were identified.

Table 8.2 lists the known radiological attributes associated with the concrete rubble, as well as data gaps showing where additional information is required to design a disposition survey. As presented, the existing information is not adequate to design a disposition survey. Preliminary surveys were designed and implemented to address the data gaps identified in Table 8.2. The results of the preliminary surveys were used to modify the conceptual site model by filling some of the data gaps.

Table 8.1 Physical Attributes of the Concrete Rubble

Attribute	Description
Dimensions	Total Mass 400 ft × 100 ft × 1 ft ≈ 40,000 ft³ 40,000 ft³ × 0.0283 m³/ft³ ≈ 1,132 m³ The approximate density of crushed concrete is 2.3×10^6 g/m³ 1,132 m³ × 2.3×10^6 g/m³ = 2.60×10^9 g = 2.60×10^6 kg Shape The concrete has been broken into chunks less than one meter in the largest dimension. The concrete is stored in three piles. Each pile is approximately 1.5 m high, 6 m wide, and 40 m long.
Complexity	Rebar used to reinforce the floor is present in the concrete rubble. The rebar will be segregated and removed, and treated as a separate waste stream.
Accessibility	The concrete rubble may require further reduction in size to ensure measurability.
Inherent Value	The concrete represents inherent value for several potential disposition options. Crushed concrete serves many useful purposes, including recyclable use as roadbed material. This option presents potential cost savings over using virgin materials in place of recycled concrete and a reuse scenario that avoids the relatively high cost for disposal.

The radionuclides of potential concern are the uranium (^{238}U) and thorium (^{232}Th) natural radioactive decay series. Based on process knowledge, radionuclide concentrations in the raw ore average between 750 and 1,100 Bq/kg for members of the uranium series, and between 200 and 400 Bq/kg for members of the thorium series. Following processing, some ^{238}U and ^{232}Th decay products may not have been in equilibrium with the parents. The amount of time since the plant ceased operations (i.e., 10 years) indicates there is a potential for the thorium series radionuclides to have re-established secular equilibrium. Preliminary survey measurements are required to determine the equilibrium status of the uranium and thorium series radionuclides.

8.2.3.3   Design and Implement Preliminary Surveys

Limited scanning of concrete rubble was performed using a GM detector. The purpose of the scanning was to determine how the radioactivity associated with the concrete was distributed. The scanning survey also included additional visual inspection of the concrete.

Intermittent staining within the concrete rubble and scanning surfaces of concrete chunks demonstrates that the radioactivity was heterogeneously deposited on the processing building floor. Higher levels of radioactivity were found in areas where spills occurred historically (i.e., discolored concrete). The staining did not appear to have penetrated more than one-quarter inch into the concrete when the floor was intact. Prior to demolition, the presence of cracks and other structural irregularities in the concrete floor provided preferential pathways for activity to penetrate to greater depths. This resulted in some variance in activity with depth of the original concrete floor.

Table 8.2 Radiological Attributes of the Concrete Rubble

Attribute	Description			Data Gaps
Radionuclides	Uranium Series Radionuclides	Principal Emission Particle	Emission Energy (MeV)	The radioactivity is likely to have come in contact with the M&E through spills of process fluids and dumping of solid tailings on the concrete floor. Equilibrium status of the decay series is unknown, although sufficient time has elapsed since site closure for the thorium series to have re-established secular equilibrium.
	^{238}U	Alpha	4.20	
	^{234}Th	Beta	0.1886	
	^{234m}Pa	Beta/Gamma	2.28/1.001	
	^{234}Pa	Beta	0.224	
	^{234}U	Alpha	4.77	
	^{230}Th	Alpha	4.688	
	^{226}Ra	Alpha/Gamma	4.78/0.186	
	^{222}Rn	Alpha	5.49	
	^{218}Po	Alpha	6.00	
	^{214}Pb	Beta/Gamma	0.67/0.352	
	^{214}Bi	Beta/Gamma	1.54/0.609	
	^{214}Po	Alpha	7.687	
	^{210}Pb	Beta	0.016	
	^{210}Bi	Beta	1.161	
	^{210}Po	Alpha	5.305	
	Thorium Series Radionuclides	Principal Emission Particle	Emission Energy (MeV)	
	^{232}Th	Alpha	4.01	
	^{228}Ra	Beta	0.0389	
	^{228}Ac	Beta/Gamma	1.17/0.911	
	^{228}Th	Alpha	5.42	
	^{224}Ra	Alpha	5.686	
	^{220}Rn	Alpha	6.288	
	^{216}Po	Alpha	6.78	
	^{212}Pb	Beta/Gamma	0.334/0.238	
	^{212}Bi	Alpha/Beta	6.05/2.246	
	^{212}Po (64%)	Alpha	8.785	
	^{208}Tl (36%)	Beta	1.80	
Activity	Activity levels range from background (approximately 40 Bq/kg) to 4,000 Bq/kg from isolated portions of the concrete rubble where spills occurred.			The expected range of activity is an estimate. Nature and extent of activity needs to be investigated to provide better estimates of average and maximum activity. Better estimates of background are needed.
Distribution	The radioactivity is heterogeneously distributed throughout the mass of concrete rubble.			No data gaps were identified. The current distribution is not a concern because the concrete will be crushed to 2–3 cm size prior to survey.
Location	The concrete rubble is considered a volumetrically impacted mass. The residual radioactivity that is present is a combination of fixed and removable.			The distribution of radioactivity with depth may provide useful information for selecting measurement methods because it can impact the total measurement efficiency.

Samples were collected from the crushed concrete from the processing mill floor to determine concentrations of residual radioactivity using alpha spectrometry and gamma spectroscopy. Concrete samples were collected from four biased locations, including two areas of elevated gross activity within the concrete rubble with GM readings as high as 250 cpm and visible staining (Samples 1 and 2), and two samples with readings consistent with the average readings observed during scanning (40 to 45 cpm) (Samples 3 and 4). Process knowledge and limited historical site information indicates that radiological materials were never used or stored within the on-site administrative building. Reference Samples 1 and 2 were collected from the concrete floor of the onsite administrative building to provide information on background activities in non-impacted concrete for the uranium and thorium decay series for the conceptual model. The six samples were sent to a radioanalytical laboratory for analysis, and the results of the analyses are provided in Tables 8.3 through 8.6.

Table 8.3 Preliminary Alpha Spectrometry Results for Uranium Series Radionuclides

Sample ID	^{234}U	CSU[1]	MDC[2]	^{235}U	CSU[1]	MDC[2]	^{238}U	CSU[1]	MDC[2]
Sample 1	7,000	± 2,100	1,900	340	± 1,900	1,600	7,600	± 2,400	1,900
Sample 2	7,200	± 2,300	1,900	320	± 1,700	1,600	7,000	± 2,100	1,900
Sample 3	21	± 7.4	3.7	0.74	± 1.9	0.74	21	± 7.0	3.7
Sample 4	25	± 8.1	3.7	0.74	± 3.0	0.74	21	± 7.0	3.7
Reference Sample 1	19	± 5.2	3.7	0.37	± 0.74	0.74	20	± 5.6	3.7
Reference Sample 2	13	± 3.7	3.7	0.37	± 0.74	0.74	11	± 3.3	3.7

All units in Bq/kg
[1] CSU is the combined standard uncertainty of the measurement result reported by the analytical laboratory.
[2] MDC is the minimum detectable concentration reported by the analytical laboratory.

Table 8.4 Preliminary Alpha Spectrometry Results for Thorium Series Radionuclides

Sample ID	^{232}Th	CSU[1]	MDC[2]	^{228}Th	CSU[1]	MDC[2]
Sample 1	1,400	± 110	110	1,300	± 150	110
Sample 2	1,200	± 130	110	1,500	± 190	110
Sample 3	21	± 1.5	1.1	23	± 1.5	1.1
Sample 4	26	± 1.1	1.1	24	± 1.1	1.1
Reference Sample 1	21	± 1.1	1.1	22	± 1.1	1.1
Reference Sample 2	23	± 1.1	1.1	23	± 1.1	1.1

All units in Bq/kg
[1] CSU is the combined standard uncertainty of the measurement result reported by the analytical laboratory.
[2] MDC is the minimum detectable concentration reported by the analytical laboratory.

Table 8.5 Preliminary Gamma Spectroscopy Results for Uranium Series Radionuclides

Sample ID	^{214}Bi	CSU[1]	MDC[2]	^{214}Pb	CSU[1]	MDC[2]	^{226}Ra	CSU[1]	MDC[2]
Sample 1	93	± 920	1,400	530	± 780	1,300	47	± 1,100	1,500
Sample 2	740	± 1,000	1,300	1,000	± 870	1,200	192	± 1,200	1,400
Sample 3	21	± 1.1	3.6	21	± 1.1	6.3	64	± 9.6	16
Sample 4	22	± 1.1	4.1	23	± 1.1	7.0	68	± 8.5	19
Reference Sample 1	17	± 1.1	3.1	17	± 1.1	7.0	36	± 6.3	18
Reference Sample 2	20	± 1.1	3.4	20	± 1.1	5.6	52	± 7.1	17

All units in Bq/kg
[1] CSU is the combined standard uncertainty of the measurement result reported by the analytical laboratory.
[2] MDC is the minimum detectable concentration reported by the analytical laboratory.

Table 8.6 Preliminary Gamma Spectroscopy Results for Thorium Series Radionuclides

Sample ID	^{228}Ac	CSU[1]	MDC[2]
Sample 1	1,600	± 180	52
Sample 2	1,400	± 130	41
Sample 3	14	± 2.6	4.4
Sample 4	21	± 3.1	6.3
Reference Sample 1	15	± 3.3	5.9
Reference Sample 2	16	± 3.4	3.4

All units in Bq/kg
[1] CSU is the combined standard uncertainty of the measurement result reported by the analytical laboratory.
[2] MDC is the minimum detectable concentration reported by the analytical laboratory.

Note the results provided in Tables 8.3 through 8.6 are from actual samples collected from a real site. However, the sample results included as part of this illustrative example were selected to provide specific information supporting the application of MARSAME guidance and represent a portion of the total amount of information available. The number and type of samples collected as part of preliminary survey should be determined using the DQO Process as discussed in Section 2.3.

8.2.3.4   Select a Disposition Option

The preferred disposition of the concrete rubble is clearance. It is expected that the concrete will be reused as roadbed or disposed of in a municipal landfill. If the activity levels exceed the project action levels, then the concrete may need to be disposed of as discrete naturally occurring or accelerator-produced (NARM) waste. If the activity is below the alternate action levels, the concrete may either be reused or disposed of as diffuse NARM waste.

8.2.3.5 Document the Results of the Initial Assessment

The results of the IA were documented in a letter report. The purpose of the letter report was to document the categorization decision and all supporting information. The letter report was reviewed and finalized by the facility owner. Detailed results of the IA will be included in the final documentation of the survey design.

8.2.4 Develop a Decision Rule

Following completion of the IA, additional information was needed to develop the disposition survey design.

8.2.4.1 Select Radionuclides or Radiations of Concern

The list of radionuclides of concern was finalized based on the preliminary survey results. Uranium-238, ^{234}U, and ^{226}Ra are the radionuclides of concern for the uranium natural decay series. The alpha spectrometry results indicate that ^{238}U and ^{234}U are in equilibrium (i.e., have equal concentrations). Because alpha spectrometry for uranium isotopes provides results for both ^{238}U and ^{234}U, both isotopes (and their decay products with half-lives less than six months) will be kept as radionuclides of concern. There is no indication of enrichment or depletion of uranium as a result of site activities based on the uranium alpha spectrometry results listed in Table 8.3. Radium-226 decay products, including ^{210}Pb, are assumed to be out of secular equilibrium with the other uranium series radionuclides (e.g., ^{238}U and ^{234}U) because process knowledge shows the chemical processing at the plant would separate uranium from radium. Bismuth-214 and ^{214}Pb can be used as beta or gamma emission surrogates for ^{226}Ra, because the decay products of ^{226}Ra should be in secular equilibrium with one another. However, a 21-day ingrowth period may be required to confirm this assumption. The planning team determined an ingrowth study was not required for this project following discussions with the regulators.

Thorium-232 is the radionuclide of concern for the thorium natural decay series. Based on the alpha spectrometry and gamma spectroscopy results shown in Table 8.3, all members of the thorium natural decay series are in secular equilibrium. Actinium-228 emits gamma rays that are easy to quantify using gamma spectroscopy, and can be used as a surrogate for the members of the thorium series.

8.2.4.2 Identify Action Levels

For the purposes of this illustrative example, an action level of 0.01 mSv/y was selected based on discussions with the planning team. Using information provided in NUREG-1640 (NRC 2003), the action levels were converted into concentration units based on clearance as the disposition option. Incorporating the concrete rubble into roadbed material would provide the highest potential doses following clearance. The mean values from NUREG-1640 (NRC 2003), Table I1.13 ("Normalized effective dose equivalents from all pathways: Driving on road [µSv/y per Bq/g]"), are the basis for the action levels.

Radionuclide of concern	^{238}U	^{234}U	^{232}Th	^{226}Ra
Mass-based EDE mean values (µSv/y per Bq/g)	0.26	8.2×10^{-4}	30	22

The action levels from Table I1.13, NUREG-1640 (NRC 2003) are expressed in units of µSv/y per Bq/g, but the preliminary survey measurement results are in Bq/kg. To make a direct comparison, the action levels were converted to units of Bq/kg. Note that a hypothetical dose for clearance was selected only for the purpose of showing example calculations for this illustrative example. Clearance criteria should be provided by the regulator for actual applications of this guidance. The action levels were converted to concentrations by inverting the action levels and multiplying by the hypothetical dose limit (i.e., the inverted action levels in units of Bq/g per µSv/y are multiplied by 0.01 mSv/y, 1,000 g/kg, and 1,000 µSv/mSv providing action levels in Bq/kg). Table 8.7 lists the action levels in concentration units of Bq/kg.

Table 8.7 Radionuclide-Specific Action Levels

Radionuclide	Mass-Based EDE Mean Values (Bq/g per µSv/y)	Action Level (Bq/kg)
^{238}U	$\dfrac{1\,\text{Bq/g}}{0.26\,\mu\text{Sv/y}} \times 0.01\,\text{mSv/y} \times 1 \times 10^{6} = 38{,}000$	38,000
^{234}U	$\dfrac{1\,\text{Bq/g}}{8.2 \times 10^{-4}\,\mu\text{Sv/y}} \times 0.01\,\text{mSv/y} \times 1 \times 10^{6} = 12{,}000{,}000$	12,000,000
^{232}Th	$\dfrac{1\,\text{Bq/g}}{3.0 \times 10^{1}\,\mu\text{Sv/y}} \times 0.01\,\text{mSv/y} \times 1 \times 10^{6} = 330$	330
^{226}Ra	$\dfrac{1\,\text{Bq/g}}{2.2 \times 10^{1}\,\mu\text{Sv/y}} \times 0.01\,\text{mSv/y} \times 1 \times 10^{6} = 450$	450

The unity rule (Equation 8-1) is used to account for the individual radionuclide action levels. The unity rule is satisfied when the summed analyses of each radionuclide against its respective action level yields a value less than one:

$$\text{The Unity Rule} = \frac{C_1}{AL_1} + \frac{C_2}{AL_2} + \dots \frac{C_n}{AL_n} \leq 1 \qquad (8\text{-}1)$$

Where:

C = concentration of each individual radionuclide (1, 2, … n)

AL = action level value for each individual radionuclide (1, 2, … n)

Equation 8-1 is used to calculate the sum of fractions for each of the preliminary survey results:

$$\text{The Unity Rule} = \frac{C_{^{238}\text{U}}}{AL_{^{238}\text{U}}} + \frac{C_{^{234}\text{U}}}{AL_{^{234}\text{U}}} + \frac{C_{^{232}\text{Th}}}{AL_{^{232}\text{Th}}} + \frac{C_{^{226}\text{Ra}}}{AL_{^{226}\text{Ra}}} \leq 1$$

$$\text{Sample 1} = \frac{7,600 \text{ Bq/kg}}{38,000 \text{ Bq/kg}} + \frac{7,000 \text{ Bq/kg}}{12,000,000 \text{ Bq/kg}} + \frac{1,400 \text{ Bq/kg}}{330 \text{ Bq/kg}} + \frac{47 \text{ Bq/kg}}{450 \text{ Bq/kg}} = 4.5$$

$$\text{Sample 2} = \frac{6,900 \text{ Bq/kg}}{38,000 \text{ Bq/kg}} + \frac{7,200 \text{ Bq/g}}{12,000,000 \text{ Bq/kg}} + \frac{1,230 \text{ Bq/kg}}{330 \text{ Bq/kg}} + \frac{192 \text{ Bq/kg}}{450 \text{ Bq/g}} = 4.2$$

$$\text{Sample 3} = \frac{21 \text{ Bq/kg}}{38,000 \text{ Bq/kg}} + \frac{21 \text{ Bq/kg}}{12,000,000 \text{ Bq/kg}} + \frac{21 \text{ Bq/kg}}{330 \text{ Bq/kg}} + \frac{64 \text{ Bq/kg}}{450 \text{ Bq/g}} = 0.21$$

$$\text{Sample 4} = \frac{21 \text{ Bq/kg}}{38,000 \text{ Bq/kg}} + \frac{25 \text{ Bq/kg}}{12,000,000 \text{ Bq/kg}} + \frac{26 \text{ Bq/kg}}{330 \text{ Bq/kg}} + \frac{68 \text{ Bq/kg}}{450 \text{ Bq/g}} = 0.23$$

The results of the calculations for Samples 1 and 2 exceed a sum of fractions of 1.0, and indicate the presence of small volumes of concrete with elevated activity. Note that the reported MDCs for gamma spectroscopy for ^{226}Ra in Samples 1 and 2 would not meet the MQOs for clearance (i.e., the MDC exceeds the action level). However, the radionuclide concentrations in these two samples clearly exceed the action level. Therefore, the quality of these results is acceptable to support the disposition survey design.

The results of the calculations for Samples 3 and 4 indicate that, on average, the concrete rubble is expected to have radionuclide concentrations below the action levels. Therefore, the average activity in the concrete rubble is expected to be below the action level. Large blocks containing elevated levels of radioactivity may be visually identified via staining, verified with a GM detector, and segregated prior to removal of the rebar.

8.2.4.3   Modify the Action Levels to Account for Multiple Radionuclides

Radionuclide-specific action levels need to be combined into a single gross gamma action level for evaluating the field instrument for detection of low-energy radiation (FIDLER) scan measurements. The information in Section 3.3.3.1 requires an estimate of the relative fraction of the total activity contributed by each radionuclide. A consistent relationship between ^{238}U and ^{232}Th concentrations is not expected based on the IA, because different ore bodies could contain different ratios of these radionuclides. Rather than develop a preliminary survey attempting to develop this relationship, a conservative approach was adopted for this project.

Assuming the entire radioactivity detected by the FIDLER results from the presence of the most restrictive radionuclide will provide the most conservative gross gamma action level. The ratios of exposure rate to radionuclide concentration (µR/h per Bq/kg) and instrument response to exposure rate (cpm per µR/h) were developed in Section 7.11 during development of the scan MDC for both ^{238}U and ^{232}Th. These ratios can be used to calculate the count rate above background associated with a radionuclide activity equal to the action level as shown in Equation 8-2.

$$GG_{AL} = AL \times \left(\frac{\mu R/h}{Bq/kg} \right) \times \left(\frac{cpm}{\mu R/h} \right) \tag{8-2}$$

Where:

GG_{AL} = gross gamma action level (cpm)

AL = action level value for each individual radionuclide (Bq/kg)

Equation 8-2 was used to calculate a gross gamma count rate above background for the FIDLER assuming each radionuclide of concern was present at a concentration equal to the action level. The gross gamma count rates were divided by two to account for uncertainty associated with the detector efficiency calculation and added to the background count rate. The result is a gross gamma action level for the FIDLER to identify locations with unexpectedly high gamma activity that could result in doses near the action level of 0.01 mSv/y. The results of the calculations are shown in Table 8.8. The ^{232}Th gross gamma action level of 30,000 cpm is more conservative than the ^{238}U gross gamma action level of 140,000 cpm, so 30,000 cpm was selected as the gross gamma action level.

Table 8.8 Calculation of the Gross Gamma Action Level

Action Level (Bq/kg)	µR/h per Bq/kg	cpm per µR/h	Gross Gamma Count Rate (cpm)	Adjusted Gross Gamma Count Rate (cpm)	Background Count Rate (cpm)	Gross Gamma Action Level (cpm)
^{238}U 38,000	1.413×10^{-4}	45,593	244,807	122,404	12,870	140,000
^{232}Th 330	2.619×10^{-2}	3,923	33,905	16,953	12,870	30,000

FIDLER readings that exceed the ^{232}Th gross gamma action level indicate locations where radionuclide concentrations could result in doses exceeding the 0.01 mSv/y used for this illustrative example if all of the activity results from ^{232}Th.

Because ^{232}Th has decay products in secular equilibrium that can be used to estimate the ^{232}Th activity, gamma spectroscopy can be used to quantify ^{232}Th concentrations. FIDLER readings that exceed 140,000 cpm identify locations where radionuclide concentrations could result in doses exceeding 0.01 mSv/y if all of the activity results from ^{238}U. Alpha spectrometry is required to quantify ^{238}U concentrations.

8.2.4.4   Describe the Parameter of Interest

Because the disposition option is stated in terms of dose, the parameter of interest is the mean radionuclide concentration. The target population is all of the possible measurement results that could be obtained within a survey unit. This means the target population will be defined by the survey unit boundaries (Section 8.2.4.6) and the selected measurement method (Section 8.2.4.8).

8.2.4.5   Identify Alternative Actions

The alternative actions identify the results of decisions based on the measurement results. If the radionuclide concentrations do not result in a dose that exceeds the action level, the material is cleared. If the dose exceeds the action level, materials exceeding the action level will be segregated and investigated for disposal as NARM waste.

8.2.4.6   Identify Survey Units

Survey unit boundaries are based primarily on the modeling assumptions used to develop the action levels. The volume of concrete used to model exposures for building a road is 83 m^3 (NUREG-1640 [NRC 2003] Volume 2, Appendix B, Tables B-8 and B-11). Each survey unit will consist of approximately 80 m^3 of crushed concrete (approximately 25 m $\times$ 22 m $\times$ 0.15 m).

The volume of concrete poured to create the floor of the processing mill was approximately 1,100 m^3. Crushing the concrete and removing the rebar is expected to result in approximately a 25% increase in volume due to air gaps, for a total volume of 1,400 m^3 of crushed concrete. Using these calculations, there will therefore be a total of 18 survey units plus one reference area.

The concrete rubble can be spread into a relatively uniform layer approximately 15 cm thick and scanned. This adapts an approach used in MARSSIM to survey the top 15 cm of surface soil as a two-dimensional object.

8.2.4.7   Define the Decision Rules

MARSSIM-type surveys are designed to evaluate the average radionuclide concentration in a survey unit using samples or direct measurements, as well as small areas of elevated activity using scans. Small areas of elevated activity receive additional investigation. Because there are multiple action levels and multiple decisions to be made, there are multiple decision rules for the disposition survey. The first two decision rules address how small areas of elevated activity are identified by scans and what investigations will be performed. The third decision rule evaluates the results of the investigations of small areas of elevated activity. The fourth decision rule evaluates the average activity in each survey unit.

1. If any FIDLER scanning measurement result exceeds the gross gamma action level of 30,000 cpm (see Section 8.2.5.4), a biased sample will be collected for laboratory analysis by gamma spectroscopy, otherwise no biased samples will be collected.
2. If any FIDLER scanning measurement exceeds 140,000 cpm, the biased sample collected for gamma spectroscopy analysis will also be analyzed by alpha spectrometry for uranium and thorium isotopes, otherwise the concrete will be held awaiting the results of the gamma spectroscopy analysis.
3. If the results from a biased sample result in a sum of fractions for ^{238}U, ^{234}U, ^{226}Ra, and ^{232}Th exceeding 1.0, the concrete will be segregated and investigated for disposal as NARM waste. Otherwise, the survey unit will be evaluated based on the WRS test results for the samples taken over a systematic grid.

4. If the mean sum of fractions in a survey unit exceeds 1.0, the concrete will be segregated and investigated for disposal as NARM waste. Otherwise, the WRS test will be performed to support the final disposition decision for that survey unit.

8.2.4.8 Develop Inputs for Selection of Provisional Measurement Methods

The selected measurement method will be required, at a minimum, to detect radionuclide concentrations at or below the action levels in Table 8.8 (page 8-11). The survey planners considered each of the possible measurement techniques (Section 5.9.1).

Scan-only techniques have the ability to detect surface activity at concentrations below the action levels. In situ measurement techniques are also expected to have the ability to measure radionuclide concentrations at the action levels. However, uncertainties associated with the efficiency for both techniques will be large. In order to reduce these uncertainties to a level where the radionuclide concentrations are measurable, the concrete would need to be pulverized and mixed rather than just crushed to 2–3 cm size. Because the cost of processing the concrete this way would be a major cost associated with the disposition survey, a MARSSIM-type survey design was selected for the disposition survey. No method-based survey designs were identified that matched the description of the M&E, so no method-based survey designs were considered.

Concrete samples will be analyzed in a laboratory using alpha spectrometry for uranium isotopes (i.e., ^{234}U and ^{238}U) as well as gamma spectroscopy for other radionuclides of concern (i.e., ^{214}Bi, ^{214}Pb, and ^{228}Ac). Sample sizes must be sufficient to allow quantification of radionuclide concentrations at the action levels. By convention, the MQC for each radionuclide of concern is selected so the measurement method uncertainty at concentrations equal to the action levels in Table 8.7 is 10%. Alternatively, the samples can be sealed in airtight containers for at least twenty-one days to allow secular equilibrium to be reestablished prior to analysis by gamma spectroscopy so decay products can be used as surrogate radionuclides.

Due to the rough, irregular shape of the concrete rubble, alpha radiation is attenuated easily and is difficult to measure. Beta and gamma measurements typically provide a more accurate assessment of thorium and uranium activity on most building surfaces because surface conditions cause significantly less attenuation of beta and gamma particles than alpha particles. For this reason, scanning will be performed using instruments that detect beta or gamma radiation. Surface scans, using a 12.7-cm by 0.16-cm FIDLER sodium iodide (NaI[Tl]) scintillation detector, are used to scan for gamma emissions. The approximate detection sensitivity of the FIDLER is 300 Bq/kg for natural uranium and 20 Bq/kg for natural thorium when activity is present at the surface. The FIDLER is a large detector and can detect gammas from a greater height above the crushed concrete than alpha or beta detection equipment, making it a more practical choice for surveying large volumes of material. The selection of the FIDLER over more conventional NaI(Tl) detectors (e.g., a three-inch by three-inch gamma scintillation detector) is primarily based on the FIDLER's ability to detect low-energy gamma radiation, which comprises the majority of the gamma radiation from the radionuclides of concern.

8.2.4.9   Identify Reference Materials

Concrete from the administrative building contains non-impacted materials, as established by the process knowledge discussed in Section 8.2.3.1. The reference material measurements will be performed on the floor in the administrative building. The geometry of the floor is similar enough to the concrete rubble (after crushing to 2–3 cm size and arrangement into a 15-cm thick layer) that modifications to the building are not required.

8.2.5   Develop a Survey Design

The concrete rubble from the mineral processing facility is surveyed for clearance using a MARSSIM-type disposition survey. The survey includes scanning to identify small areas of elevated activity combined with collection and analysis of samples to evaluate the average activity in the concrete rubble.

Scenario A is used to design the survey, because decisions will be made based on average radionuclide concentrations and radioactivity levels in each survey unit. The null hypothesis is that the radionuclide concentrations in the concrete rubble will result in a dose that exceeds 0.01 mSv/y. There are two decisions for MARSSIM-type surveys. The first decision is based on the average radionuclide concentrations in the survey unit, and the second decision is based on the scanning survey results and subsequent biased sample results from flagged locations. The same null hypothesis applies to both decisions.

A Type I decision error would occur if the decision-maker decided the activity levels in the concrete rubble were below the action level when they actually exceeded the action level. The consequence of making this decision error could result in increased doses to members of the public and failing to identify small areas of elevated radionuclide concentrations. The members of the planning team agreed to a Type I decision error rate of 5% based on the consequence of making this decision error. This Type I error rate applies to both the scanning portion of the survey design as well as sampling on a systematic grid.

A Type II decision error would occur if the decision-maker decided the activity levels in the concrete rubble exceeded the action level when they were actually below the action level. The consequence of making this decision error could result in increased disposal costs. The members of the planning team agreed to a Type II decision error rate of 10% based on the consequence of making this decision error for sampling. However, during scanning the consequence of making this decision error is the need to perform additional investigation., As such, a Type II decision error rate of 40% is selected for the scanning surveys.

8.2.5.1   Classify the M&E

All of the concrete rubble from the floor of the processing facility has the potential to exceed one or more of the action levels. The concrete rubble is classified as Class 1 M&E.

8.2.5.2   Design the Scanning Survey

The concrete must be crushed prior to performing the scanning survey to reduce the size of individual particles to less than 2–3 cm in diameter. This provides a uniform matrix of material ensuring a representative sample can be collected, and also allows the rebar to be removed. The crushed concrete is distributed in a layer approximately 15 cm thick, and surveyed using a FIDLER at a height of 10 cm above the surface. The scan speed is 0.25 m/s, which is consistent with the scan MDC calculations. One hundred percent of the concrete rubble is scanned with readings in excess of 30,000 cpm flagged for additional investigation. The additional investigations include collection and analysis of samples using gamma spectroscopy to quantify activity levels for the radionuclides of concern. Samples collected from locations with readings in excess of 140,000 cpm are also analyzed for uranium and thorium isotopes by alpha spectrometry.

8.2.5.3   Design the Sample Collection Survey

The concrete rubble is divided into survey units and a statistically based number of samples are collected from each survey unit. Because multiple radionuclides are present, the unity rule is used to evaluate the sample results. Because the radionuclides are present in background, the Wilcoxon Rank Sum (WRS) test is used to evaluate the survey results.

The upper bound of the gray region (UBGR) is set equal to the action level, which is a sum of fractions of 1.0 above background. The lower bound of the gray region (LBGR) is set equal to the expected sum of fractions based on results from the preliminary survey. The expected average activity in the concrete rubble is close to background, even though isolated areas have results more than four times the action level. An LBGR value of 0.15 is selected, which is consistent with results reported in Tables 8.3 through 8.6 for the two randomly selected samples (i.e., samples 3 and 4). Because the values are not corrected for background, this value is considered conservative. The shift (UBGR – LBGR) is 0.85.

The variability in the activity levels for the concrete rubble, σ_S, is not well defined. To be conservative, the variability in the results should be large for results near the LBGR. A value of 0.15 was selected for the variability. This value is equal to the LBGR, and represents 100% variability in results that are at or near background. The relative shift equals 5.6 (0.85 divided by 0.15 and rounded down). Because relative shifts greater than 4.0 do not result in significantly smaller numbers of samples, a relative shift of 4.0 was used to determine the number of samples and also help to ensure adequate statistical power.

Table A.2b (Appendix A) lists the number of samples required for each survey unit and reference area for use with the WRS test. Seven samples are required for each survey unit and reference area using a relative shift of 4.0, Type I decision error rate of five percent, and Type II decision error rate of 10 percent. The radionuclide or radioactivity concentrations derived from the dose-based action level are based on an average radionuclide concentration or level of radioactivity over the entire survey unit. No adjustments need to be made to the number of measurements to account for the scan MDC, because the scan MDC is less than the action level for both ^{238}U and ^{232}Th.

Seven samples of approximately 1,000 g of concrete rubble are collected from each survey unit. This mass corresponds to a cylinder with a diameter of approximately 6 cm (2.5 in) to a depth of 15 cm (6 in). This disposition survey design will be applied to all of the concrete rubble, including the concrete segregated based on visual inspection and elevated scanning results with a GM detector during the preliminary surveys (Section 8.2.3.3).

8.2.5.4   Develop an Operational Decision Rule

The action level is stated in terms of incremental dose above background. In a MARSSIM survey, there are requirements for both sample measurements and scanning results. Samples will be collected from non-impacted concrete to represent background radionuclide concentrations. The WRS test will be used to evaluate the survey results. If the test statistic for the WRS test is less than or equal to 65 ($n = m = 7$, $\alpha = 0.05$), decide that the dose from that survey unit exceeds 0.01 mSv/y and the concrete will not be cleared.

For the scanning results, if any FIDLER measurement exceeds 30,000 cpm, collect a biased concrete sample at the location of the elevated measurement for analysis by gamma spectroscopy. If any FIDLER measurement exceeds 140,000 cpm, analyze the biased concrete sample by alpha spectrometry as well. If the sum of fractions for any biased sample exceeds 1.0, decide that the dose from that survey unit exceeds the 0.01 mSv/y used for this illustrative example and the concrete will not be cleared.

8.2.5.5   Document the Survey Design

The final survey design was documented in a detailed work plan. The work plan provided the results of the IA, as well as all of the assumptions used to develop the survey design. The DQOs and MQOs for the survey design were also included.

The draft work plan was submitted to the planning team for review. Comments were received, and responses to comments developed and approved. The approved responses to comments were incorporated into a final work plan documenting the disposition survey design.

8.2.6   Implement the Survey Design

8.2.6.1   Ensure Protection of Health and Safety

A job safety analysis (JSA) was performed based on the tasks defined in the work plan documenting the disposition survey design. Table 8.9 shows the results of the JSA. Potential health and safety hazards identified by the JSA are addressed in a site-specific health and safety plan. No hazards associated with the concrete rubble will notably affect how the disposition survey is implemented.

Table 8.9 Job Safety Analysis for Surveying Concrete Rubble

Sequence of Basic Job Steps	Potential Hazards	Recommended Action or Procedure
1. Dividing rubble into manageable survey units	Use of front end loader by untrained personnel	Ensure equipment operators are adequately trained
	Personnel in area could be struck by heavy equipment	Area workers must maintain eye contact with equipment operators
		Reflective vests will be worn to improve visibility
	Exposure to silica	Use of a real-time dust monitor will document dust levels. Respiratory protection will be used if dust levels exceed established action levels (dependent on silica content of concrete)
	Lower back strain from lifting	Proper lifting techniques will be used
		Loads will be sized so as not to create unreasonable weights for manual lifting
	Exposure to radiological contamination	PPE including booties, Tyveks, and gloves will be used
2. Establish exclusion zone for survey area	None anticipated	
3. Use hand-held survey instruments to perform survey measurements on the crushed concrete	Unstable footing may result in slips, trips, or falls	Spread out rubble in a way to minimize tripping hazards by creating clear rows between rows of concrete
4. Physical handling of larger pieces of concrete debris to expose underside for gamma surveying	Rough surfaces may cut and scrape skin on hands	Wear a set of work gloves to protect hands when handling concrete pieces
5. Entering Exclusion Zone (EZ) to perform survey	Tripping	Maintain good housekeeping in survey area
	Exposure to radiological contamination	PPE including booties, Tyveks, and gloves will be used
	Spread of radiological contamination outside EZ	Establish step-off area outside of EZ
6. Moving contaminated or clean material to appropriate disposal containers	Use of front end loader by untrained personnel	Ensure equipment operators are adequately trained
	Lower back strain from lifting	Proper lifting techniques will be used. Loads will be sized so as not to create unreasonable weights for manual lifting
	Exposure to radiological contamination	PPE including booties, Tyveks, and gloves will be used
	Exposure to silica	Use of a real-time dust monitor will document dust levels. Respiratory protection will be used if dust levels exceed established action levels (dependent on silica content of concrete)

8.2.6.2   Consider Issues for Handling the M&E

The concrete rubble must be crushed to a uniform size of less than one inch to implement the disposition survey design and meet the MQOs. The crushing process will generate dust potentially containing radioactive material. Controls to limit dust generation were implemented during concrete crushing activities. Equipment involved in handling the concrete during crushing activities (e.g., front loader, crusher, rebar separator, conveyor belts, dump trucks) is categorized as impacted and will require a disposition survey before the equipment can be released. Surveys of the front loader are discussed in Sections 8.3 and 8.4.

8.2.6.3   Segregate the M&E

Concrete rubble with visible stains and pitting on the floor surface is segregated as having higher activity concentrations. Stained and unstained concrete were grouped into separate survey units. Following segregation, the concrete was crushed to 2–3-cm diameter pieces, and the rebar was removed.

8.2.6.4   Set Measurement Quality Objectives

The two most important MQOs for this survey design are the required measurement method uncertainty, u_{MR}, for the scan MDCs for the FIDLER measurements (Section 8.2.6.5) and the concrete samples collected on the systematic grid (Section 8.2.6.6). Other MQOs were established during the development of the survey design to support selection of measurement methods. These included setting the MQC for each radionuclide of concern so the relative measurement method uncertainty, φ_{MR}, at concentrations equal to the action levels in Table 8.7 is 10% (Section 8.2.4.8) and calculating the scan MDCs for the FIDLER (Section 7.11).

8.2.6.5   Determine Measurement Uncertainty for the Scan MDC

This section describes the calculation of the uncertainty for the scan MDC measurements performed as part of this survey using the FIDLER. An upper bound for an expanded uncertainty for the scan MDC calculation is derived to reduce the probability that the scan MDC has been underestimated. The result is used as the investigation level for evaluating the results of the scan survey.

The uncertainty calculations presented in this section may be performed using commercially available statistical software (Section 5.6). Detailed solutions for this illustrative example are provided below.

The scan MDCs for the FIDLER measurements, y, are calculated in Section 7.11. The scan MDC for natural uranium, y_U, is approximately 400 Bq/kg. The scan MDC for natural thorium, y_{Th}, is approximately 25 Bq/kg. Both scan MDCs are less than their respective action levels of 38,000 and 330 Bq/kg. The values used to calculate the scan MDCs for the FIDLER measurements are:

b_i = average number of counts in the background interval (214.5 counts). Here b_i is assumed to have a triangular distribution with a half-width of 30% or 64 counts, so the mean value of b_i is 215 and $u(b_i) = 64/\sqrt{6} = 26$.

i = observation interval length (one second). Here i is assumed to have a triangular distribution with a half-width of 0.5, so the mean value of $i = 1.0$ and $u(i) = 0.5/\sqrt{6} = 0.2$.

p = efficiency of a less than ideal surveyor, range of 0.5 to 0.75 from NUREG-1507 (NRC 1998b); a value 0.5 was chosen as a conservative value. Here p is assumed to have a rectangular distribution with a half-width of 0.125, so the mean value of $p = 0.625$ and $u(p) = 0.125/\sqrt{3} = 0.072$.

d' = detectability index from Table 6.1 of NUREG-1507 (NRC 1998b); a value of 1.90 was selected and treated as a constant.

W_T = total weighted instrument sensitivity (cpm per µR/h)

= 44,923 for natural uranium from Table 7.10 and

= 3,881 for natural thorium from Table 7.11.

R_T = total exposure rate with buildup (µR/h)

= 1.413×10^{-4} for natural uranium from Table 7.10 and

= 2.619×10^{-2} for natural thorium from Table 7.11.

C = concentration of source term (set at 1 Bq/kg and treated as a constant).

y = Scan MDC (in Bq/kg) introduced here for simplicity of notation.

Because we are assuming there are no correlations among the input variables, the combined standard uncertainty of y can be calculated using Equation 7.33 from Section 7.8.1.6:

$$u_c^2(y) = \sum_{i=1}^{N} \left(\frac{\partial y}{\partial x_i} \right)^2 u^2(x_i) = \sum_{i=1}^{N} c_i^2 u^2(x_i)$$

The concentration of the source term, C, and the detectability index, d', are treated as constants with no associated uncertainty, so this expands to:

$$u_c^2(y) = \left(\frac{\partial y}{\partial b_i} \right)^2 u^2(b_i) + \left(\frac{\partial y}{\partial i} \right)^2 u^2(i) + \left(\frac{\partial y}{\partial p} \right)^2 u^2(p) + \left(\frac{\partial y}{\partial R_T} \right)^2 u^2(R_T) + \left(\frac{\partial y}{\partial W_T} \right)^2 u^2(W_T)$$

The sensitivity coefficients, c_i^2, are calculated as follows:

$$\left(\frac{\partial y}{\partial b_i} \right) = \frac{\partial \left(\frac{60 C d' \sqrt{b_i}}{i W_T R_T \sqrt{p}} \right)}{\partial b_i} = \left(\frac{1}{2} \right) \frac{60 C d'}{i W_T R_T \sqrt{p} \sqrt{b_i}} = \left(\frac{1}{2} \right) \frac{y}{b_i}$$

$$\left(\frac{\partial y}{\partial p}\right) = \frac{\partial\left(\frac{60Cd'\sqrt{b_i}}{iW_T R_T \sqrt{p}}\right)}{\partial p} = \left(-\frac{1}{2}\right)\frac{60Cd'\sqrt{b_i}}{iW_T R_T p^{3/2}} = \left(-\frac{1}{2}\right)\frac{y}{p}$$

$$\left(\frac{\partial y}{\partial i}\right) = \frac{\partial\left(\frac{60Cd'\sqrt{b_i}}{iW_T R_T \sqrt{p}}\right)}{\partial i} = -\frac{60Cd'\sqrt{b_i}}{i^2 W_T R_T \sqrt{p}} = -\frac{y}{i}$$

$$\left(\frac{\partial y}{\partial R_T}\right) = \frac{\partial\left(\frac{60Cd'\sqrt{b_i}}{iW_T R_T \sqrt{p}}\right)}{\partial R_T} = -\frac{60Cd'\sqrt{b_i}}{iW_T R_T^2 \sqrt{p}} = -\frac{y}{R_T}$$

$$\left(\frac{\partial y}{\partial W_T}\right) = \frac{\partial\left(\frac{60Cd'\sqrt{b_i}}{iW_T R_T \sqrt{p}}\right)}{\partial W_T} = -\frac{60Cd'\sqrt{b_i}}{iW_T^2 R_T \sqrt{p}} = -\frac{y}{W_T}$$

Therefore,

$$u_c^2(y) = \left(\frac{y}{2b_i}\right)^2 u^2(b_i) + \left(\frac{-y}{i}\right)^2 u^2(i) + \left(\frac{-y}{2p}\right)^2 u^2(p) + \left(\frac{-y}{R_T}\right)^2 u^2(R_T) + \left(\frac{-y}{W_T}\right)^2 u^2(W_T)$$

$$= y^2 \left[\left(\frac{u(b_i)}{2b_i}\right)^2 + \left(\frac{u(i)}{i}\right)^2 + \left(\frac{u(p)}{2p}\right)^2 + \left(\frac{u(R_T)}{R_T}\right)^2 + \left(\frac{u(W_T)}{W_T}\right)^2\right]$$

$$= y^2 \left[\left(\frac{26}{2(215)}\right)^2 + \left(\frac{0.2}{1}\right)^2 + \left(\frac{0.072}{2(0.625)}\right)^2 + \left(\frac{u(R_T)}{R_T}\right)^2 + \left(\frac{u(W_T)}{W_T}\right)^2\right].$$

The most notable sources of uncertainty associated with W_T and R_T are the modeling assumptions for the source-to-detector separation distance during scanning and the depth distribution of the radioactivity in the crushed concrete. To calculate uncertainties, the same basic modeling assumptions as those for the MDC calculations were applied, though with variations to both the source-to-detector separation distance during scanning and the distribution of the radioactivity in the crushed concrete. While the MDC calculation assumes a source-to-detector distance of 10 cm and that the activity is uniformly distributed within a cylindrical volume of crushed concrete 15 cm thick with a radius of 28 cm, several other calculations were made using source-to-detector separation distances during scanning of 8, 10, and 12 cm, and by varying the distribution of the radioactivity in the crushed concrete from uniform to uniformly distributed within both the top and bottom 7.5 cm of the cylindrical volume of crushed concrete, to assess the potential variability in the MDC. In each calculation the total activity was the same, only the distribution with depth was changed. The extreme cases were for a source-to-detector distance of 8 cm with the activity uniformly distributed within the top 7.5 cm of the concrete versus a source-to-detector distance of 12 cm with the activity uniformly distributed within the bottom 7.5 cm of the concrete. While more extreme conditions might be imagined, the foregoing were considered to represent reasonable bounds on the source-to-detector distance and the activity distribution with depth. The other assumptions used in the calculations were the same as used in Section 7.11. Therefore, there are three values each to describe the distribution of the possible values of W_T and R_T: The estimated mean value calculated for a uniform distribution of radioactivity in the 15 cm of concrete surveyed at 10 cm above; an estimated lower bound calculated for a uniform distribution of radioactivity in the bottom 7.5 cm of concrete surveyed at 12 cm above; and an estimated upper bound calculated for a uniform distribution of radioactivity in the top 7.5 cm of concrete surveyed at 8 cm above.

The values for W_T and R_T at the extremes considered were not equally distant from the mean, i.e., their distribution was not symmetric. However the GUM suggests that in the absence of more information the simplest approximation is a symmetric rectangular distribution of the same total width. With this approximation, $u(W_T) = 6673$ and $u(R_T) = 4.638 \times 10^{-5}$ for natural uranium and $u(W_T) = 539$ and $u(R_T) = 7.315 \times 10^{-3}$ for natural thorium.

Using the information for natural uranium in Equation 7-34 we find:

$$u_c^2(y_U) = y_U^2 \left[\left(\frac{26}{2(215)} \right)^2 + \left(\frac{0.2}{1} \right)^2 + \left(\frac{0.072}{2(0.625)} \right)^2 + \left(\frac{u(R_T)}{R_T} \right)^2 + \left(\frac{u(W_T)}{W_T} \right)^2 \right]$$

$$= (238)^2 \left[\left(\frac{26}{2(215)} \right)^2 + \left(\frac{0.2}{1} \right)^2 + \left(\frac{0.072}{2(0.625)} \right)^2 + \left(\frac{4.638 \times 10^{-5}}{1.413 \times 10^{-4}} \right)^2 + \left(\frac{6673}{44{,}923} \right)^2 \right]$$

$$= 10{,}013 \ (\text{Bq/kg})^2.$$

So, taking the square root of the variance and rounding the result, $u_c(y_U) = 100$ Bq/kg.

Therefore the FIDLER Scan MDC for natural uranium, y_U, is 400 Bq/kg with an expanded uncertainty of 200 Bq/kg, using a coverage factor of 2 and an estimated coverage probability of 95%. The upper bound of the Scan MDC using this interval is 600 Bq/kg.

Similarly substituting the information for natural thorium into Equation 7-34 we find:

$$u_c^2(y_{Th}) = y_{Th}^2 \left[\left(\frac{26}{2(215)} \right)^2 + \left(\frac{0.2}{1} \right)^2 + \left(\frac{0.072}{2(0.625)} \right)^2 + \left(\frac{u(R_T)}{R_T} \right)^2 + \left(\frac{u(W_T)}{W_T} \right)^2 \right]$$

$$= (15)^2 \left[\left(\frac{26}{2(215)} \right)^2 + \left(\frac{0.2}{1} \right)^2 + \left(\frac{0.072}{2(0.625)} \right)^2 + \left(\frac{7.315 \times 10^{-3}}{2.619 \times 10^{-2}} \right)^2 + \left(\frac{539}{3{,}881} \right)^2 \right]$$

$$= 32 \ (\text{Bq/kg})^2.$$

So, taking the square root of the variance and rounding the result, $u_c(y_{Th}) = 6$ Bq/kg.

Therefore the FIDLER Scan MDC for natural thorium, y_{Th}, is 25 Bq/kg with an expanded uncertainty of 12 Bq/kg, using a coverage factor of 2 and an estimated coverage probability of 95%. The upper bound of the Scan MDC using this interval is 37 Bq/kg.

The upper bound of the scan MDCs of approximately 600 Bq/kg for natural uranium and 37 Bq/kg for natural thorium are both less than their respective action levels of 38,000 and 330 Bq/kg. Therefore, the FIDLER is an acceptable instrument for performing the scan measurements.

8.2.6.6   Determine Measurement Uncertainty for Concrete Samples

The primary measurement quality objective is the required measurement method uncertainty at the action level. MARSAME recommends $u_{MR} \leq \Delta/10$ by default when decisions are being made about the mean of a sampled population.

For this illustrative example, the Unity Rule, $\dfrac{C_{238_U}}{AL_{238_U}} + \dfrac{C_{234_U}}{AL_{234_U}} + \dfrac{C_{232_{Th}}}{AL_{232_{Th}}} + \dfrac{C_{226_{Ra}}}{AL_{226_{Ra}}} \leq 1$, is used to compare the sum of the ratios of the radionuclide concentrations to their respective action levels.[1] Because the results of the survey are used to calculate a sum of fractions, the action level is normalized to 1. The required measurement method uncertainty at this action level is $\Delta/10 =$ (UBGR − LBGR)/10. Because the LBGR was chosen to be 0.15, then $u_{MR} \leq \Delta/10 =$ (UBGR − LBGR)/10 = (1.0 − 0.15)/10 = 0.085.

Therefore, we require that:

$$u_c\left(\frac{C_{238_U}}{AL_{238_U}} + \frac{C_{234_U}}{AL_{234_U}} + \frac{C_{232_{Th}}}{AL_{232_{Th}}} + \frac{C_{226_{Ra}}}{AL_{226_{Ra}}}\right) \leq 0.085 \text{ when } \left(\frac{C_{238_U}}{AL_{238_U}} + \frac{C_{234_U}}{AL_{234_U}} + \frac{C_{232_{Th}}}{AL_{232_{Th}}} + \frac{C_{226_{Ra}}}{AL_{226_{Ra}}}\right) = 1.0.$$

Clearly, if each of the four terms in the sum is constrained to a fourth of its limit, the unity rule will be satisfied.

If the concentrations of the radionuclides of concern are independent, then:

$$u_{MR}^2\left(\frac{C_{238_U}}{0.25\,AL_{238_U}} + \frac{C_{234_U}}{0.25\,AL_{234_U}} + \frac{C_{232_{Th}}}{0.25\,AL_{232_{Th}}} + \frac{C_{226_{Ra}}}{0.25\,AL_{226_{Ra}}}\right)$$

$$= \left(\frac{u\left(C_{238_U}\right)}{0.25\,AL_{238_U}}\right)^2 + \left(\frac{u\left(C_{234_U}\right)}{0.25\,AL_{234_U}}\right)^2 + \left(\frac{u\left(C_{232_{Th}}\right)}{0.25\,AL_{232_{Th}}}\right)^2 + \left(\frac{u\left(C_{226_{Ra}}\right)}{0.25\,AL_{226_{Ra}}}\right)^2 \leq (0.085)^2$$

If the required relative measurement method uncertainty is the same for each radionuclide, therefore, the required relative method uncertainty for each individual radionuclide is:

---

[1] MARSSIM Section 4.3.3 and MARSSIM Appendix I.11 provide information on applying the unity rule.

$$\left(\frac{u\left(C_{238_U}\right)}{0.25AL_{238_U}}\right)^2 = \left(\frac{u\left(C_{234_U}\right)}{0.25AL_{234_U}}\right)^2 = \left(\frac{u\left(C_{232_{Th}}\right)}{0.25AL_{232_{Th}}}\right)^2 = \left(\frac{u\left(C_{226_{Ra}}\right)}{0.25AL_{226_{Ra}}}\right)^2 \leq \frac{(0.085)^2}{4} = (0.0425)^2$$

The required relative measurement method uncertainties for each radionuclide are provided in Table 8.10.

Table 8.10 Radionuclide-Specific Required Relative Measurement Method Uncertainties

Radionuclide	Modified Action Level (Bq/kg)	Required Relative Measurement Method Uncertainty, φ_{MR}
^{238}U	$38,000/4 = 9500$	4.25%
^{234}U	$12,000,000/4 = 3,000,000$	4.25%
^{232}Th	$330/4 = 82.5$	4.25%
^{226}Ra	$450/4 = 112.5$	4.25%

The required measurement method uncertainty for each radionuclide was provided to the analytical laboratory. The analytical laboratory used this information to specify sample volumes required to ensure this MQO was achieved.

8.2.6.7   Collect Survey Data

As the concrete is removed from the crusher, it is placed in a wooden frame (measuring 8 m by 10 m by 15 cm) on a concrete pad. The wooden frame's volume (12 m^3) corresponds to the volume associated with each sample from the survey design (i.e., 83 m^3 divided by 7 samples). Therefore, 7 batches of concrete equal 1 survey unit. One sample is collected from the center of the concrete rubble residing in the wooden form for each batch of crushed concrete. One hundred percent of the surface is scanned to identify locations with count rates greater than 30,000 cpm to investigate for areas of elevated activity and establish biased sampling points. A sample is collected at each location exceeding 30,000 cpm.

If no scan results exceed 30,000 cpm, the concrete is removed from the form and placed in the non-impacted concrete staging area awaiting laboratory analysis of the samples. If the scan survey identifies areas exceeding 30,000 cpm, the concrete is transferred to a holding container to control access to the concrete until the laboratory analyses are completed. A total of 126 batches of concrete are scanned (7 batches for each of the 18 survey units). Seventeen batches of concrete are segregated as potentially containing elevated levels of radioactivity based on the scan survey results, and one additional sample is collected from each batch as part of the investigation. No areas exceeding 100,000 cpm are identified during implementation of the disposition survey.

Five additional samples are collected from random locations on the floor of the administrative building to provide a total of seven reference area samples. The results of the two samples collected from the administrative building during the preliminary surveys are reviewed and determined to be of adequate quality for the disposition survey.

All of the concrete samples collected during implementation of the disposition survey are sent to a laboratory for analysis by gamma spectroscopy and alpha spectrometry for uranium isotopes. Thorium-232 is quantified based on the ^{228}Ac gamma spectroscopy results. Radium-226 is quantified based on the ^{214}Bi gamma spectroscopy results. A total of 150 samples are analyzed, including seven samples from the reference area. The 17 biased-sample locations identified by the scan survey were analyzed by gamma spectroscopy.

Performance checks of the FIDLER were made at the beginning and end of collection activities for each survey unit. These performance checks included a blank measurement in an area away from potential sources of radioactivity and a source check. Control charts were constructed to monitor the performance of the FIDLER throughout the survey. One FIDLER was dropped while performing a scan survey and the window was damaged. The instrument was removed from service and all scan measurements were repeated using a replacement FIDLER for that survey unit. No quality related problems were identified during the performance of the scan surveys.

The offsite laboratory provided the results of the laboratory analyses. The quality control measurements specified in the work plan were performed. All of the QC results were within the limits specified in the work plan. No quality related issues were identified during the performance of the sampling surveys.

8.2.7   Evaluate the Survey Results

8.2.7.1   Conduct a Data Quality Assessment

The disposition survey design for the concrete rubble is verified as having been executed very closely to the survey design, with the appropriate number of measurements collected for each of the survey units.

The quality control sample results from the laboratory are reviewed and the data are deemed acceptable. An exploratory data analysis of the entire data set is performed to gain an understanding of the structure of the data.

The sum of fractions for each sample is calculated using the results for ^{238}U, ^{234}U, ^{232}Th (^{228}Ac), and ^{226}Ra (^{214}Bi) and the radionuclide specific action levels. Only two samples result in sums of fractions greater than 1.0 without correcting for background. Both of these samples came from batches that were segregated prior to crushing based on visual evidence of staining within the concrete rubble; these were also the two locations with the highest scan survey results. A frequency plot (Figure 8.1) and normal cumulative frequency plot (Figure 8.2) were constructed to provide visual representations of the data.

8.2.7.2   Conduct the Wilcoxon Rank Sum Test

The Wilcoxon Rank Sum test was used to compare the reference area data to the survey unit data. In each case the test statistic exceeded the critical value of 65, so the null hypothesis was rejected for all 17 survey units. It was concluded that the average activity in all the crushed concrete exceeds background by less than a sum of fractions of 1.0.

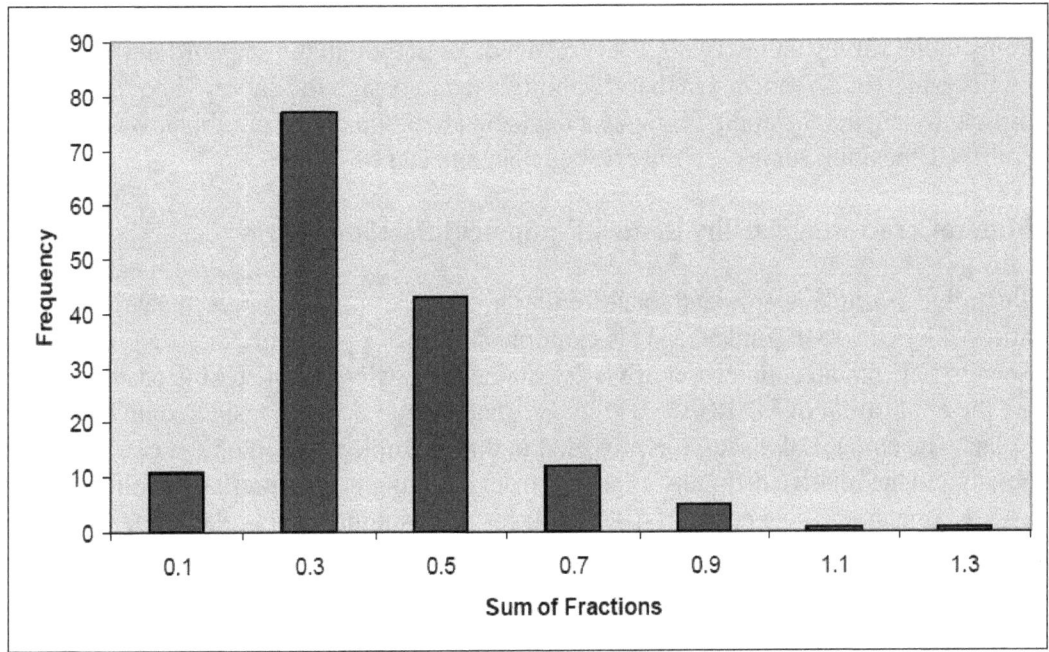

Figure 8.1 Frequency Plot of Illustrative Example Data

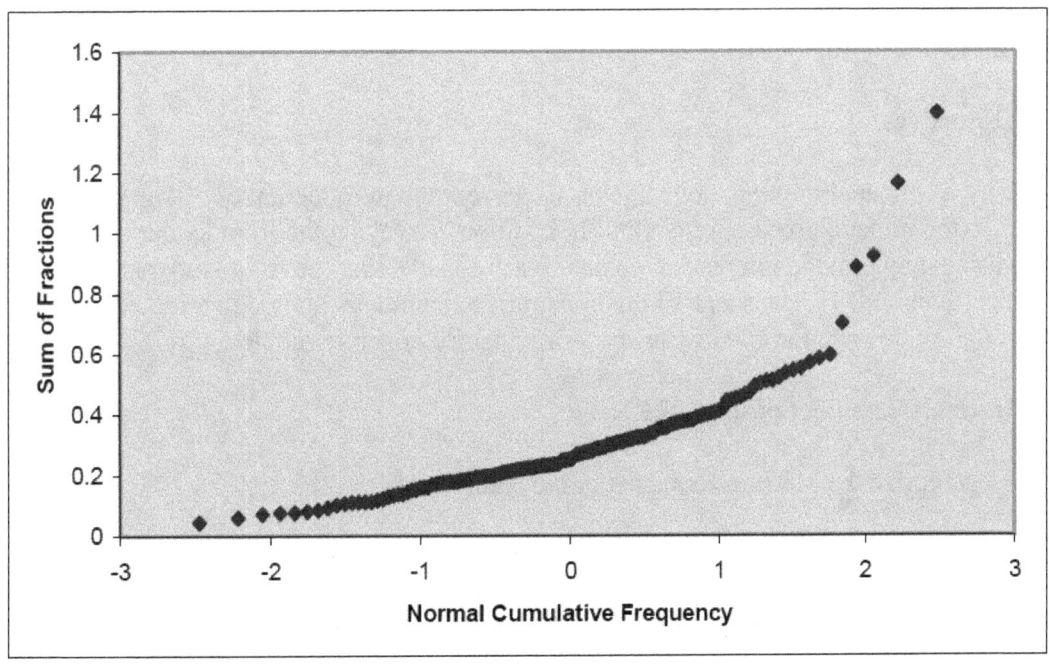

Figure 8.2 Cumulative Frequency Plot of Illustrative Example Data

8.2.8   Evaluate the Results: The Decision

In every survey unit, including those with stained concrete, the test statistic for the WRS test exceeded the critical value in Table A.4 in Appendix A. The null hypothesis that the mean sum of fractions in the survey unit exceeds 1.0 is rejected. Even though the standard deviation of the survey unit results (0.287) exceeded the variability used to design the survey (i.e., 0.15), it did not significantly impact the ability to make a decision about the concrete rubble. Based on the results of the disposition survey, all the crushed concrete can be cleared.

8.3   Mineral Processing Facility Rented Equipment Baseline Survey

This illustrative example is provided for information purposes only and presents a theoretical application of MARSAME guidance. This example describes a scan-only interdiction survey using Scenario B with an action level of no detectable radioactivity. The text is provided to illustrate the application of MARSAME guidance, and should not be considered an example survey plan. The amount of discussion provided in this example is based on the complexity of the problem and the relative difficulty expected from applying or interpreting specific portions of MARSAME guidance. The amount of discussion for this example is not related to, and should not be used as an estimate of, the level of effort associated with planning, implementing, or assessing an actual disposition survey.

8.3.1   Description

Heavy equipment is required to move the piles of concrete rubble at the mineral processing facility discussed in Section 8.2. A front loader is rented to assist with the work. The radiological history of the rented front loader is unknown.

8.3.2   Objectives

The objective is to apply interdiction controls to prevent the introduction of offsite radioactive materials to the mineral processing facility. In addition, surveying the front loader before it enters the site may provide reference area data for use in clearing the front loader at the end of the project (Section 8.4). The scope of this illustrative example is limited to a rented front loader being brought to the site for on-site transport of impacted concrete rubble.

8.3.3   Initial Assessment of the M&E

8.3.3.1   Categorize the M&E as Impacted or Non-Impacted

The material to be assessed is a rented front loader (Figure 8.3). A review of the existing information shows it is not adequate to categorize the front loader (see Figure 2.1 in Chapter 2). A visual inspection of the front loader as it is delivered to the site shows the equipment has been used, but there are no notable quantities of soil. No detailed historical records pertaining to the usage history of the front loader are available for review, other than that available from the rental company pertaining to the types of sites where heavy equipment is rented and used. Natural radionuclides are present in or commingled with soil, sediment, rubble, debris, and water. Heavy equipment is in direct contact with natural uranium and thorium during operations. Because there is a possibility the M&E may contain radionuclide concentrations or radioactivity exceeding the background at the mineral processing facility, the front loader is categorized as impacted.

Figure 8.3 Front Loader

Sentinel measurements were performed to provide information on whether the difficult-to-measure portions of the front loader, specifically the engine, were impacted by activities conducted prior to arrival at the site. The existing air filter was removed and a sentinel measurement of the used air filter was performed to determine if any radioactivity was associated with the air filter. A smear sample was taken from the air intake beyond the air filter to test for removable radioactivity. A second smear sample was taken inside the exhaust pipe to test for removable radioactivity exiting the difficult-to-measure engine areas. Measurements were performed using a hand-held gas proportional detector with an effective probe area of 100 cm^2, a detection limit less than 1,000 dpm per 100 cm^2 (Section 8.3.5.2), and counting for 1 minute. Smear measurements were made using a dual phosphor detector with a detection limit less than 1,000 dpm per 100 cm^2 (Section 8.3.5.2), and counting for 2 minutes. The results of the sentinel measurements are shown in Table 8.11.

Table 8.11 Sentinel Measurement Results

Sample Description	Reference Material Counts (Before Use, N_B)		Sample Counts (After Use, N_S)		Net Count ($N_S - N_B$)		Critical Value of the Net Instrument Signal (S_C, Table 7.5)	
	α	β	α	β	α	β	α	β
Air Filter	2	154	5	156	3	11	4.96	28.1
Air Intake Smear	0	66	1	74	1	8	2.82	18.9
Exhaust Smear	0	66	0	68	0	2	2.82	18.9

The sentinel measurement results are below the critical value of the net instrument signal, so no radioactivity was detected by the sentinel measurements. As long as the results of the interdiction survey do not detect any radioactivity, the decision will be that the difficult-to-measure areas of the front loader are non-impacted.

8.3.3.2   Describe the M&E

The information available after categorizing the front loader is not adequate to select a disposition option (see Figure 2.2 in Chapter 2). The data gaps for the front loader are associated with describing the physical and radiological attributes of the front loader. The scoping survey design includes scanning external and easily measurable areas of the front loader that have the highest potential to contact radioactive materials.

A description of the physical attributes of the front loader is listed in Table 8.12 (per Table 2.1). The front loader is a large, complicated piece of machinery. It incorporates four wheels that are 50 centimeters (cm) (1 feet [ft], 8 inches [in]) wide and 150 cm (5 ft) tall, a wheelbase of 345 cm (11 ft, 4 in), an additional section of 246 cm (8 ft, 1 in) behind the rear wheels for the engine housing, and a height of 363 cm (11 ft, 9 in) to the top of the operator cab.

Table 8.12 Physical Attributes Used to Describe the Front Loader

Attribute	Description
Dimensions	Size: Total Mass ≈25,490 kg (56,196 lbs) Shape: Total Surface Area ≈180 m^2
Complexity	The front loader is composed of multiple materials. Most external components are painted steel. However, the tires are rubber, the cab is comprised of large sections of glass, hydraulic fluid hoses are composed of high-pressure silicon, and the joints are coated with grease. Disassembly would ideally be avoided for the considerable time and expense it adds to performing disposition surveys on the equipment. Options for surveying interior surfaces include surveying of the engine air filters and interior surfaces of the exhaust plumbing to determine whether it is likely radioactive materials have spread into the engine.
Accessibility	The inside corners of the bucket and portions of each tire and wheel are difficult to measure using conventional hand-held measurements, even with a relatively small hand-held GM detector. The large height of the front loader, the underside of the front loader, and the varying orientation of surfaces associated with the equipment represent a scenario that makes accessibility difficult. There are only a few porous surfaces that allow permeation of radioactivity, such as the grease used on external hinges and joints. Air inlets, grease used on external hinges and joints, and air vents in the external panels represent areas where radioactivity could penetrate to difficult-to-measure areas.
Inherent Value	The front loader can be decontaminated, reused, or recycled. The costs associated with either replacing impacted portions of the front loader, or disposing of the front loader and replacing it, are very high. As long as only exterior surfaces of the front loader become impacted, the cost of decontamination to allow unrestricted release and reuse elsewhere will probably not be substantial.

The front loader uses a 320-cm wide (10 ft, 6 in), 4.7-m^3) capacity bucket (6 yd^3). The overall length with the bucket is 914 cm (30 ft, 0 in).

The surface area was estimated by dividing the front loader into components with regular geometric shapes and rounding to the nearest square meter. For example, the tires were modeled as cylinders and the cab was modeled as a box. The bucket has a surface area of 13.5 m^2, which is applied to the inside and outside surfaces for a total of 27 m^2. The exterior surfaces of the body have a surface area of approximately 76 m^2. The tires have a surface area of 24 m^2, and the inside of the cab is estimated at 16 m^2. Because the surfaces are not actually regular geometric shapes, a contingency factor of 25% (35 m^2) was used to account for irregular surfaces, hoses, etc. This contingency factor was based on professional judgment and approved through discussions with the regulators. The rounded total surface area is 180 m^2.

The front loader is composed of multiple materials. Most external components are painted steel. However, the tires are rubber, the cab is comprised of large sections of glass, hydraulic fluid hoses are composed of high-pressure silicon, and the joints are coated with grease. The front loader is deemed accessible, as the areas most likely to contain radioactivity are all accessible (though some portions of the front loader are more accessible than others) for conducting measurements with hand-held instruments. Internal areas of the front loader are inaccessible without disassembly.

The radiological attributes of the front loader are listed in Table 8.13 (per Table 2.2). Radionuclides of potential concern include any radionuclides that may be present. Members of the uranium and thorium radioactive decay series are used as a preliminary list of radionuclides because these are the radionuclides of concern for the site (Appendix C lists types of sites where uranium and thorium series radionuclides may be present). These are the radionuclides that are known to be present at the mineral processing facility. Radioactivity associated with the front loader is anticipated to be present at near-background concentrations. Materials may have built up in specific locations on the front loader (e.g., joints with external grease, tires, corners of the bucket) resulting in small areas of elevated radioactivity. The distribution of radioactive material is expected to be concentrated on the underside and lower edges of the front loader. Horizontal surfaces also present areas for the potential deposition of airborne radioactivity (angled and vertical surfaces also present areas for the potential deposition of airborne radioactivity but deposition of radioactivity is less likely in these areas due to surface orientation).

Table 8.13 Radiological Attributes Used to Describe the Front Loader

Attribute	Description
Radionuclides	Radionuclides of potential concern are any radionuclides that can be identified. The uranium and thorium series radionuclides are used as a preliminary list, because these are the radionuclides of concern for the mineral processing facility.
Activity	Radionuclide concentrations are expected to be close to background or zero.
Distribution	Radioactivity is expected to be associated with materials that have come in contact with the front loader. These materials will likely build up in specific locations resulting in small areas of elevated activity that can be visually identified.
Location	Radioactivity associated with the front loader is expected to be surficial and removable.

Given the unknown use history of the front loader, professional judgment and process knowledge are used to develop a likely scenario for the potential distribution of radioactivity. Radioactivity associated with the front loader is expected to be surficial only. Because the radioactivity is expected to be associated with materials from the site, the radioactivity is also expected to be removable.

Process knowledge does not provide a likely scenario for activation or other method for volumetrically impacting the front loader.

8.3.3.3 Design and Implement Preliminary Surveys

A Geiger-Mueller (GM) meter is used to collect initial scanning survey data to help address data gaps on the bucket and tires (i.e., external and easily measurable areas of the front loader that have the highest potential for residual radioactivity). The maximum reading from the bucket was 80 counts per minute (cpm), and the maximum reading from the tires was 65 cpm. A collimated in situ gamma spectrum made of the front loader showed no gamma lines other than those associated with natural uranium, potassium, and thorium. Although one might expect some trace amounts of ^{137}Cs from atmospheric fallout, there was not enough to show up in the spectrum. A non-impacted section of steel I-beam approximately one foot long (which resembles the majority of the surfaces of the front loader) is used as a reference material to establish the GM's background count rate. Scanning measurements are collected from flat surfaces, edges, and inside corners of the I-beam; count rates of 30 to 35 cpm are observed. Daily quality control checks were performed to ensure the instruments were operating properly.

8.3.3.4 Select a Disposition Option

The disposition options for the front loader are to accept it for use at the mineral processing facility following an interdiction survey, or to return it to the rental company.

8.3.3.5 Document the Results of the Initial Assessment

The results of the IA were documented in a letter report to the project manager. The decision to categorize the front loader as impacted was included in the report, along with the descriptions of the physical and radiological attributes of the front loader. The letter report described the scoping survey and listed the results of the measurements.

8.3.4 Develop a Decision Rule

Following completion of the IA, additional information needed to develop the disposition survey design is collected.

8.3.4.1 Select Radionuclides or Radiations of Concern

The initial assessment indicates that natural uranium and natural thorium are the radionuclides of potential concern.

8.3.4.2 Identify Action Levels

The action level selected for the interdiction survey is no detectable surface radioactivity above background. Because there are multiple radionuclides to be evaluated during the interdiction survey, additional discussion of action levels may be necessary.

8.3.4.3 Describe the Parameter of Interest

The parameter of interest for an interdiction survey with an action level of no detectable activity is the level of radioactivity above background reported for each measurement. Any measurement that detects the presence of radioactivity above background indicates the action level has been exceeded.

8.3.4.4 Identify Alternative Actions

The alternative actions are determined by the disposition option. If the front loader is refused access to the site, it will be returned to the rental company. If the front loader is granted access to the site, it will be used to transport concrete rubble.

8.3.4.5 Develop a Decision Rule

The decision rule incorporates the action level, parameter of interest, and alternative actions into an "if…then" statement.

If the results of any measurement identify surface radioactivity in excess of background, then the front loader will be refused access to the site. If no surface radioactivity in excess of background is detected, then the front loader will be granted access to the site.

8.3.4.6 Identify Survey Units

A survey unit is defined as the quantity of M&E for which a separate disposition decision will be made. The front loader is the survey unit. The decision rule will be applied by comparing individual measurement results to the critical value for detection. All measurements must be below the critical value (i.e., no surface radioactivity in excess of background detected) in order to accept the front loader.

8.3.4.7 Develop Inputs for Selection of Provisional Measurement Methods

The selection of a measurement method depends on the list of radionuclides or radiations of concern and will affect the survey unit boundaries. Establishing performance characteristics for the measurement method (i.e., measurement quality objectives [MQOs]) will help ensure the measurement results are adequate to support the disposition decision. Three provisional measurement methods were identified by the planning team for consideration; scan-only, in situ, or a combination of both methods in a MARSSIM-type survey design. No method-based survey designs were identified that matched the description of the M&E, so no method-based survey designs were considered.

Detection Capability

Because the action level is stated in terms of detection capability, the detection capability is critical in selecting an acceptable measurement method. The detection capability is defined as the minimum detectable concentration (MDC). The survey design will need to specify how hard to look (i.e., select an appropriate discrimination limit) before the MQO for detection capability can be established. The MDC for the selected measurement method must be less than or equal to the discrimination limit.

Measurement Method Uncertainty

The measurement method uncertainty is also important in selecting a measurement method. The MQO for detection capability will determine the acceptability of a measurement method, but it will also include information on the measurement method uncertainty. The measurement method uncertainty at background concentrations is used to calculate the MDC, as well as the critical value for the detection decision.

Range

The selected measurement method must be able to detect radionuclide concentrations or radioactivity at the discrimination limit. However, the measurement method must also be able to operate and quantify radionuclide concentrations or radioactivity at levels equal to those identified in the M&E at the site.

Specificity

The requirement for specificity will be tied to the list of radionuclides and radiations of concern. If radionuclide specific measurements are required, the measurement method must be able to identify radioactivity associated with specific radionuclides. If radionuclide specific measurements are not required, methods that measure gross activity may be acceptable.

Ruggedness

Ruggedness is not expected to be a major concern for selecting a measurement method. Because only surficial radioactivity is expected, in situ measurements of front loader surfaces will be used to collect data for comparison to the action levels. The selected measurement method must be able to perform these surface measurements in the field where the front loader is located. The environmental conditions will depend on the site location (e.g. northeast versus southwest) and the time of the year (e.g., winter versus summer).

8.3.4.8   Reference Materials

The majority of the surfaces on the front loader are metal (e.g., steel), although there are several rubber surfaces as well (e.g., tires, hoses). The small steel I-beam used to estimate background during the preliminary surveys will be used as the reference materials for the disposition survey. There is no inherent radioactivity from the uranium or thorium decay series expected in steel or rubber, so the selection of the reference material is not expected to result in any bias during interpretation of the results.

8.3.5 Develop a Survey Design

8.3.5.1 Select a Null Hypothesis

The hypotheses being tested are:

- Null Hypothesis: The front loader contains no detectable radionuclide concentrations or radioactivity above background levels (i.e., indistinguishable from background).
- Alternative Hypothesis: The front loader contains detectable radionuclide concentrations or radioactivity above background levels.

MARSAME processes require the use of Scenario B when the action level is zero, which is the case for indistinguishable from background.

8.3.5.2 Set the Discrimination Limit

The discrimination limit is the radionuclide concentration or level of radioactivity that can be reliably distinguished from the action level by performing measurements. Under Scenario B, the discrimination limit determines how hard the surveyor needs to look to determine there is no detectable radioactivity.

Acceptable surface activity levels derived from the relevant regulatory agency were selected as the discrimination limits for radionuclides of potential concern. Table 8.14 lists the potential discrimination limits based on the preliminary list of radionuclides of concern.

Table 8.14 Potential Discrimination Limits

Radionuclide of Potential Concern	Natural U	Natural Th
Average (dpm/100 cm^2)	5,000	1,000
Maximum (dpm/100 cm^2)	15,000	3,000

Based on the preliminary selection of radionuclides of potential concern, the discrimination limits for natural thorium represent the limiting case.

8.3.5.3 Specify the Limits on Decision Errors

A Type I decision error occurs when the null hypothesis is rejected when it is true. For this survey, a Type I decision error would be refusing to allow the front loader onto the site even though there is no radioactivity present that exceeds background. The consequences of this decision error may include unnecessarily returning the front loader and taking additional time to locate a replacement, or possibly deciding to decontaminate the front loader prior to use on the site. During scanning, the consequence of making a Type I decision error is the need to perform an investigation to determine the reason for the elevated reading. A Type I decision error rate of 25% is selected for the scanning survey to balance the potential of additional rental costs for the front loader while additional investigations are performed and evaluated against the additional time required to scan at slower speeds to achieve this DQO.

A Type II decision error occurs when the null hypothesis is not rejected when it is false. For this survey, a Type II decision error would be allowing the front loader to be used on the site when there is radioactivity above background. The consequences of a Type II decision error may include introducing additional radionuclides on to the site and slightly increased exposures to workers. It may also make it difficult to clear the front loader and return it to the rental company when the work is complete. For this reason a Type II decision error rate of 5% is selected for the scanning.

8.3.5.4   Select a Measurement Technique

At this point in the survey design process, the planning team decides to evaluate each of the three provisional measurement methods from Section 8.3.4.7 to determine what might be feasible for surveying the front loader. Final selection of a measurement technique will help determine the final survey design and decide between the multiple options currently available for the survey.

A scan-only survey approach requires that the measurement method be capable of detecting radioactivity at the discrimination limit. Any results exceeding the critical value would provide evidence of radioactivity levels exceeding background. There would be no need to record individual measurement results, because every result would be compared to the critical value. The calculation of the total efficiency is expected to be a major source of measurement method uncertainty. Additional measurements or assumptions are required to select a source term as the basis for the efficiency calculations. Scanning can be performed for alpha, beta, gamma, or some combination of the types of radiation. The amount of the front loader requiring scanning (i.e. 10 to 100%) would be determined by the classification. It is unknown if any scan-only measurement methods are available that meet the MQOs.

In situ survey approaches also require that the measurement method be capable of detecting radioactivity at the discrimination limit. In situ techniques allow identification of specific radionuclides, if necessary. The major source of measurement method uncertainty will likely be the model used to calculate the efficiency. Additional measurements or assumptions are required to select a source term as the basis for the efficiency calculations. The amount of the front loader requiring measurement (i.e., 10 to 100%) would be determined by the classification. The final number of measurements will be linked to the field of view of the detector. For example, a detector with a 1-m^2 field of view would require more than 180 measurements to measure 100% of the external surfaces of the front loader. An instrument such as the GM detector used during the scoping survey with a field of view of less than 100 cm^2 would require thousands of measurements to measure the minimum 10% of the front loader.

A MARSSIM-type approach would use a combination of direct measurements or samples with scanning to support a disposition decision. Sampling could damage the front loader, so direct measurements would be preferred. Locating measurements on the surface of the front loader will be problematic. Similar to scan-only and in situ designs, the scanning and direct measurements should be capable of detecting radioactivity at the discrimination limit. The MARSSIM-type survey design would require the most resources to implement.

Based on the evaluation of measurement techniques, a scan-only survey design is the preferred approach. Assumptions about the radionuclides of concern will need to be established and the availability of scan-only measurement methods needs to be verified.

8.3.5.5   Finalize Selection of Radiations to be Measured

Scan-only measurement methods are available for alpha, beta, and gamma radiations. The higher background associated with scanning for gamma radiation makes it unlikely that the measurement method could detect radioactivity at the discrimination limit. Alpha particles are attenuated more than beta particles, increasing the uncertainty caused by variations in source to detector distance. Scan-only measurement methods for beta radiation should provide the optimum survey design. However, the lower detection limits associated with alpha measurements may be required to meet the detection capability MQO. Any radioactivity in excess of background is assumed to result from natural thorium, which is the limiting radionuclide.

8.3.5.6   Develop an Operational Decision Rule

A scan-only survey will be performed for beta (and possibly alpha) radiation. Any result that exceeds the critical value associated with the MDC set at the discrimination limit will result in rejection of the null hypothesis, and the front loader will not be allowed on the site. Additional constraints on data collection activities include that the front loader be clean and dry when the measurements are performed.

8.3.5.7   Classify the M&E

The expected levels of radioactivity are background (see Table 8.13). No radioactivity in excess of background is expected, so the front loader is classified as Class 3.

8.3.5.8   Select a Measurement Method

The planning team decided to verify the availability of an acceptable measurement method prior to finalizing the survey design. The GM detector used to perform the preliminary survey is evaluated first. The expected range of radioactivity based on the reference material and preliminary survey data is approximately 35 cpm (i.e., background) to 80 cpm.

Based on the scanning survey data collected using the GM detector during the preliminary surveys, the anticipated Scan MDC of the GM detector may not be capable of detecting radioactivity at the discrimination limit of 1000 dpm/100 cm^2 (see Table 8.14).

An alpha-beta gas proportional detector utilizing a larger effective probe area will help achieve a lower scan MDC. The maximum reading for measurements from the bucket is 250 cpm; and the maximum reading from the tires is 220 cpm. Measurements collected from flat surfaces, edges, and inside corners of the reference material I-beam provide count rates between 180 and 190 cpm. The maximum background count rate is converted to scan MDC using NUREG-1761 (NRC 2002a) Equations 4-3 and 4-4.

$$s_i = d'\sqrt{b_i} = 2.32 \times \sqrt{8.3} = 6.7 \; counts$$

$$MDCR = s_i \times \frac{60}{i} = 6.7 \times \frac{60}{2} = 201 \; cpm$$

$$Scan \; MDC = \frac{MDCR}{\sqrt{p\,\varepsilon_i\varepsilon_s}} = \frac{201}{\sqrt{0.5} \times 1.29} = 220 \; dpm/100 \; cm^2$$

Where:

b_i	=	average number of background counts in the observation interval 2(250/60) = 8.3 counts)
i	=	the interval length (2 s) based on a scan speed of 5 cm/s
p	=	efficiency of a less than ideal surveyor, range of 0.5 to 0.75 from NUREG-1507 (NRC 1998b); a value 0.5 was chosen as a conservative value
d'	=	detectability index from Table 6.1 of NUREG-1507 (NRC 1998b); a value of 2.32 was selected, which represents a true positive detection rate of 95% and a false positive detection rate of 25%
s_i	=	minimum detectable number of net source counts in the observation interval (counts)
MDCR	=	minimum detectable count rate (cpm)
$\varepsilon_i\varepsilon_s$	=	weighted total alpha-beta efficiency for natural thorium in equilibrium with its progeny on the surveyed media (1.29, see Table 8.15)

The scan MDC for activity is now below 1,000 dpm/ 100 cm^2 and is good enough to detect radioactivity at the ^{232}Th discrimination limit.

Table 8.15 Detector Efficiency for the Mineral Processing Facility (^{232}Th in Complete Equilibrium with its Progeny) using a Gas Proportional Detector

Radionuclide	Average Energy (keV)	Fraction	Instrument Efficiency	Surface Efficiency	Weighted Efficiency
^{232}Th	alpha	1	0.40	0.25	0.1
^{228}Ra	7.2 keV beta	1	0	0	0
^{228}Ac	377 keV beta	1	0.54	0.50	0.27
^{228}Th	alpha	1	0.40	0.25	0.1
^{224}Ra	alpha	1	0.40	0.25	0.1
^{220}Rn	alpha	1	0.40	0.25	0.1
^{216}Po	alpha	1	0.40	0.25	0.1
^{212}Pb	102 keV beta	1	0.40	0.25	0.1
^{212}Bi	770 keV beta	0.64	0.66	0.50	0.211
^{212}Bi	alpha	0.36	0.40	0.25	0.036
^{212}Po	alpha	0.64	0.40	0.25	0.064
^{208}Tl	557 keV beta	0.36	0.58	0.50	0.104
			Total efficiency =		1.29

From NUREG-1761 (NRC 2002a), Table 4.3

8.3.5.9   Optimize the Disposition Survey Design

A scan-only interdiction survey will be performed of the exterior surfaces of the front loader. Because the front loader is Class 3, approximately 10% of the external surface area will be surveyed. Professional judgment will be used to select the locations for the scans in the locations with the highest potential for radioactivity (i.e., the bucket, tires, and floor of the cab). Approximately 50% of each of these areas will be surveyed, for a total of approximately 18 m^2 (7 m^2 of the bucket, 10 m^2 of the tires, and 1 m^2 of the cab floor). Experienced technicians will be used to perform the surveys. The scan speed will be 5 cm per second, so the scan should take approximately one man-hour to complete. The scans will be performed using a 100 cm^2 active probe area alpha-beta gas-proportional detector.

If while scanning, an area is perceived to exceed background (i.e., exceeds the scan MDC), the surveyor will suspend the scan survey and perform an investigation survey consisting of a one-minute measurement to verify the result of the scan measurement. The one-minute time interval was chosen to meet the DQOs and MQOs for this measurement. If the results of the one-minute verification measurement exceed the critical value calculated in 8.3.6.5, the radioactivity at that location exceeds background and should be recorded on a log sheet. The location of any one-minute verification measurement that exceeds the critical value will be clearly marked.

Quality control (QC) measurements will be performed prior to the start of the survey and at the completion of the survey. These QC measurements will demonstrate that the instruments were working properly while the survey was being performed. In addition, approximately 5% of the survey will be repeated using a different surveyor to confirm the results of the initial survey.

8.3.5.10 Document the Disposition Survey Design

The interdiction survey design was documented in a letter report to the project manager. The results of the IA were also included in this letter report.

8.3.6   Implement the Survey Design

8.3.6.1   Ensure Protection of Health and Safety

Protection of health and safety was performed as part of the survey implementation, but is not included in this illustrative example (see Section 8.2.6.1 for an example Job Safety Analysis.)

8.3.6.2   Consider Issues for Handling M&E

Because only a portion of the front loader needs to be accessed to implement the survey design, the front loader does not need to be moved to provide access to additional areas during the survey (e.g., bottom of tires, underside of bucket). Areas included in the survey do not need to be marked, outside of the small area that will be re-surveyed as part of the QC checks and locations of measurements exceeding the critical value. The front loader will not be parked adjacent to areas known to contain radionuclide concentrations or radioactivity in excess of background (e.g., piles of concrete rubble) while the survey is performed.

8.3.6.3 Segregate the M&E

No segregation of the front loader is required to implement the survey design.

8.3.6.4 Determine the Measurement Detectability for the Scan Survey

Section 8.3.4.7 established the MQO for the measurement detectability. The scan MDC must be less than or equal to the discrimination limit.

8.3.6.5 Determine the Measurement Detectability for the Investigation Survey

As indicated in Section 8.3.5.9, an investigation survey will be performed for any result that exceeds the scan survey investigation level (i.e. scan MDC). Both Type I and Type II errors that might occur during the investigation survey are equally undesirable. The consequence of incorrectly alleging that the front loader contains radioactivity in excess of background (Type I error) may raise unnecessary regulatory concerns. On the other hand, accepting a front loader that has radioactivity detectable above facility background (Type II error) may make it difficult to clear when the work is finished. Thus it is desirable to initially set $\alpha = \beta = 0.01$. The critical value for the one-minute measurement may be calculated from the equation in line 1 of Table 7.5:

$$S_C = z_{1-\alpha}\sqrt{N_B\frac{t_S}{t_B}\left(1+\frac{t_S}{t_B}\right)} = 2.326\sqrt{2\times 250} = 2.326\sqrt{500} = 52 \text{ net counts},$$

Where:

S_C	=	the critical value
N_B	=	the mean background count (250 counts)
t_S	=	the count time for the test source (one minute)
t_B	=	the count time for the background (one minute)
$z_{1-\alpha}$	=	the $(1-\alpha)$-quantile of the standard normal distribution (2.326 when α =0.01).

The minimum detectable net count can be calculated from the equation in line 1 of Table 7.6:

$$S_D = S_C + \frac{z_{1-\beta}^2}{2} + z_{1-\beta}\sqrt{\frac{z_{1-\beta}^2}{4} + S_C + N_B\frac{t_S}{t_B}\left(1+\frac{t_S}{t_B}\right)}$$

$$= 52 + \frac{2.326^2}{2} + 2.326\sqrt{\frac{2.326^2}{4} + 52 + 250(2)} = 109 \text{ net counts},$$

Where:

$z_{1-\beta}$	=	the $(1-\beta)$-quantile of the standard normal distribution (2.326 when β =0.01)
S_D	=	the minimum detectable value of the net instrument signal (discrimination limit, 7 cpm)

The MDC can be calculated from Equation 4-1 in NUREG-1761 (NRC 2002a):

$$MDC = \frac{\text{detection limit}}{\text{total efficiency} \times \text{sample size}} = \frac{S_D}{\varepsilon_i \varepsilon_s \times \dfrac{\text{Probe Area}}{100}} = \frac{(109)}{(1.29) \times \dfrac{100}{100}} = \frac{109}{1.29} = 84.5 \text{ dpm/100 cm}^2$$

of natural thorium.

8.3.6.6   Determine Measurement Uncertainty for the Investigation Survey MDC

$$MDC = \frac{S_D}{(\varepsilon_i \varepsilon_s) \times \dfrac{\text{Probe Area}}{100}}$$

Assuming a negligible uncertainty in the probe area, the combined standard uncertainty of the MDC is (see Equation 7-33):

$$u_c^2(MDC) = \left(\frac{\partial MDC}{\partial S_D}\right)^2 u^2(S_D) + \left(\frac{\partial MDC}{\partial \varepsilon_i \varepsilon_s}\right)^2 u^2(\varepsilon_i \varepsilon_s).$$

Note that $\varepsilon_i \varepsilon_s$ is treated as a single input variable because it is the weighted total alpha-beta efficiency for natural thorium in equilibrium with its progeny on the surveyed media.

Because the MDC is of the form of a ratio of products, Equation 7-34 may be used:

$$u_c^2(MDC) = MDC^2 \left(\frac{u^2(S_D)}{S_D^2} + \frac{u^2(\varepsilon_i \varepsilon_s)}{\varepsilon_i^2 \varepsilon_s^2}\right)$$

$$S_D = S_C + \frac{z_{1-\beta}^2}{2} + z_{1-\beta}\sqrt{\frac{z_{1-\beta}^2}{4} + S_C + N_B \frac{t_S}{t_B}\left(1 + \frac{t_S}{t_B}\right)}$$

$$= S_C + \frac{2.326^2}{2} + 2.326\sqrt{\frac{2.326^2}{4} + S_C + N_B(2)}$$

$$= z_{1-\alpha}\sqrt{N_B \frac{t_S}{t_B}\left(1 + \frac{t_S}{t_B}\right)} + \frac{2.326^2}{2} + 2.326\sqrt{\frac{2.326^2}{4} + \left(z_{1-\alpha}\sqrt{N_B \frac{t_S}{t_B}\left(1 + \frac{t_S}{t_B}\right)}\right) + 2N_B}$$

$$= 2.326\sqrt{N_B(2)} + \frac{2.326^2}{2} + 2.326\sqrt{\frac{2.326^2}{4} + \left(2.326\sqrt{N_B(2)}\right) + 2N_B}$$

Where the formula for S_C and the values of the constants have been inserted. The uncertainties in the times are assumed to be negligible, so these have also been treated as constants. Thus, the uncertainty in S_D will be due entirely to the uncertainty in the background count:

$$u^2(S_D) = \left(\frac{\partial S_D}{\partial N_B}\right)^2 u^2(N_B)$$

The sensitivity coefficient for S_D at $N_B = 250$ is

$$\left(\frac{\partial S_D}{\partial N_B}\right) = \left(\frac{\partial\left(2.326\sqrt{N_B(2)}+\frac{2.326^2}{2}+2.326\sqrt{\frac{2.326^2}{4}+\left(2.326\sqrt{N_B(2)}\right)+2N_B}\right)}{\partial N_B}\right)$$

$$= \left(\frac{\partial\left(2.326\sqrt{N_B(2)}\right)}{\partial N_B}\right) + \frac{\partial\left(\frac{2.326^2}{2}\right)}{\partial N_B} + \frac{\partial\left(2.326\sqrt{\frac{2.326^2}{4}+\left(2.326\sqrt{N_B(2)}\right)+2N_B}\right)}{\partial N_B}$$

$$= \left(\frac{\left(2.326\sqrt{2}\right)}{2\sqrt{N_B}}\right)+0+2.326\left[\frac{\partial\left(\sqrt{\frac{2.326^2}{4}+\left(2.326\sqrt{N_B(2)}\right)+2N_B}\right)}{\partial\left(\frac{2.326^2}{4}+\left(2.326\sqrt{N_B(2)}\right)+2N_B\right)}\right]\left[\frac{\partial\left(\frac{2.326^2}{4}+\left(2.326\sqrt{N_B(2)}\right)+2N_B\right)}{\partial N_B}\right]$$

$$= \left(\frac{1.6447}{\sqrt{N_B}}\right)+2.326\left[\frac{\partial\left(\sqrt{0.5815+3.289\sqrt{N_B}+2N_B}\right)}{\partial\left(0.5815+3.289\sqrt{N_B}+2N_B\right)}\right]\left[\frac{\partial\left(0.5815+3.289\sqrt{N_B}+2N_B\right)}{\partial N_B}\right]$$

$$= \frac{1.6447}{\sqrt{N_B}}+\frac{1.163\left(\frac{1.6447}{\sqrt{N_B}}+2\right)}{\sqrt{\left(0.5815+3.289\sqrt{N_B}+2N_B\right)}}$$

$$= 0.104+\frac{2.447}{23.5} = 0.208$$

Suppose the spatial variability in N_B can be described by a triangular distribution with a mean of 250 and a half-width of 50, then,

$$u(N_B) = 50/\sqrt{6} = 20.4$$

and

$$u(S_D) = \left(\frac{\partial S_D}{\partial N_B}\right)u(N_B) = (0.208)(20.4) = 4.2$$

A complete analysis of the uncertainty in $\varepsilon_i \varepsilon_s$, the weighted total alpha-beta efficiency for natural thorium in equilibrium with its progeny on the surveyed media involves propagation of uncertainty through all of the input quantities in Table 8.15. The uncertainty in the weighted total alpha-beta efficiency is

$$u(\varepsilon_i \varepsilon_s) = 0.5/\sqrt{6} = 0.20.$$

Putting this information together into Equation 7-34 for the combined total variance of the MDC we have:

$$u_c^2(\text{MDC}) = \text{MDC}^2 \left(\frac{u^2(S_D)}{S_D^2} + \frac{u^2(\varepsilon_i \varepsilon_s)}{\varepsilon_i^2 \varepsilon_s^2} \right)$$

$$= 84.5^2 \left(\frac{4.2^2}{109^2} + \frac{0.20^2}{1.29^2} \right)$$

$$= 7,140 \left(0.000148 + .024 \right)$$

$$= 172.4$$

So the estimated combined standard uncertainty in the MDC is $u_c(\text{MDC}) = 13.1$.

8.3.6.7   Perform Quality Control Measurements

The required QC measurements are performed as described in the survey design.

8.3.6.8   Collect Survey Data

Data from the survey of the front loader is collected consistent with the survey design and provides a complete record of the data collected. Thirty-seven locations were flagged during the survey for investigations using one-minute measurements. None of the one-minute measurement results exceeded the critical value.

8.3.7   Evaluate the Survey Results

8.3.7.1   Conduct a Data Quality Assessment

The surveying procedure utilized for the front loader was verified as having been executed very closely to the survey design, with the appropriate survey coverage. The results of the QC measurements demonstrated that the instruments were working properly and a different surveyor could duplicate the results of the survey. Control charts used to check the performance of the survey instruments did not identify any potential problems with the instruments.

8.3.7.2   Conduct a Preliminary Data Review

The preliminary data review for this baseline survey does not yield identifying patterns, relationships, or potential anomalies. The locations of the additional investigations appear to be randomly located based on visual inspection of the front loader.

8.3.7.3   Conduct the Statistical Tests

The statistical test selected for this scanning survey is direct comparison to the critical level. If all the results are below the critical level associated with the discrimination limit, there is no detectable radioactivity above background. All of the scanning results that exceeded the critical value were subjected to additional investigation. All of the results of the additional investigations were below the critical value.

8.3.8   Evaluate the Results: The Decision

Based on the results of the baseline survey, the front loader is determined to have no detectable radioactivity above background and is therefore allowed to enter the site.

8.4   Mineral Processing Facility Rented Equipment Disposition Survey

This illustrative example is provided for information purposes only and presents a theoretical application of MARSAME guidance. This example describes a scan-only disposition survey using Scenario A. Because this example uses the same M&E and the same survey design used in Section 8.3, it points out the similarities and differences between interdiction and release surveys. The examples in Sections 8.3 and 8.4 also point out the similarities and differences between surveys designed using Scenario A and surveys designed using Scenario B. The text is provided to illustrate the application of MARSAME guidance, and should not be considered an example survey plan. The amount of discussion provided in this example is based on the complexity of the problem and the relative difficulty expected from applying or interpreting specific portions of MARSAME guidance. The amount of discussion for this example is not related to, and should not be used as an estimate of, the level of effort associated with planning, implementing, or assessing an actual disposition survey.

8.4.1   Description

The radiological surveys at the mineral processing facility described in Section 8.2 have been completed. The front loader that was brought on site to assist with handling the concrete rubble (Section 8.3) is no longer being used. The front loader must be cleared before it can be returned to the rental company.

8.4.2   Objectives

The objective is to demonstrate the front loader can be cleared. The scope of this illustrative example is limited to the rented front loader used for the on-site transport of impacted concrete rubble.

An interdiction survey was performed to demonstrate there was no detectable radioactivity above background associated with the front loader when it entered the site. This illustrative example provides a comparison between interdiction and clearance surveys performed on the same piece of equipment.

8.4.3   Initial Assessment of the M&E

8.4.3.1   Categorize the M&E as Impacted or Non-Impacted

The existing information is adequate to categorize the front loader. The front loader was used to transport concrete rubble containing radionuclides with concentrations exceeding background. The front loader is impacted. Following use, the front loader was steam cleaned to remove loose dirt and grease (together with any associated radioactivity) for acceptance by the rental company.

Sentinel measurements were performed to provide information on whether the difficult-to-measure portions of the front loader, specifically the engine, were impacted by site activities. In addition to sentinel measurements, dust control measures were used to minimize the potential for airborne radioactivity from soil particulates throughout the project. Air monitoring of the work zone and the breathing zone of the front loader operator was performed throughout the project to estimate inhalation exposure for project workers.

A new air filter was installed at the beginning of the project and a single measurement of radioactivity associated with the air filter was performed prior to use to provide an estimate for background. Following completion of soil handling activities the air filter was removed and stored for 72 hours to allow for decay of short-lived radon decay products. A sentinel measurement of the used air filter was performed following storage to determine if any radioactivity was associated with the air filter after being used at the site. A smear sample was taken from the air intake beyond the air filter to determine if there was any removable radioactivity. Measurements were performed using a hand-held gas proportional detector with an effective probe area of 100 cm^2, a detection limit less than 1,000 dpm per 100 cm^2 (see Section 8.3.5.2), and counting for 1 minute. Smear measurements were made using dual phosphor detector with a detection limit less than 1,000 dpm per 100 cm^2 (see Section 8.3.5.2), and counting for 2 minutes. The results of the sentinel measurements are shown in Table 8.16.

Table 8.16 Sentinel Measurement Results

Sample Description	Reference Material Counts (Before Use, N_B)		Sample Counts (After Use, N_S)		Net Count ($N_S - N_B$)		Critical Net Signal (S_C, Table 7.5)	
	α	β	α	β	α	β	α	β
Air Filter	2	145	4	168	2	23	4.96	28.1
Air Intake Smear	0	66	1	79	1	13	2.82	18.9

The engineering controls minimized the potential for airborne contamination. The work zone and breathing zone air monitoring results reported no detectable radioactivity with detection limits below the acceptable derived air concentrations (DACs). The sentinel measurement results are below the critical net signal, so no radioactivity was detected by the sentinel measurements. The combination of engineering controls, air monitoring measurements, and sentinel measurements support categorization of the difficult-to-measure portions of the front loader as non-impacted.

8.4.3.2   Describe the M&E

The description of the physical attributes associated with the front loader has not changed (see Table 8.7). The uranium series and thorium series radionuclides listed in Table 8.2 are the radionuclides of potential concern for the front loader. The existing information is adequate to select a disposition option, and there are no data gaps.

8.4.3.3   Select a Disposition Option

The preferred disposition option for the front loader is clearance. The existing interdiction survey design used to allow the front loader access to the site will be evaluated for applicability as a clearance survey (Section 8.4.4.2).

8.4.3.4   Document the Results of the Initial Assessment

The decision to categorize the front loader as impacted will be documented with the results of the survey. The planning team determined that no other documentation is necessary.

8.4.4   Develop a Decision Rule

8.4.4.1   Identify Action Levels

The action level selected for the interdiction survey was no detectable surface radioactivity above background. The action levels in this case are the limits shown in Table 8.13 The limiting value is 1000 dpm/100 cm^2 for natural thorium.

8.4.4.2   Evaluate an Existing Survey Design

Because the same front loader is being surveyed, the measurement method is still adequate. The scan MDC of 132 dpm/100 cm^2 for natural thorium is well below the action level. There were no problems identified during the interdiction survey that would prevent using the measurement method for a clearance survey. The population parameter of interest and the survey unit boundaries are linked to the measurement method (see Sections 8.3.4.3 and 8.3.4.6).

The alternative actions are different for the clearance survey. If the front loader is cleared, it will be returned to the rental company. If the front loader is not cleared, it will remain on site. This results in a change to the decision rule. If the results of any measurement identify surface radioactivity in excess of background, the front loader will remain on site and radiological controls will remain in place. If no surface radioactivity in excess of 1,000 dpm/100 cm^2 over background is detected, the front loader will be cleared and returned to the rental company.

8.4.5   Develop a Survey Design

8.4.5.1   Select the Null Hypothesis

Scenario A is being used, so the hypotheses being tested are:

- Null Hypothesis: The front loader contains detectable radionuclide concentrations or radioactivity equal to or in excess of 1,000 dpm/100 cm^2 above background levels
- Alternative Hypothesis: The front loader contains radionuclide concentrations or radioactivity less than 1,000 dpm/100 cm^2 above background levels.

8.4.5.2   Set the Discrimination Limit

The discrimination limit is the radionuclide concentration or level of radioactivity that can be reliably distinguished from the action level by performing measurements. Under Scenario A, the discrimination limit should represent a prudently conservative estimate of any amount of natural thorium that may be present on the front loader in excess of background.

8.4.5.3   Specify Limits on Decision Errors

A Type I decision error occurs when the null hypothesis is rejected when it is true. For this survey, a Type I decision error would be clearing the front loader when there is radioactivity detectable more than 1,000 dpm/100 cm^2 above background. The consequence of a Type I decision error may include releasing radionuclides from the site and increased exposures to members of the public. The existing survey design specifies a Type I decision error rate of 5% for scanning measurements for this decision error.

A Type II decision error occurs when the null hypothesis is not rejected when it is false. For this survey, a Type II decision error would be refusing to clear the front loader even though the radioactivity present exceeds background by less than 1,000 dpm/100 cm^2. The consequence of this decision error may include the need to perform an investigation to determine the reason for the elevated reading, unnecessarily remediating the front loader, incurring additional costs for extra rental time, or even purchasing the front loader and disposing of it as low-level radioactive waste. The existing survey design specifies a Type II decision error rate of 25% for the scanning measurements for this decision error. Note that the definitions of Type I and Type II decision errors are reversed compared to the existing survey design from Section 8.3.

8.4.5.4   Classify the M&E

The potential for radioactivity exceeding background has increased because the front loader is known to have contacted concrete rubble containing radionuclides at concentrations that exceed background. This increased potential for radioactivity exceeding background results in a higher classification for portions of the front loader for the clearance survey. The inside of the bucket is now classified as Class 1. The remaining external surfaces are considered Class 3 so professional judgment can still be used to determine where surveys will be performed.

8.4.5.5   Optimize the Existing Survey Design

The front loader will be scanned with an alpha-beta gas proportional detector. Experienced technicians will perform the surveys. If while scanning, an area is perceived to exceed background, a one-minute measurement will be performed at that location to verify the scan results. If the results of the one-minute count exceed 1,000 dpm/100 cm^2 above background the front loader will require further remediation before it can be released. The results of all one-minute verification counts will be recorded on a log sheet. The location of any one-minute count that exceeds the critical value will be clearly marked.

Based on the classification of the inside of the bucket as Class 1, 100% of the inside of the bucket will be surveyed. In addition, 25% of the outside surface of the bucket will be surveyed, concentrating on the bottom where the bucket frequently came in contact with the concrete rubble. Similar to the interdiction survey, 50% of the tires and the floor of the cab will be surveyed. In addition, 10% of the bottom and 5% the top (i.e., horizontal surfaces) will be included in the clearance survey. Areas to be scanned will be biased to locations where residual dirt or grease is visible. The increased surface area to be scanned is expected to increase the scan time to approximately three man-hours. Based on professional judgment, four times as many investigations are expected for the clearance survey, or approximately 150 one-minute measurements. The additional investigations are expected to require an additional three man-hours.

Implementation of this survey design will likely identify locations on the front loader bucket with radioactivity levels exceeding 1,000 dpm/100 cm^2 above background. To minimize these occurrences, the front loader will be steam cleaned and dried prior to implementing the survey design. Locations on the bucket (which is a Class 1 survey unit) where the additional measurement exceeds the action level will be delineated using scanning techniques, scrubbed clean to remove any surface radioactivity, and re-surveyed (i.e., clean-as-you-go). Locations with radioactivity exceeding 1,000 dpm/100 cm^2 above background are not expected anywhere else on the front loader.

8.4.5.6   Document the Disposition Survey Design

The modified survey design was documented in a letter report to the project manager. The letter report included the results of the categorization decision (Section 8.4.3.1).

8.4.6   Implement the Survey Design

The front loader was positioned on a concrete pad during steam cleaning operations. The water was collected and containerized for survey prior to release. The bucket was lifted off the ground and supported with wooden beams to provide access to the bottom of the bucket.

The survey was implemented as described in the survey design. The beta background in the area underneath the bucket was higher than expected (i.e., 350 cpm instead of the 250 cpm used to design the survey). The bucket was lifted higher off the ground (i.e., 1.5 meters instead of 15 cm) and the scan survey was repeated with a lower background. The survey results were documented in a letter report to the project manager.

8.4.7   Evaluate the Survey Results

8.4.7.1   Conduct a Data Quality Assessment

The surveying procedure utilized for the front loader was verified as having been executed very closely to the survey design. The surveys included the appropriate scan coverage and number of additional investigations. The preliminary data review for this baseline survey does not yield identifying patterns, relationships, or potential anomalies. Control charts documenting the results of quantitative QC checks and performance checks indicate the DQOs have been achieved for this clearance survey.

8.4.7.2   Conduct the Statistical Tests

The statistical test selected for this scanning survey is direct comparison to the action level of 1,000 dpm/100 cm^2 above background. If all of the measurement results are below the action level, the average natural thorium above background cannot exceed 1,000 dpm/100 cm^2 above background.

At 83 locations the scan MDC of 132 dpm/100 cm^2 above background appeared to be exceeded. However, none of the one-minute follow up counts at those locations exceeded 500 dpm/100 cm^2 above background.

8.4.8   Evaluate the Results: The Decision

Based on the results of the disposition survey, the front loader is determined to have no radioactivity above the action level and so can be cleared.

A. STATISTICAL TABLES AND PROCEDURES

A.1 Normal Distribution

Table A.1 Cumulative Normal Distribution Function $\Phi(z)$

z	0.00	0.01	0.02	0.03	0.04	0.05	0.06	0.07	0.08	0.09
0.00	0.5000	0.5040	0.5080	0.5120	0.5160	0.5199	0.5239	0.5279	0.5319	0.5359
0.10	0.5398	0.5438	0.5478	0.5517	0.5557	0.5596	0.5636	0.5674	0.5714	0.5753
0.20	0.5793	0.5832	0.5871	0.5910	0.5948	0.5987	0.6026	0.6064	0.6103	0.6141
0.30	0.6179	0.6217	0.6255	0.6293	0.6331	0.6368	0.6406	0.6443	0.6480	0.6517
0.40	0.6554	0.6591	0.6628	0.6664	0.6700	0.6736	0.6772	0.6808	0.6844	0.6879
0.50	0.6915	0.6950	0.6985	0.7019	0.7054	0.7088	0.7123	0.7157	0.7190	0.7224
0.60	0.7257	0.7291	0.7324	0.7357	0.7389	0.7422	0.7454	0.7486	0.7517	0.7549
0.70	0.7580	0.7611	0.7642	0.7673	0.7704	0.7734	0.7764	0.7794	0.7823	0.7852
0.80	0.7881	0.7910	0.7939	0.7967	0.7995	0.8023	0.8051	0.8078	0.8106	0.8133
0.90	0.8159	0.8186	0.8212	0.8238	0.8264	0.8289	0.6315	0.8340	0.8365	0.8389
1.00	0.8413	0.8438	0.8461	0.8485	0.8508	0.8531	0.8554	0.8577	0.8599	0.8621
1.10	0.8643	0.8665	0.8686	0.8708	0.8729	0.8749	0.8770	0.8790	0.8810	0.8830
1.20	0.8849	0.8869	0.8888	0.8907	0.8925	0.8944	0.8962	0.8980	0.8997	0.9015
1.30	0.9032	0.9049	0.9066	0.9082	0.9099	0.9115	0.9131	0.9147	0.9162	0.9177
1.40	0.9192	0.9207	0.9222	0.9236	0.9251	0.9265	0.9279	0.9292	0.9306	0.9319
1.50	0.9332	0.9345	0.9357	0.9370	0.9382	0.9394	0.9406	0.9418	0.9429	0.9441
1.60	0.9452	0.9463	0.9474	0.9484	0.9495	0.9505	0.9515	0.9525	0.9535	0.9545
1.70	0.9554	0.9564	0.9573	0.9582	0.9591	0.9599	0.9608	0.9616	0.9625	0.9633
1.80	0.9641	0.9649	0.9656	0.9664	0.9671	0.9678	0.9686	0.9693	0.9699	0.9706
1.90	0.9713	0.9719	0.9726	0.9732	0.9738	0.9744	0.9750	0.9756	0.9761	0.9767
2.00	0.9772	0.9778	0.9783	0.9788	0.9793	0.9798	0.9803	0.9808	0.9812	0.9817
2.10	0.9821	0.9826	0.9830	0.9834	0.9838	0.9842	0.9846	0.9850	0.9854	0.9857
2.20	0.9861	0.9864	0.9868	0.9871	0.9875	0.9878	0.9881	0.9884	0.9887	0.9890
2.30	0.9893	0.9896	0.9898	0.9901	0.9904	0.9906	0.9909	0.9911	0.9913	0.9916
2.40	0.9918	0.9920	0.9922	0.9925	0.9927	0.9929	0.9931	0.9932	0.9934	0.9936
2.50	0.9938	0.9940	0.9941	0.9943	0.9945	0.9946	0.9948	0.9949	0.9951	0.9952
2.60	0.9953	0.9955	0.9956	0.9957	0.9959	0.9960	0.9961	0.9962	0.9963	0.9964
2.70	0.9965	0.9966	0.9967	0.9968	0.9969	0.9970	0.9971	0.9972	0.9973	0.9974
2.80	0.9974	0.9975	0.9976	0.9977	0.9977	0.9978	0.9979	0.9979	0.9980	0.9981
2.90	0.9981	0.9982	0.9982	0.9983	0.9984	0.9984	0.9985	0.9985	0.9986	0.9986
3.00	0.9987	0.9987	0.9987	0.9988	0.9988	0.9989	0.9989	0.9989	0.9990	0.9990
3.10	0.9990	0.9991	0.9991	0.9991	0.9992	0.9992	0.9992	0.9992	0.9993	0.9993
3.20	0.9993	0.9993	0.9994	0.9994	0.9994	0.9994	0.9994	0.9995	0.9995	0.9995
3.30	0.9995	0.9995	0.9995	0.9996	0.9996	0.9996	0.9996	0.9996	0.9996	0.9997
3.40	0.9997	0.9997	0.9997	0.9997	0.9997	0.9997	0.9997	0.9997	0.9997	0.9998

Negative values of z can be obtained from the relationship $\Phi(-z) = 1 - \Phi(z)$

A.2   Sample Sizes for Statistical Tests

Table A.2a Sample Sizes for Sign Test
(Number of measurements to be performed in each survey unit)

Δ/σ	\multicolumn{15}{c}{(α,β) or (β,α)}														
	0.01 0.01	0.01 0.025	0.01 0.05	0.01 0.1	0.01 0.25	0.025 0.025	0.025 0.05	0.025 0.1	0.025 0.25	0.05 0.05	0.05 0.1	0.05 0.25	0.1 0.1	0.1 0.25	0.25 0.25
0.1	4,095	3,476	2,984	2,463	1,704	2,907	2,459	1,989	1,313	2,048	1,620	1,018	1,244	725	345
0.2	1,035	879	754	623	431	735	622	503	333	518	410	258	315	184	88
0.3	468	398	341	282	195	333	281	227	150	234	185	117	143	83	40
0.4	270	230	197	162	113	192	162	131	87	136	107	68	82	48	23
0.5	178	152	130	107	75	126	107	87	58	89	71	45	54	33	16
0.6	129	110	94	77	54	92	77	63	42	65	52	33	40	23	11
0.7	99	83	72	59	41	70	59	48	33	50	40	26	30	18	9
0.8	80	68	58	48	34	57	48	39	26	40	32	21	24	15	8
0.9	66	57	48	40	28	47	40	33	22	34	27	17	21	12	6
1.0	57	48	41	34	24	40	34	28	18	29	23	15	18	11	5
1.1	50	42	36	30	21	35	30	24	17	26	21	14	16	10	5
1.2	45	38	33	27	20	32	27	22	15	23	18	12	15	9	5
1.3	41	35	30	26	17	29	24	21	14	21	17	11	14	8	4
1.4	38	33	28	23	16	27	23	18	12	20	16	10	12	8	4
1.5	35	30	27	22	15	26	22	17	12	18	15	10	11	8	4
1.6	34	29	24	21	15	24	21	17	11	17	14	9	11	6	4
1.7	33	28	24	20	14	23	20	16	11	17	14	9	10	6	4
1.8	32	27	23	20	14	22	20	16	11	16	12	9	10	6	4
1.9	30	26	22	18	14	22	18	15	10	16	12	9	10	6	4
2.0	29	26	22	18	12	21	18	15	10	15	12	8	10	6	3
2.5	28	23	21	17	12	20	17	14	10	15	11	8	9	5	3
3.0	27	23	20	17	12	20	17	14	9	14	11	8	9	5	3

Table A.2b Sample Sizes for Wilcoxon Rank Sum Test
(Number of measurements to be performed on the reference material and for each survey unit)

Δ/σ	(α,β) or (β,α)														
	0.01 0.01	0.01 0.025	0.01 0.05	0.01 0.1	0.01 0.25	0.025 0.025	0.025 0.05	0.025 0.1	0.025 0.25	0.05 0.05	0.05 0.1	0.05 0.25	0.1 0.1	0.1 0.25	0.25 0.25
0.1	5,452	4,627	3,972	3,278	2,268	3,870	3,273	2,646	1,748	2,726	2,157	1,355	1,655	964	459
0.2	1,370	1,163	998	824	570	973	823	665	440	685	542	341	416	243	116
0.3	614	521	448	370	256	436	369	298	197	307	243	153	187	109	52
0.4	350	297	255	211	146	248	210	170	112	175	139	87	106	62	30
0.5	227	193	166	137	95	162	137	111	73	114	90	57	69	41	20
0.6	161	137	117	97	67	114	97	78	52	81	64	40	49	29	14
0.7	121	103	88	73	51	86	73	59	39	61	48	30	37	22	11
0.8	95	81	69	57	40	68	57	46	31	48	38	24	29	17	8
0.9	77	66	56	47	32	55	46	38	25	39	31	20	24	14	7
1.0	64	55	47	39	27	46	39	32	21	32	26	16	20	12	6
1.1	55	47	40	33	23	39	33	27	18	28	22	14	17	10	5
1.2	48	41	35	29	20	34	29	24	16	24	19	12	15	9	4
1.3	43	36	31	26	18	30	26	21	14	22	17	11	13	8	4
1.4	38	32	28	23	16	27	23	19	13	19	15	10	12	7	4
1.5	35	30	25	21	15	25	21	17	11	18	14	9	11	7	3
1.6	32	27	23	19	14	23	19	16	11	16	13	8	10	6	3
1.7	30	25	22	18	13	21	18	15	10	15	12	8	9	6	3
1.8	28	24	20	17	12	20	17	14	9	14	11	7	9	5	3
1.9	26	22	19	16	11	19	16	13	9	13	11	7	8	5	3
2.0	25	21	18	15	11	18	15	12	8	13	10	7	8	5	3
2.25	22	19	16	14	10	16	14	11	8	11	9	6	7	4	2
2.5	21	18	15	13	9	15	13	10	7	11	9	6	7	4	2
2.75	20	17	15	12	9	14	12	10	7	10	8	5	6	4	2
3.0	19	16	14	12	8	14	12	10	6	10	8	5	6	4	2
3.5	18	16	13	11	8	13	11	9	6	9	8	5	6	4	2
4.0	18	15	13	11	8	13	11	9	6	9	7	5	6	4	2

A.3   Critical Values for the Sign Test

Table A.3 Critical Values for the Sign Test Statistic, $S+$

N	Alpha								
	0.005	0.01	0.025	0.05	0.1	0.2	0.3	0.4	0.5
4	4	4	4	4	3	3	3	2	2
5	5	5	5	4	4	3	3	3	2
6	6	6	5	5	5	4	4	3	3
7	7	6	6	6	5	5	4	4	3
8	7	7	7	6	6	5	5	4	4
9	8	8	7	7	6	6	5	5	4
10	9	9	8	8	7	6	6	5	5
11	10	9	9	8	8	7	6	6	5
12	10	10	9	9	8	7	7	6	6
13	11	11	10	9	9	8	7	7	6
14	12	11	11	10	9	9	8	7	7
15	12	12	11	11	10	9	9	8	7
16	13	13	12	11	11	10	9	9	8
17	14	13	12	12	11	10	10	9	8
18	14	14	13	12	12	11	10	10	9
19	15	14	14	13	12	11	11	10	9
20	16	15	14	14	13	12	11	11	10
21	16	16	15	14	13	12	12	11	10
22	17	16	16	15	14	13	12	12	11
23	18	17	16	15	15	14	13	12	11
24	18	18	17	16	15	14	13	13	12
25	19	18	17	17	16	15	14	13	12
26	19	19	18	17	16	15	14	14	13
27	20	19	19	18	17	16	15	14	13
28	21	20	19	18	17	16	15	15	14
29	21	21	20	19	18	17	16	15	14
30	22	21	20	19	19	17	16	16	15

Table A.3 Critical Values for the Sign Test Statistic, $S+$ (continued)

N	Alpha								
	0.005	0.01	0.025	0.05	0.1	0.2	0.3	0.4	0.5
31	23	22	21	20	19	18	17	16	15
32	23	23	22	21	20	18	17	17	16
33	24	23	22	21	20	19	18	17	16
34	24	24	23	22	21	19	19	18	17
35	25	24	23	22	21	20	19	18	17
36	26	25	24	23	22	21	20	19	18
37	26	26	24	23	22	21	20	19	18
38	27	26	25	24	23	22	21	20	19
39	27	27	26	25	23	22	21	20	19
40	28	27	26	25	24	23	22	21	20
41	29	28	27	26	25	23	22	21	20
42	29	28	27	26	25	24	23	22	21
43	30	29	28	27	26	24	23	22	21
44	30	30	28	27	26	25	24	23	22
45	31	30	29	28	27	25	24	23	22
46	32	31	30	29	27	26	25	24	23
47	32	31	30	29	28	26	25	24	23
48	33	32	31	30	28	27	26	25	24
49	33	33	31	30	29	27	26	25	24
50	34	33	32	31	30	28	27	26	25

For N greater than 50, the table (critical) value can be calculated from:

$$\frac{N}{2} + \frac{z_{1-\alpha}}{2}\sqrt{N}$$

(A-1)

where:

$z_{1-\alpha}$ = $(1-\alpha)$ percentile of a standard normal distribution (page A-9)

A.4   Critical Values for the WRS Test

The parameter, m, is the number of reference area samples and the parameter, n, is the number of survey unit samples. When using this table under Scenario A, m is the number of reference area samples and n is the number of survey unit samples. When using this table for Scenario B, the roles of m and n in this table are reversed.

Table A.4 Critical Values for the WRS Test

m	α	2	3	4	5	6	7	8	9	10	11	12	13	14	15	16	17	18	19	20
																				n
2	0.001	7	9	11	13	15	17	19	21	23	25	27	29	31	33	35	37	39	41	43
	0.005	7	9	11	13	15	17	19	21	23	25	27	29	31	33	35	37	39	40	42
	0.01	7	9	11	13	15	17	19	21	23	25	27	28	30	32	34	36	38	39	41
	0.025	7	9	11	13	15	17	18	20	22	23	25	27	29	31	33	34	36	38	40
	0.05	7	9	11	12	14	16	17	19	21	23	24	26	27	29	31	33	34	36	38
	0.1	7	8	10	11	13	15	16	18	19	21	22	24	26	27	29	30	32	33	35
3	0.001	12	15	18	21	24	27	30	33	36	39	42	45	48	51	54	56	59	62	65
	0.005	12	15	18	21	24	27	30	32	35	38	40	43	46	48	51	54	57	59	62
	0.01	12	15	18	21	24	26	29	31	34	37	39	42	45	47	50	52	55	58	60
	0.025	12	15	18	20	22	25	27	30	32	35	37	40	42	45	47	50	52	55	57
	0.05	12	14	17	19	21	24	26	28	31	33	36	38	40	43	45	47	50	52	54
	0.1	11	13	16	18	20	22	24	27	29	31	33	35	37	40	42	44	46	48	50
4	0.001	18	22	26	30	34	38	42	46	49	53	57	60	64	68	71	75	78	82	86
	0.005	18	22	26	30	33	37	40	44	47	51	54	58	61	64	68	71	75	78	81
	0.01	18	22	26	29	32	36	39	42	46	49	52	56	59	62	66	69	72	76	79
	0.025	18	22	25	28	31	34	37	41	44	47	50	53	56	59	62	66	69	72	75
	0.05	18	21	24	27	30	33	36	39	42	45	48	51	54	57	59	62	65	68	71
	0.1	17	20	22	25	28	31	34	36	39	42	45	48	50	53	56	59	61	64	67
5	0.001	25	30	35	40	45	50	54	58	63	67	72	76	81	85	89	94	98	102	107
	0.005	25	30	35	39	43	48	52	56	60	64	68	72	77	81	85	89	93	97	101
	0.01	25	30	34	38	42	46	50	54	58	62	66	70	74	78	82	86	90	94	98
	0.025	25	29	33	37	41	44	48	52	56	60	63	67	71	75	79	82	86	90	94
	0.05	24	28	32	35	39	43	46	50	53	57	61	64	68	71	75	79	82	86	89
	0.1	23	27	30	34	37	41	44	47	51	54	57	61	64	67	71	74	77	81	84
6	0.001	33	39	45	51	57	63	67	72	77	82	88	93	98	103	108	113	118	123	128
	0.005	33	39	44	49	54	59	64	69	74	79	83	88	93	98	103	107	112	117	122
	0.01	33	39	43	48	53	58	62	67	72	77	81	86	91	95	100	104	109	114	118
	0.025	33	37	42	47	51	56	60	64	69	73	78	82	87	91	95	100	104	109	113
	0.05	32	36	41	45	49	54	58	62	66	70	75	79	83	87	91	96	100	104	108
	0.1	31	35	39	43	47	51	55	59	63	67	71	75	79	83	87	91	94	98	102

Table A.4 Critical Values for the WRS Test (continued)

m	α	2	3	4	5	6	7	8	9	10	11	12	13	14	15	16	17	18	19	20
7	0.001	42	49	56	63	69	75	81	87	92	98	104	110	116	122	128	133	139	145	151
	0.005	42	49	55	61	66	72	77	83	88	94	99	105	110	116	121	127	132	138	143
	0.01	42	48	54	59	65	70	76	81	86	92	97	102	108	113	118	123	129	134	139
	0.025	42	47	52	57	63	68	73	78	83	88	93	98	103	108	113	118	123	128	133
	0.05	41	46	51	56	61	65	70	75	80	85	90	94	99	104	109	113	118	123	128
	0.1	40	44	49	54	58	63	67	72	76	81	85	90	94	99	103	108	112	117	121
8	0.001	52	60	68	75	82	89	95	102	109	115	122	128	135	141	148	154	161	167	174
	0.005	52	60	66	73	79	85	92	98	104	110	116	122	129	135	141	147	153	159	165
	0.01	52	59	65	71	77	84	90	96	102	108	114	120	125	131	137	143	149	155	161
	0.025	51	57	63	69	75	81	86	92	98	104	109	115	121	126	132	137	143	149	154
	0.05	50	56	62	67	73	78	84	89	95	100	105	111	116	122	127	132	138	143	148
	0.1	49	54	60	65	70	75	80	85	91	96	101	106	111	116	121	126	131	136	141
9	0.001	63	72	81	88	96	104	111	118	126	133	140	147	155	162	169	176	183	190	198
	0.005	63	71	79	86	93	100	107	114	121	127	134	141	148	155	161	168	175	182	188
	0.01	63	70	77	84	91	98	105	111	118	125	131	138	144	151	157	164	170	177	184
	0.025	62	69	76	82	88	95	101	108	114	120	126	133	139	145	151	158	164	170	176
	0.05	61	67	74	80	86	92	98	104	110	116	122	128	134	140	146	152	158	164	170
	0.1	60	66	71	77	83	89	94	100	106	112	117	123	129	134	140	145	151	157	162
10	0.001	75	85	94	103	111	119	128	136	144	152	160	167	175	183	191	199	207	215	222
	0.005	75	84	92	100	108	115	123	131	138	146	153	160	168	175	183	190	197	205	212
	0.01	75	83	91	98	106	113	121	128	135	142	150	157	164	171	178	186	193	200	207
	0.025	74	81	89	96	103	110	117	124	131	138	145	151	158	165	172	179	186	192	199
	0.05	73	80	87	93	100	107	114	120	127	133	140	147	153	160	166	173	179	186	192
	0.1	71	78	84	91	97	103	110	116	122	128	135	141	147	153	160	166	172	178	184
11	0.001	88	99	109	118	127	136	145	154	163	171	180	188	197	206	214	223	231	240	248
	0.005	88	98	107	115	124	132	140	148	157	165	173	181	189	197	205	213	221	229	237
	0.01	88	97	105	113	122	130	138	146	153	161	169	177	185	193	200	208	216	224	232
	0.025	87	95	103	111	118	126	134	141	149	156	164	171	179	186	194	201	208	216	223
	0.05	86	93	101	108	115	123	130	137	144	152	159	166	173	180	187	195	202	209	216
	0.1	84	91	98	105	112	119	126	133	139	146	153	160	167	173	180	187	194	201	207
12	0.001	102	114	125	135	145	154	164	173	183	192	202	210	220	230	238	247	256	266	275
	0.005	102	112	122	131	140	149	158	167	176	185	194	202	211	220	228	237	246	254	263
	0.01	102	111	120	129	138	147	156	164	173	181	190	198	207	215	223	232	240	249	257
	0.025	100	109	118	126	135	143	151	159	168	176	184	192	200	208	216	224	232	240	248
	0.05	99	108	116	124	132	140	147	155	165	171	179	186	194	202	209	217	225	233	240
	0.1	97	105	113	120	128	135	143	150	158	165	172	180	187	194	202	209	216	224	231
13	0.001	117	130	141	152	163	173	183	193	203	213	223	233	243	253	263	273	282	292	302
	0.005	117	128	139	148	158	168	177	187	196	206	215	225	234	243	253	262	271	280	290
	0.01	116	127	137	146	156	165	174	184	193	202	211	220	229	238	247	256	265	274	283
	0.025	115	125	134	143	152	161	170	179	187	196	205	214	222	231	239	248	257	265	274
	0.05	114	123	132	140	149	157	166	174	183	191	199	208	216	224	233	241	249	257	266
	0.1	112	120	129	137	145	153	161	169	177	185	193	201	209	217	224	232	240	248	256

Table A.4 Critical Values for the WRS Test (continued)

m	α	2	3	4	5	6	7	8	9	10	11	12	13	14	15	16	17	18	19	20
14	0.001	133	147	159	171	182	193	204	215	225	236	247	257	268	278	289	299	310	320	330
	0.005	133	145	156	167	177	187	198	208	218	228	238	248	258	268	278	288	298	307	317
	0.01	132	144	154	164	175	185	194	204	214	224	234	243	253	263	272	282	291	301	311
	0.025	131	141	151	161	171	180	190	199	208	218	227	236	245	255	264	273	282	292	301
	0.05	129	139	149	158	167	176	185	194	203	212	221	230	239	248	257	265	274	283	292
	0.1	128	136	145	154	163	171	180	189	197	206	214	223	231	240	248	257	265	273	282
15	0.001	150	165	178	190	202	212	225	237	248	260	271	282	293	304	316	327	338	349	360
	0.005	150	162	174	186	197	208	219	230	240	251	262	272	283	293	304	314	325	335	346
	0.01	149	161	172	183	194	205	215	226	236	247	257	267	278	288	298	308	319	329	339
	0.025	148	159	169	180	190	200	210	220	230	240	250	260	270	280	289	299	309	319	329
	0.05	146	157	167	176	186	196	206	215	225	234	244	253	263	272	282	291	301	310	319
	0.1	144	154	163	172	182	191	200	209	218	227	236	246	255	264	273	282	291	300	309
16	0.001	168	184	197	210	223	236	248	260	272	284	296	308	320	332	343	355	367	379	390
	0.005	168	181	194	206	218	229	241	252	264	275	286	298	309	320	331	342	353	365	376
	0.01	167	180	192	203	215	226	237	248	259	270	281	292	303	314	325	336	347	357	368
	0.025	166	177	188	200	210	221	232	242	253	264	274	284	295	305	316	326	337	347	357
	0.05	164	175	185	196	206	217	227	237	247	257	267	278	288	298	308	318	328	338	348
	0.1	162	172	182	192	202	211	221	231	241	250	260	269	279	289	298	308	317	327	336
17	0.001	187	203	218	232	245	258	271	284	297	310	322	335	347	360	372	384	397	409	422
	0.005	187	201	214	227	239	252	264	276	288	300	312	324	336	347	359	371	383	394	406
	0.01	186	199	212	224	236	248	260	272	284	295	307	318	330	341	353	364	376	387	399
	0.025	184	197	209	220	232	243	254	266	277	288	299	310	321	332	343	354	365	376	387
	0.05	183	194	205	217	228	238	249	260	271	282	292	303	313	324	335	345	356	366	377
	0.1	180	191	202	212	223	233	243	253	264	274	284	294	305	315	325	335	345	355	365
18	0.001	207	224	239	254	268	282	296	309	323	336	349	362	376	389	402	415	428	441	454
	0.005	207	222	236	249	262	275	288	301	313	326	339	351	364	376	388	401	413	425	438
	0.01	206	220	233	246	259	272	284	296	309	321	333	345	357	370	382	394	406	418	430
	0.025	204	217	230	242	254	266	278	290	302	313	325	337	348	360	372	383	395	406	418
	0.05	202	215	226	238	250	261	273	284	295	307	318	329	340	352	363	374	385	396	407
	0.1	200	211	222	233	244	255	266	277	288	299	309	320	331	342	352	363	374	384	395
19	0.001	228	246	262	277	292	307	321	335	350	364	377	391	405	419	433	446	460	473	487
	0.005	227	243	258	272	286	300	313	327	340	353	366	379	392	405	419	431	444	457	470
	0.01	226	242	256	269	283	296	309	322	335	348	361	373	386	399	411	424	437	449	462
	0.025	225	239	252	265	278	290	303	315	327	340	352	364	377	389	401	413	425	437	450
	0.05	223	236	248	261	273	285	297	309	321	333	345	356	368	380	392	403	415	427	439
	0.1	220	232	244	256	267	279	290	302	313	325	336	347	358	370	381	392	403	415	426
20	0.001	250	269	286	302	317	333	348	363	377	392	407	421	435	450	464	479	493	507	521
	0.005	249	266	281	296	311	325	339	353	367	381	395	409	422	436	450	463	477	490	504
	0.01	248	264	279	293	307	321	335	349	362	376	389	402	416	429	442	456	469	482	495
	0.025	247	261	275	289	302	315	329	341	354	367	380	393	406	419	431	444	457	470	482
	0.05	245	258	271	284	297	310	322	335	347	360	372	385	397	409	422	434	446	459	471
	0.1	242	254	267	279	291	303	315	327	339	351	363	375	387	399	410	422	434	446	458

Reject the null hypothesis if the test statistic (W_r) is greater than the table (critical) value.
For n or m greater than 20 with few or no ties, the table (critical) value can be calculated from:

$$Critical\ Value = \frac{m(n+m+1)}{2} + z\sqrt{\frac{nm(n+m+1)}{12}} \qquad (A-2)$$

If there are ties, the critical value can be calculated from:

$$Critical\ Value = \frac{m(n+m+1)}{2} + z\sqrt{\frac{nm}{12}\left[(n+m+1) - \sum_{j=1}^{g}\frac{t_j(t_j^2-1)}{(n+m)(n+m+1)}\right]} \qquad (A-3)$$

Where:

g	=	number of groups of tied measurements
t_j	=	number of tied measurements in the jth group
z	=	$(1-\alpha)$ percentile of a standard normal distribution (see list below)

α	z
0.001	3.090
0.005	2.575
0.01	2.326
0.025	1.960
0.05	1.645
0.1	1.282

Other values for z can be obtained from Table A.1.

A.5 Critical Values for the Quantile Test

Tables A.5a–d contain values of the parameters r and k needed for the Quantile test calculated by Gilbert and Simpson (Gilbert 1992) for certain combinations of m (the number of measurements in the reference area) and n (the number of measurements in the survey unit). The value of α listed is that obtained from simulation studies.

Table A.5a Values of r and k for the Quantile Test When α Is Approximately 0.01

m		5	10	15	20	25	30	35	40	45	50	55	60	65	70	75	80	85	90	95	100
5	r,k	r,k		11,11	13,13	16,16	19,19	22,22	25,25	28,28											r,k
	α			0.008	0.015	0.014	0.013	0.013	0.013	0.012											α
10			6,6	7,7	9,9	11,11	13,13	14,14	16,16	18,18	19,19	21,21	23,23	25,25	26,26	28,28	30,30				
			0.005	0.013	0.012	0.011	0.01	0.014	0.013	0.012	0.015	0.014	0.013	0.012	0.015	0.014	0.013				
15		3,3	7,6	6,6	7,7	8,8	10,10	11,11	12,12	13,13	15,15	16,16	17,17	18,18	19,19	21,21	22,22	23,23	24,24	26,26	27,27
		0.009	0.007	0.008	0.011	0.014	0.009	0.011	0.013	0.014	0.011	0.012	0.013	0.014	0.015	0.012	0.013	0.014	0.015	0.013	0.013
20		6,4	4,4	5,5	6,6	7,7	8,8	9,9	10,10	11,11	12,12	13,13	14,14	15,15	16,16	17,17	18,18	19,19	19,19	20,20	21,21
		0.005	0.008	0.009	0.01	0.011	0.011	0.011	0.011	0.011	0.011	0.011	0.012	0.012	0.012	0.012	0.012	0.012	0.015	0.015	0.015
25		4,3	7,5	4,4	5,5	6,6	7,7	8,8	9,9	9,9	10,10	11,11	12,12	12,12	13,13	14,14	15,15	16,16	16,16	17,17	18,18
		0.009	0.012	0.015	0.013	0.011	0.01	0.009	0.009	0.014	0.012	0.011	0.011	0.015	0.014	0.013	0.012	0.011	0.014	0.014	0.013
30		4,3	3,3	4,4	5,5	6,6	6,6	7,7	8,8	8,8	9,9	10,10	10,10	11,11	1211	12,12	13,13	14,14	14,14	15,15	15,15
		0.006	0.012	0.009	0.007	0.006	0.012	0.01	0.008	0.013	0.011	0.009	0.013	0.011	0.014	0.013	0.012	0.011	0.014	0.012	0.015
35		2,2	3,3	4,4	4,4	5,5	6,6	6,6	7,7	7,7	8,8	9,9	9,9	10,10	10,10	11,11	11,11	12,12	13,13	13,13	14,14
		0.013	0.008	0.006	0.014	0.01	0.007	0.012	0.009	0.014	0.011	0.009	0.013	0.01	0.014	0.011	0.015	0.012	0.011	0.013	0.012
40		2,2	3,3	7,5	4,4	5,5	5,5	6,6	6,6	7,7	7,7	8,8	8,8	9,9	9,9	10,10	10,10	11,11	11,11	12,12	12,12
		0.01	0.006	0.013	0.01	0.006	0.012	0.008	0.013	0.009	0.013	0.01	0.014	0.011	0.014	0.011	0.014	0.012	0.014	0.012	0.014
45		2,2	6,4	3,3	4,4	4,4	5,5	5,5	6,6	6,6	7,7	7,7	8,8	8,8	9,9	9,9	10,10	10,10	10,10	11,11	11,11
		0.008	0.008	0.013	0.007	0.014	0.008	0.014	0.009	0.013	0.009	0.013	0.009	0.012	0.009	0.012	0.009	0.012	0.015	0.012	0.014
50			4,3	3,3	4,4	4,4	5,5	5,5	5,5	6,6	6,6	7,7	7,7	8,8	8,8	8,8	9,9	9,9	10,10	10,10	10,10
			0.013	0.01	0.005	0.01	0.006	0.01	0.015	0.009	0.013	0.009	0.012	0.009	0.011	0.014	0.011	0.013	0.01	0.012	0.015
55			4,3	3,3	7,5	4,4	4,4	5,5	5,5	6,6	6,6	6,6	7,7	7,7	8,8	8,8	8,8	9,9	9,9	9,9	10,10
			0.01	0.008	0.013	0.008	0.014	0.007	0.011	0.007	0.01	0.014	0.009	0.012	0.008	0.01	0.013	0.009	0.012	0.014	0.011
60			4,3	3,3	3,3	4,4	4,4	5,5	5,5	5,5	6,6	6,6	6,6	7,7	7,7	7,7	8,8	8,8	8,8	9,9	9,9
			0.008	0.007	0.014	0.006	0.011	0.006	0.009	0.013	0.007	0.01	0.014	0.009	0.011	0.014	0.01	0.012	0.015	0.01	0.013
65			4,3	3,3	3,3	6,5	4,4	4,4	5,5	5,5	5,5	6,6	6,6	6,6	7,7	7,7	7,7	8,8	8,8	8,8	9,9
			0.007	0.006	0.012	0.006	0.009	0.013	0.007	0.01	0.014	0.008	0.011	0.014	0.009	0.011	0.014	0.009	0.011	0.014	0.01
70			2,2	6,4	3,3	7,5	4,4	4,4	5,5	5,5	5,5	5,5	6,6	6,6	6,6	7,7	7,7	7,7	8,8	8,8	8,8
			0.014	0.008	0.01	0.013	0.007	0.011	0.005	0.008	0.011	0.015	0.008	0.011	0.014	0.009	0.011	0.013	0.009	0.011	0.013
75			2,2	4,3	3,3	3,3	4,4	4,4	4,4	5,5	5,5	5,5	6,6	6,6	6,6	6,6	7,7	7,7	7,7	8,8.	8,8
			0.013	0.014	0.008	0.014	0.006	0.009	0.013	0.006	0.009	0.012	0.007	0.009	0.011	0.014	0.009	0.011	0.013	0.008	0.01
80			2,2	4,3	3,3	3,3	6,5	4,4	4,4	5,5	5,5	5,5	5,5	6,6	6,6	6,6	6,6	7,7	7,7	7,7	7,7
			0.011	0.012	0.007	0.012	0.006	0.008	0.011	0.005	0.007	0.01	0.013	0.007	0.009	0.012	0.014	0.009	0.01	0.013	0.015
85			2,2	4,3	3,3	3,3	7,5	4,4	4,4	4,4	5,5	5,5	5,5	5,5	6,6	6,6	6,6	6,6	7,7	7,7	7,7
			0.01	0.01	0.006	0.011	0.013	0.006	0.009	0.013	0.006	0.008	0.011	0.014	0.008	0.01	0.012	0.014	0.008	0.01	0.012
90				4,3	3,3	3,3	3,3	4,4	4,4	4,4	5,5	5,5	5,5	5,5	5,5	6,6	6,6	6,6	6,6	7,7	7,7
				0.009	0.005	0.009	0.014	0.005	0.008	0.011	0.005	0.007	0.009	0.012	0.015	0.008	0.01	0.012	0.014	0.008	0.019
95				4,3	6,4	3,3	3,3	6,5	4,4	4,4	4,4	5,5	5,5	5,5	5,5	6,6	6,6	6,6	6,6	6,6	7,7
				0.008	0.008	0.008	0.013	0.005	0.007	0.01	0.013	0.006	0.008	0.01	0.013	0.007	0.008	0.01	0.012	0.014	0.008
100	r,k	r,k		4,3	4,3	3,3	3,3	7,5	4,4	4,4	4,4	4,4	5,5	5,5	5,5	5,5	6,6	6,6	6,6	6,6	6,6
	α	α		0.007	0.014	0.007	0.011	0.013	0.006	0.008	0.011	0.015	0.007	0.009	0.011	0.013	0.007	0.008	0.01	0.012	0.014

Table A.5b Values of *r* and *k* for the Quantile Test When α Is Approximately 0.025

m	5	10	15	20	25	30	35	40	45	50	55	60	65	70	75	80	85	90	95	100
5 r,k	r,k		9,9	12,12	15,15	17,17	20,20	22,22	25,25											r,k
α	α		0.03	0.024	0.021	0.026	0.024	0.028	0.025											α
10 r,k		7,6	6,6	8,8	9,9	11,11	12,12	14,14	17,17	18,18	20,20	21,21	23,23	24,24	26,26	27,27				
α		0.029	0.028	0.022	0.029	0.024	0.029	0.025	0.025	0.029	0.026	0.029	0.026	0.029	0.026	0.029				
15 r,k	11,5	6,5	5,5	6,6	7,7	8,8	9,9	10,10	11,11	13,13	15,15	14,14	16,16	17,17	18,18	19,19	21,21	21,21	22,22	23,23
α	0.03	0.023	0.021	0.024	0.026	0.027	0.028	0.029	0.03	0.022	0.023	0.023	0.024	0.025	0.025	0.026	0.021	0.027	0.027	0.027
20 r,k	8,4	3,3	4,4	5,5	6,6	7,7	12,11	13,12	9,9	10,10	11,11	12,12	13,13	13,13	14,14	15,15	16 16	17,17	17,17	18,18
α	0.023	0.03	0.026	0.024	0.022	0.02	0.021	0.024	0.028	0.026	0.024	0.023	0.022	0.029	0.027	0.026	0.025	0.024	0.029	0.028
25 r,k	2,2	8,5	6,5	7,6	5,5	6,6	10,9	7,7	8,8	13,12	9,9	10,10	11,11	11,11	12,12	13,13	13,13	14,14	15,15	15,15
α	0.023	0.027	0.021	6.023	0.025	0.02	0.026	0.027	0.023	0.027	0.027	0.024	0.022	0.028	0.025	0.823	0.628	0.025	0.023	0.028
30 r,k	6,3	6,4	9,6	4,4	7,6	5,5	9,8	6,6	7,7	12,11	8,8	9,9	9,9	10,10	10,10	11,11	11,11	12,12	13,13	13,13
α	0.026	0.026	0.026	0.021	0.029	0.026	0.024	0.029	0.023	0.021	0.025	0.021	0.027	0.023	0.029	0.025	0.03	0.026	0.023	0.027
35 r,k	7,3	4,3	3,3	6,5	4,4	10,8	5,5	9,8	6,6	7,7	7,7	8,8	8,8	9,9	9,9	10,10	10,10	11,11	11,11	12,12
α	0.03	0.03	0.023	0.02	0.026	0.022	0.027	0.024	0.027	0.02	0.027	0.021	0.027	0.022	0.027	0.022	0.027	0.022	0.027	0.023
40 r,k	3,2	4,3	8,5	11,7	6,5	4,4	10,8	5,5	9,8	6,6	10,9	7,7	12,11	8,8	8,8	9,9	9,9	10,10	10,10	11,11
α	0.029	0.022	0.028	0.025	0.028	0.03	0.026	0.027	0.023	0.026	0.028	0.024	0.02	0.023	0.029	0.022	0.027	0.021	0.026	0.021
45 r,k	3,2	8,4	6,4	3,3	8,6	4,4	7,6	5,5	5,5	9,8	6,6	10,9	7,7	7,7	8,8	8,8	8,8	9,9	9,9	10,10
α	0.023	0.029	0.036	0.026	0.021	0.023	0.025	0.02	0.028	0.023	0.024	0.026	0.022	0.027	0.02	0.025	0.03	0.023	0.027	0.021
50 r,k		2,2	6,4	3,3	11,7	6,5	4,4	7,6	5,5	5,5	9,8	6,6	6,6	7,7	7,7	12,11	8,8	8,8	13,12	9,9
α		0.025	0.022	0.021	0.077	6.026	0.026	0.028	0.021	0.028	0.022	0.023	0.029	0.02	0.025	0.02	0.022	0.026	0.027	0.023
55 r,k		2,2	4,3	8,5	3,3	8,6	4,4	4,4	10,8	5,5	5,5	9,8	6,6	6,6	10,9	7,7	7,7	12,11	8,8	8,8
α		0.022	0.029	0.028	0.028	0.021	0.02	0.029	0.021	0.022	0.028	0.022	0.092	0.028	0.029	0.023	0.027	0.023	0.023	0.027
60 r,k		14,5	4,3	8,5	3,3	11,7	6,5	4,4	7,6	10,8	5,5	5,5	9,8	6,6	6,6	10,9	7,7	7,7	7,7	8,8
α		0.022	0.024	0.021	0.023	0.029	0.024	0.023	0.023	0.024	0.023	0.029	0.022	0.022	0.027	0.027	0.021	0.025	0.03	0.021
65 r,k		6,3	7,4	6,4	10,6	3,3	8,6	6,5	4,4	7,6	10,8	5,5	5,5	9,8	6,6	6,6	10,9	7,7	7,7	7,7
α		0.028	0.021	0.025	0.025	0.029	0.021	0.029	0.026	0.026	0.026	0.023	0.029	0.022	0.021	0.026	0.026	0.020	0.024	0.028
70 r,k		6,3	2,2	6,4	8,5	3,3	13,8	6,5	4,4	4,4	7,6	10,8	5,5	5,5	9,8	6,6	6,6	6,6	10,9	7,7
α		0.024	0.029	0.021	0.028	0.025	0.026	0.023	0.022	0.028	0.028	0.027	0.024	0.029	0.022	0.021	0.025	0.029	0.03	0.022
75 r,k		11,4	2,2	4,3	8,5	3,3	9,6	8,6	6,5	4,4	7,6	7,6	10,8	5,5	5,5	9,8	6,6	6,6	6,6	10,9
α		0.022	0.026	0.028	0.022	0.022	0.028	0.021	0.027	0.024	0.023	0.03	0.029	0.024	0.029	0.021	0.021	0.024	0.028	0.028
80 r,k		7,3	2,2	4,3	6,4	10,6	3,3	13,8	6,5	4,4	4,4	7,6	10,8	5,5	5,5	5,5	9,8	6,6	6,6	6,6
α		0.028	0.024	0.024	0.028	0.024	0.027	0.027	0.023	0.02	0.026	0.024	0.023	0.07	0.025	0.029	0.021	0.02	0.024	0.027
85 r,k		3,2	2,2	4,3	6,4	8,5	3,3	9,6	8,6	6,5	4,4	4,4	7,6	10,8	5,5	5,5	5,5	9,8	6,6	6,6
α		0.029	0.021	0.021	0.023	0.028	0.023	0.03	0.02	0.026	0.022	0.028	0.026	0.024	0.021	0.025	0.029	0.021	0.02	0.023
90 r,k			5,3	11,5	9,5	8,5	3,3	3,3	13,8	6,5	6,5	4,4	4,4	7,6	10,8	5,5	5,5	5,5	9,8	9,8
α			0.02	0.027	0.023	0.023	0.021	0.028	0.028	0.022	0.029	0.024	0.029	0.028	0.026	0.022	0.025	0.03	0.021	0.025
95 r,k			10,4	2,2	4,3	6,4	10,6	3,3	11,7	8,6	6,5	4,4	4,4	7,6	7,6	10,8	5,5	5,5	5,5	9,8
α			0.029	0.029	0.028	0.029	0.023	0.025	0.026	0.02	0.025	0.021	0.026	0.024	0.029	0.027	0.022	0.026	0.03	0.021
100 r,k	r,k	6,3	2,2	4,3	6,4	8,5	3,3	3,3	13,8	6,5	6,5	4,4	4,4	7,6	10,8	10,8	5,5	5,5	5,5	
α	α	0.029	0.027	0.025	0.025	0.028	0.022	0.029	0.028	0.022	0.028	0.023	0.027	0.025	0.022	0.028	0.022	0.026	0.03	

Table A.5c Values of *r* and *k* for the Quantile Test When α Is Approximately 0.05

m	5	10	15	20	25	30	35	40	45	50	55	60	65	70	75	80	85	90	95	100
						Number of Survey Unit Measurements, n														
5	r,k		8,8	10,10	13 13	15 15	17,17	19,19	21,21											r,k
	α		0.051	0.057	0.043	0.048	0.051	0.054	0.056											α
10		4,4	5,5	14,12	8,8	9,9	10,10	12,12	13,13	14,14	15,15	17,17	18,18	19,19	20,20	21,21	23,23			
		0.043	0.057	0.045	0.046	0.052	0.058	0.046	0.05	0.054	0.057	0.049	0.052	0.055	0.057	0.059	0.053			
15	2,2	3,3	4,4	5,5	6,6	7,7	8,8	9,9	9,9	10,10	11,11	12,12	13,13	14,14	15,15	16,16	16,16	17,17	18,18	19,19
	0.053	0.052	0.05	0.048	0.046	0.045	0.052	0.043	0.06	0.057	0.055	0.054	0.052	0.051	0.05	0.049	0.058	0.057	0.056	0.055
20	9,4	8,5	6,5	4,4	5,5	9,8	6,6	7,7	8,8	8,8	9,9	10,10	10,10	11,11	12,12	12,12	13,13	14,14	14,14	15,15
	0.04	0.056	0.04	0.053	0.043	0.052	0.056	48	0.043	0.057	0.051	0.046	0.057	0.052	0.048	0.057	0.053	0.049	0.057	0.054
25	6,3	6,4	3,3	6,5	4,4	5,5	5,5	6,6	11,10	7,7	8,8	8,8	9,9	9,9	10,10	11,11	11,11	11,11	12,12	12,12
	0.041	0.043	0.046	0.052	0.055	0.041	0.059	0.046	0.042	0.05	0.042	0.053	0.045	0.055	0.048	0.042	0.05	0.058	0.052	0.06
30	3,2	2,2	10,6	3,3	11,8	4,4	8,7	5,5	6,6	6,6	7,7	7,7	8,8	8,8	9,9	9,9	9,9	10,10	10,10	11,11
	0.047	0.058	0.052	0.058	0.045	0.056	0.044	0.054	0.04	0.053	0.041	0.052	0.042	0.051	0.042	0.05	0.059	0.049	0.057	0.049
35	8,3	2,2	6,4	3,3	6,5	4,4	4,4	8,7	5,5	9,8	6,6	6,6	7,7	7,7	8,8	8,8	8,8	9,9	9,9	10,10
	0.046	0.045	0.058	0.043	0.041	0.04	0.057	0.043	0.051	0.052	0.047	0.057	0.043	0.053	0.041	0.049	0.057	0.046	0.053	0.044
40	4,2	5,3	4,3	10,6	3,3	6,5	4,4	4,4	8,7	5,5	9,8	6,6	6,6	11,10	7,7	7,7	8,8	8,8	8,8	9,9
	0.055	0.048	0.057	0.059	0.053	0.048	0.043	0.058	0.042	0.048	0.047	0.042	0.051	0.042	0.045	0.053	0.041	0.048	0.055	0.043
45	4,2	9,4	2,2	8,5	3,3	8,6	6,5	4,4	4,4	8,7	5,5	5,5	9,8	6,6	6,6	11,10	7,7	7,7	8,8	8,8
	0.045	0.047	0.059	0.052	0.042	0.041	0.054	0.045	0.058	0.041	0.046	0.057	0.056	0.047	0.055	0.046	0.047	0.054	0.041	0.047
50		6,3	2,2	6,4	12,7	3,3	8,6	6,5	4,4	4,4	8,7	5,5	5,5	9,8	6,6	6,6	6,6	7,7	7,7	7,7
		0.051	0.05	0.051	0.05	0.049	0.049	0.059	0.047	0.059	0.041	0.045	0.054	0.051	0.043	0.05	0.058	0.041	0.048	0.054
55		3,2	2,2	4,3	8,5	3,3	5,4	6,5	9,7	4,4	4,4	8,7	5,5	5,5	9,8	6,6	6,6	6,6	11,10	7,7
		0.059	0.043	0.056	0.058	0.041	0.041	0.046	0.042	0.048	0.059	0.04	0.043	0.052	0.048	0.04	0.047	0.054	0.043	0.043
60		3,2	5,3	4,3	6,4	3,3	3,3	8,6	6,5	9,7	4,4	4,4	13,10	5,5	5,5	5,5	9,8	6,6	6,6	6,6
		0.052	0.052	0.046	0.059	0.035	0.047	0.043	51	0.046	0.049	0.059	0.052	0.042	0.05	0.058	0.054	0.044	0.05	0.056
65		.3,2	5,3	2,2	6,4	10,6	3,3	3,3	6,5	6,5	4,4	4,4	4,4	13,10	5,5	5,5	5,5	9,8	6,6	6,6
		0.045	0.043	0.053	0.048	0.05	0.04	0.052	0.041	0.055	0.042	0.05	0.06	0.052	0.041	0.048	0.055	0.051	0.041	0.047
70		8,3	9,4	2,2	4,3	8,5	5,4	3,3	3,3	6,5	6,5	4,4	4,4	4,4	13,10	5,5	5,5	5,5	9,8	9,8
		0.057	0.048	0.047	0.055	0.05	0.041	0.046	0.057	0.045	0.058	0.043	0.051	0.06	0.051	0.041	0.047	0.054	0.048	0.057
75		8,3	6,3	2,2	4,3	6,4	10,6	3,3	3,3	8,6	6,5	9,7	4,4	4,4	5,5	13,10	8,7	5,5	5,5	5,5
		0.049	0.056	0.043	0.047	0.054	0.053	0.04	0.051	0.044	0.049	0.041	0.044	0.052	0.06	0.051	0.047	0.046	0.052	0.058
80		4,2	6,3	5,3	2,2	6,4	8,5	5,4	3,3	3,3	6,5	6,5	9,7	4,4	4,4	7,6	13,10	8 7	5,5	5,5
		0.059	0.048	0.053	0.055	0.046	0.055	0.041	0.045	0.055	0.041	0.052	0.043	0.045	0.053	0.058	0.051	0.046	0.045	0.051
85		4,2	3,2	5,3	2,2	4,3	4,3	10,6	5,4	3,3	3,3	6,5	6,5	9,7	4,4	4,4	7,6	10,8	8,7	5,5
		0.054	0.058	0.047	0.05	0.054	0.048	0.056	0.049	0.049	0.059	0.044	0.055	0.046	0.046	0.053	0.059	0.06	0.045	0.044
90			3,2	5,3	2,2	6,4	6,4	8,5	5,4	3,3	3,3	8,6	6,5	6,5	4,4	4,4	4,4	7,6	10,8	8,7
			0.053	0.041	0.046	0.059	0.051	0.058	0.042	0.044	0.053	0.045	0.047	0.058	0.041	0.047	0.054	0.059	0.06	0.041
95			3,2	9,4	2,2	2,2	4,3	8,5	10,6	5,4	3,3	3,3	6,5	6,5	9,7	4,4	4,4	4,4	7,6	10,8
			0.048	0.048	0.042	0.056	0.059	0.05	0.058	0.048	0.048	0.056	0.041	0.05	0.040	0.042	0.048	0.054	0 59	0.059
100	r,k	3,2	6,3	5,3	2,2	4,3	6,4	10,6	5,4	3,3	3,3	3,3	6,5	6,5	9,7	4,4	4,4	4,4	7,6	
	α	0.044	0.057	0.054	0.052	0.053	0.056	0.049	0.043	0.043	0.051	0.059	0.044	0.053	0.042	0.043	0.049	0.055	0.059	

Table A.5d   Values of *r* and *k* for the Quantile Test When α Is Approximately 0.10

	Number of Survey Unit Measurements, *n*																			
m	5	10	15	20	25	30	35	40	45	50	55	60	65	70	75	80	85	90	95	100
5	r,k		7,7	8,8	10,10	12,12	14,14	15,15	17,17											r,k
	α		0.083	0.116	0.109	0.104	0.1	0.117	0.112											α
10		3,3	4,4	5,5	6,6	7,7	8,8	9,9	10,10	11,11	12,12	13,13	14,14	15,15	16,16	1712	18,18			
		0.105	0 108	0.109	0.109	0.109	0.109	0.109	0.109	0.109	0.109	0.109	0.109	0.109	0.109	0.109	0.109			
15	9,4	10,6	3,3	4,4	5,5	5,5	6,6	7,7	7,7	8,8	9,9	9,9	10,10	11,11	11,11	12,12	13,13	13,13	14,14	15,15
	0.098	0.106	0 112	0.093	0.081	0.117	0.102	0.092	0.118	0.106	0.098	0.118	0.109	0.101	0.118	0.11	0.104	0 118	0.111	0.106
20	3,2	2,2	5,4	3,3	4,4	4,4	5,5	10,9	6,6	7,7	7,7	8,8	8,8	9,9	9,9	10,10	10,11	11,11	11,11	12,12
	0.091	0.103	0.093	0.115	0.085	0.119	0.093	0.084	0.099	0.083	0.102	0.088	0.105	0.092	0.107	0.095	0.108	0.098	0.11	0.1
25	4,2	7,4	8,5	3,3	3,3	4,4	4,4	8,7	5,5	10,9	6,6	6,6	7,7	7,7	8,8	8,8	8,8	9,9	9,9	10,10
	0.119	0.084	0 112	0.08	0.117	0.08	0.107	0.108	0.101	0.088	0.096	0.114	0.093	0.108	0.091	0.104	0.117	0.1	0.112	0.098
30	4,2.	5,3	2,2	14,8	3,3	3,3	9,7	4,4	8,7	5,5	5,5	6,6	6,6	6,6	7,7	7,7	7,7	8,8	8,8	8,8
	0.089	0.089	0 106	0.111	0.088	0.119	0.116	0.1	0.093	0.088	0.106	0.08	0.095	0.11	0.087	0.1	0.113	0.092	0.103	0.115
35	5,2	3,2	2,2	6,4	5,4	3,3	3,3	9,7	4,4	4,4	8,7	5,5	5,5	6,6	6,6	6,6	6,6	7,7	7,7	7,7
	0.109	0.119	0.086	0.12	0.091	0.093	0 12	0.112	0.094	0.114	0.107	0.094	0.11	0.081	0.094	0.107	0.12	0.094	0.105	0.116
40	5,2	3,2	5,3	2,2	12,7	5,4	3,3	6,5	9,7	4,4	4,4	8,7	5,5	5,5	5,5	6,6	6,6	6,6	6,6	7,7
	0.087	0.098	0 119	0.107	0.109	0.102	0.097	0.100	0.109	0.09	0.107	0.097	0.086	0.099	0.112	0.082	0.093	0 104	0.116	0.089
45	6,2	3,2	5,3	2,2	6,4	7,5	5,4	3,3	6,5	9,7	4,4	4,4	4,4	8,7	5,5	5,5	5,5	6,6	6,6	6,6
	0.103	0.082	0.094	0.091	0.115	0.086	0.112	0.1	0.101	0.107	0.087	0.102	0.117	0.107	0.091	0.103	0.115	0.083	0.093	0.103
50		7,3	9,4	7,4	2,2	10,6	5,4	3,3	3,3	6,5	9,7	4,4	4,4	4,4	8,7	5,5	5,5	5,5	5,5	6,6
		0.083	0 115	0.097	0.108	0.112	0.09	0.084	0.103	0.102	0.105	0.084	0.098	0.112	0.099	0.084	0.95	0.105	0.116	0.083
55		4,2	3,2	5,3	2,2	6,4	14,8	5,4	3,3	3,3	6,5	9,7	4,4	4,4	4,4	4,4	8,7	5,5	5,5	5,5
		0.109	0 114	0.114	0.095	0.112	0.111	0.098	0.088	0.104	0.103	0.104	0.082	0.095	0.107	0.12	0.107	0.088	0.098	0.108
60		4,2	3,2	5,3	2,2	2,2	8,5	5,4	5,4	3,3	3,3	6,5	9,7	4,4	4,4	4,4	4,4	8,7	5,5	5,5
		0.095	0.1	0.097	0.084	0.109	0.119	0.082	0.105	0.091	0.106	0.103	0.102	0.081	0.092	0.103	0.115	0.1	0.083	0.092
65		4,2	3,2	5,3	7,4	2,2	6,4	12,7	5,4	5,4	3,3	3,3	6,5	9,7	7,6	4,4	4,4	4,4	8,7	8,7
		0.084	0.089	0.082	0.090	0.097	0 11	0.113	0.089	0.111	0.093	0.108	0.104	0.101	0.084	0.09	0.1	0.11	0.094	0.107
70		5,2	7,3	9,4	5,3	2,2	2,2	8,5	7,5	5,4	3,3	3,3	3,3	6,5	9,7	7,6	4,4	4,4	4,4	4,4
		0.115	0 101	0.106	0.112	0.088	0.109	0.114	0.081	0.096	0.083	0.096	0.109	0.104	0.191	0.082	0.088	0.097	0.107	0.117
75		5,2	7,3	3,2	5,3	7,4	2,2	2,2	10,6	5,4	5,4	3,3	3,3	3,3	6,5	9,7	7,6	4,4	4,4	4,4
		103	0.088	0.111	0.098	0.101	0.099	0.119	0.117	0.083	0.102	0.085	0.098	0.11	0.105	0.1	0.081	0.086	0.095	0.104
80		5,2	4,2	3,2	5,3	7,4	2,2	2,2	8,5	14,8	5,4	5,4	3,3	3,3	3,3	6,5	6,5	9,7	4,4	4,4
		0 093	0 116	0.101	0.086	0.086	0.09!	0.109	0.111	0.11	0.089	0.107	0.088	0.099	0.111	0.105	0.12	0.116	0.084	0.093
85		5,2	4,2	3,2	9,4	5,3	2,2	2,2	2,2	10,6	7,5	5,4	5,4	3,3	3,3	3,3	6,5	6,5	9,7	4,4
		0.084	0 106	0.092	117	0.111	0.083	0.101	0.118	0.112	0.084	0.094	0.111	0.09	0.101	0.112	0.105	0.119	0.114	0.083
90			4,2	3,2	3,2	5,3	7,4	2,2	2,2	8,5	12,7	5,4	5,4	3,3	3,3	3,3	3,3	6,5	6,5	9,7
			0.097	0.085	0.119	0.099	0.095	0.093	0.109	0.108	0.114	0.083	0.099	0.082	0.092	0.102	0.113	0.105	0.119	0.113
95			4,2	7,3	3,2	5,3	7,4	2,2	2,2	2,2	10,6	14,8	5,4	5,4	3,3	3,3	3,3	3,3	6,5	6,5
			0.089	100	0.11	0.089	0.084	0.086	0.102	0.117	0.08	0.117	0.088	0.103	0.084	0.094	0.103	0.113	0.106	0.118
100	r,k		4,2	7,3	3,2	5,3	5,3	2,2	2,2	2,2	6,4	12,7	7,5	5,4	5,4	3,3	3,3	3,3	3,3	6,5
	α		0.082	0.09	0.102	0.08	0.109	0.08	0.095	0.11	0.118	0.109	0.086	0.093	0.08	0.086	0.095	0.104	0.114	0.106

B. SOURCES OF BACKGROUND RADIOACTIVITY

B.1   Introduction

Background radioactivity can complicate the disposition decision for M&E. Background radioactivity may be the result of environmental radioactivity, inherent radioactivity, instrument noise, or some combination of the three. Special consideration is given to issues associated with technologically enhanced naturally occurring radioactive materials (TENORM) and orphan sources as contributors to background. The planning team should consider these potential sources of background activity and determine what effect, if any, they may have on the design of the disposition survey.

Information on background radioactivity can be obtained from many sources, including:

- The Nuclear Regulatory Commission (NRC) provides information concerning background radioactivity in *Background as a Residual Radioactivity Criterion for Decommissioning* NUREG-1501 (NRC 1994).
- The United Nations Scientific Committee on the Effects of Atomic Radiation (UNSCEAR) has published a report on *Sources and Effects of Ionizing Radiation* (UNSCEAR 2000) and provides a searchable version of the report on the World Wide Web at www.unscear.org.
- The National Council on Radiation Protection and Measurements (NCRP) has published reports on *Exposure of the Population in the United States and Canada from Natural Background Radiation*, NCRP Report No. 94 (NCRP 1988a) and *Radiation Exposure of the U.S. Population from Consumer Products and Miscellaneous Sources*, NCRP Report No. 95 (NCRP 1988b).

B.2   Environmental Radioactivity

Environmental radioactivity is radioactivity from the environment where the M&E is located. There are three sources contributing to environmental radioactivity; terrestrial (Section B.2.1), manmade (Section B.2.2), and cosmic and cosmogenic (Section B.2.3). Although background radiation is present everywhere, the component radionuclide concentrations and distributions are not constant. Certain materials have higher concentrations of background radiation, and varying environmental and physical conditions can result in accumulations of background radiation. Information on environmental radioactivity is usually available from historic measurements identified during the initial assessment (IA).

If high levels of environmental radioactivity interfere with the disposition decision (e.g., action level less than environmental background, variability in environmental radioactivity determines level of survey effort), the planning team may consider moving the M&E being investigated to a location with less environmental radioactivity (see Sections 3.3.1.3 and 5.3). If the level of environmental radioactivity is unknown, it may be necessary to collect data during a preliminary survey (see Section 2.3) to provide this information.

B.2.1   Terrestrial Radioactivity

The naturally occurring forms of radioactive elements incorporated into the Earth during its formation that is still present are referred to as "terrestrial radionuclides." The most significant terrestrial radionuclides include the uranium and thorium decay series, potassium-40 and rubidium-87. Virtually all materials found in nature contain some concentration of terrestrial radionuclides. Table B.1 lists average and typical ranges of concentrations of terrestrial radionuclides. Although the ranges in the table are typical, larger variations exist in certain areas (e.g., Colorado).

Bulk materials containing elevated concentrations of terrestrial radionuclides as well as equipment used to handle or process these materials should be identified during the IA even if these materials and equipment were not impacted by site activities.

Radon is an element that occurs as a gas in nature. Isotopes of radon are members of both the uranium and thorium natural decay series. These radon isotopes decay to produce additional radioactive isotopes, which are collectively called radon progeny.

Table B.1 Typical Average Concentration Ranges of Terrestrial Radionuclides

Material	Radium-226 (Bq/kg)[a]	Uranium-238 (Bq/kg)[a]	Thorium-232 (Bq/kg)[a]	Potassium-40 (Bq/kg)[a]
Soil, U.S.	40 (8-160)[b]	35 (4-140)[b]	35 (4-130)[b]	370 (100-700)[b]
Phosphate Fertilizer	200[c] - 100,000[d]	200-1,500[b]	20[b]	--
Concrete	(19-89)[e]	(19-89)[f]	(15-120)[f]	(260-1,100)[f]
Concrete Block	(41-780)[e]	(41-780)[f]	(37-81)[f]	(290-1,100)[f]
Brick	(4-180)[e]	(4-180)[f]	(1-140)[f]	(7-1,200)[f]
Coal Tar Fly Ash-Bottom Ash	(100-300)[e] 200[e]	(100-300)[b] 200[b]	-- 200[b]	-- --
Coal, U.S.	--	18 (1-540)[g]	21 (2-320)[g]	52 (1-710)[g]
Tile	--	(550-810)[h]	650[h]	--
Porcelain, Glazed	--	(180-37,000)[h, i]		--
Ceramic, Glazed[b]		(79-1,200)[h, i]		

a   To convert Bq/kg to pCi/g, multiply by 0.027.
b   UNSCEAR, Sources and Effects of Ionizing Radiation (UNSCEAR 2000).
c   U.S. Environmental Protection Agency (EPA), 2000. *Evaluation of EPA's Guidelines for Technologically Enhanced Naturally Occurring Radioactive Materials (TENORM)*.
d   National Academy of Sciences (NAS). 1999. *Evaluation of Guidelines for Technologically Enhanced Naturally Occurring Radioactive Materials (TENORM)*, Committee on Evaluation of EPA Guidelines for Exposure to Naturally Occurring Radioactive Materials Board on Radiation Effects Research Commission on Life Sciences National Research Council, National Academy Press, p. 72.
e ^{226}Ra is assumed to be in secular equilibrium with ^{238}U.
f   Eicholz G.G., Clarke F.J., and Kahn, B., 1980. *Radiation Exposure From Building Materials*, in "Natural Radiation Environment III," U.S. Department of Energy CONF-780422.
g   Beck H.L., Gogolak C.V., Miller K.M., and Lowder W.M., 1980. *Perturbations on the Natural Radiation Environment Due to the Utilization of Coal as an Energy Source*, in "Natural Radiation Environment III," U.S. Department of Energy CONF-780422.
h   Hobbs T.G., 2000. *Radioactivity Measurements on Glazed Ceramic Surfaces*, J. Res. Natl. Inst. Stand. Technol. **105**, 275-283.
i   Values reported as total radioactivity without identification of specific radionuclides.

Radon emissions vary significantly over time based on a wide variety of factors. For example, relatively small changes in the relative pressure between the source material and the atmosphere (indoor or outdoor) can result in large changes in radon concentrations in the air. Soil moisture content also has an affect on the radon emanation rate.

Radon progeny tend to become fixed to solid particles in the air. These particles can become attached to surfaces as a result of electrostatic charge or gravitational settling. Air flow through ventilation ducts can produce an electrostatic charge that will attract these particles. A decrease in atmospheric pressure often precedes a rainstorm, which increases the radon emanation rate. Immediately prior to an electrical storm, an electrostatic charge can build up on equipment resulting in elevated radiation levels from radon progeny. Rainfall acts to scavenge these particles from the air, potentially resulting in elevated dose rates and surface activities during and immediately following rainfall.

Pb-210 is a decay product of ^{222}Rn and ^{238}U. The 22-year half-life provides opportunities for buildup ^{210}Pb and progeny in sediments and low-lying areas. As mentioned previously, rain acts to scavenge radon progeny from the air. Areas where rain collects and concentrates can result in elevated levels of ^{210}Pb and progeny over time. In addition, lead is easily oxidized and can become fixed to surfaces through corrosion processes. Rust or oxide films on equipment can be indicators of locations with a potential for elevated background radioactivity.

B.2.2 Anthropogenic Radioactive Materials

Nuclear weapons testing and nuclear power reactors have produced large quantities of radionuclides through the fissioning of uranium and other heavy elements and the activation of various elements. Examples of anthropogenic radionuclides that could be in the environment are ^{137}Cs, ^{90}Sr, and various isotopes of plutonium.

Prior to the 1963 Limited Test Ban Treaty, fallout from atmospheric nuclear tests distributed large quantities of anthropogenic radionuclides around the globe. Following the 1963 treaty most nuclear weapons tests were conducted underground, although China and France continued atmospheric testing of nuclear weapons into the late 1970s. In 1996 a Comprehensive Test Ban Treaty was negotiated with the help of the United Nations. The Comprehensive Test Ban Treaty has not been ratified by China or the United States and was broken by Pakistan and India in 1998. However, worldwide fallout concentrations have been declining since the mid 1960s. In 1964 a Department of Defense weather satellite containing a radiation source failed to achieve orbit. The Space Nuclear Auxiliary Power (SNAP) 9-A Radioisotopic Thermoelectric Generator (RTG) burned up on re-entry and dispersed the nuclear inventory (primarily plutonium-238) into the atmosphere. Incidents involving Soviet satellites with radioisotopes or nuclear reactors occurred in 1969, 1973, 1978, and 1983. In April 1986 there was a non-nuclear steam explosion and fire at the number four reactor at the nuclear power plant in Chernobyl in north-central Ukraine. Large quantities of radioactive material were released into the environment as a result of the catastrophe. The radionuclides from these incidents have been inhomogenously deposited around the world.

Isolated pockets with elevated concentrations of anthropogenic radionuclides can still be found. For example, ventilation systems that were installed prior to 1963 collected fallout radionuclides. If these systems are still in use and the ducts have not been thoroughly cleaned, there is a potential for elevated background radiation. Another potential source of elevated background radiation from anthropogenic radionuclides is wood ash. Trees filter and store some airborne pollutants, including ^{137}Cs from fallout. When the wood is burned the ^{137}Cs is concentrated in the wood ash. Materials or equipment associated with the ash could have elevated levels of background radiation.

B.2.3 Cosmic Radiation and Cosmogenic Radionuclides

Cosmic radiation consists of highly energetic particles that are believed to originate from phenomena such as solar flares and supernova explosions. The Earth's atmosphere serves as a shield for these particles, although on rare occasions a solar flare is strong enough to produce a significant radiation dose in the lower reaches of the atmosphere.

Cosmic radiation is also responsible for the production of radioactive elements called cosmogenic radionuclides. These radionuclides are produced from collisions between the highly energetic cosmic radiation with stable elements in the atmosphere. Cosmogenic radionuclides include ^{3}H, ^{7}Be, ^{14}C, and ^{22}Na. Background concentrations of cosmic radiation and cosmogenic radionuclides generally do not impact disposition surveys.

B.3 Inherent Radioactivity

Inherent radioactivity, or intrinsic radioactivity, is radioactivity that is an integral part of the M&E being investigated. For example, concrete is made from materials that contain terrestrial radionuclides and is inherently radioactive. Some equipment is constructed from radioactive components, such as electron tubes or night vision goggles containing thorium components. Information on inherent radioactivity is usually obtained from process knowledge or historical measurements identified during the IA. Manufacturers of equipment that incorporates radioactive components can usually provide the radionuclide and the activity incorporated into the equipment. Information on radionuclides and activity levels for other types of equipment or bulk materials that are inherently radioactive is usually more generic. Table B.1 lists ranges of terrestrial radionuclide concentrations in some common materials (e.g., concrete, soil, brick). The wide range of radionuclide concentrations observed in these materials prevents establishing any general rules of thumb, so it is usually necessary to obtain project-specific information. For release scenarios, it is strongly recommended that all M&E be surveyed before it enters a controlled area. This provides project-specific information on inherent radioactivity and minimizes complications when designing the disposition survey. For interdiction scenarios, it is important to understand the types of M&E being investigated and the potential for inherent radioactivity. It may be necessary to establish an administrative action level that defines the upper end of acceptable inherent radioactivity for different types of M&E (see Section 3.2).

B.4   Instrument Background

Instrument background is a combination of radioactivity in the constituent materials of the detector, ancillary equipment, and shielding, and electronic noise contributing to the instrument response. Instruments designed to measure low levels of radioactivity generally are constructed from materials with very low levels of inherent radioactivity to minimize instrument background. The electronics in radiation instruments are also designed to minimize the signal-to-noise ratio, also reducing instrument background. Instrument background becomes the primary contributor to background only for radionuclide-specific measurements for radionuclides not contributing to environmental or inherent background (e.g., ^{60}Co in bulk soil measured by gamma spectroscopy). Note that radiation from M&E can interact with instrument shielding to produce secondary effects that may contribute to instrument background (e.g., Compton backscatter, generation of secondary photons and characteristic x rays, photoelectric absorption). Additional information on instrument background is available in Chapter 20 of *Radiation Detection and Measurement* (Knoll 1999).

B.5   Technologically Enhanced Naturally Occurring Radioactive Material

Technologically Enhanced Naturally Occurring Radioactive Material (TENORM) is any naturally occurring material not subject to regulation under the Atomic Energy Act whose radionuclide concentrations or potential for human exposure have been increased above levels encountered in the natural state by human activities (NAS 1999). Some industrial processes involving natural resources concentrate naturally occurring radionuclides, producing TENORM. Much TENORM contains only trace amounts of radioactivity and is part of our everyday landscape. Some TENORM, however, contains very high concentrations of radionuclides. The majority of radionuclides in TENORM are found in the uranium and thorium natural decay series. Potassium-40 is also associated with TENORM. Radium and radon typically are measured as indicators for TENORM in the environment. TENORM is found in many industrial waste streams (e.g., scrap metal, sludges, slags, fluids) and is being discovered in industries traditionally not thought of as being affected by radiation. Examples of products and processes affected by TENORM include:

- Uranium overburden and mine spoils,
- Phosphate industrial wastes,
- Phosphate fertilizers and potash,
- Coal ash, slag, cinders,
- Oil and gas production scale and sludge,
- Sludge and other waste materials from treatment of drinking water and waste water,
- Metal mining and processing waste,
- Geothermal energy production waste,
- Paper and pulp,
- Scrap metal recycling,
- Slag from industrial processes (metal and non-metal),
- Abrasive mineral sands, and
- Cement production.

Radon and radon progeny are concerns when dealing with TENORM. Radon-222 is a decay product of ^{238}U. The 3.8-day half-life means that ^{222}Rn is capable of migrating through several decimeters of soil or building materials and reaching the atmosphere before it decays. The radioactive progeny of unsupported ^{222}Rn have short half-lives (e.g., 27 minutes for ^{214}Pb) and usually decay to background levels within a few hours. ^{220}Rn, which has a 55-second half-life, is a decay product of ^{232}Th. The short half-life limits the mobility of ^{220}Rn since it decays before it can migrate to the atmosphere. However, ^{232}Th activity that is located on or very near the surface can produce significant quantities of ^{220}Rn in the air. The radioactive progeny of unsupported ^{220}Rn can result in elevated levels of surface radioactivity for materials and equipment used or stored in these areas. The 10.6-hour half-life of ^{212}Pb means that this surface radioactivity could take a week or longer to decay to background levels.

B.6   Orphan Sources

Radiation sources are found in certain types of specialized industrial devices, such as those used for measuring the moisture content of soil and for measuring density or thickness of materials. Usually, a small quantity of the radioactive material is sealed in a metal casing and enclosed in a housing that prevents the escape of radiation. These sources present no health risk from radioactivity as long as the sources remain sealed, the housing remains intact, and the devices are handled and used properly.

If equipment containing a sealed source is disposed of improperly or sent for recycling as scrap metal, the sealed source may be "lost" and end up in a metal recycling facility or in the possession of someone who is not licensed to handle the source. Specially licensed sources bear identifying markings that can be used to trace these sources to their original owners. However, some sources do not have these markings or the markings become obliterated. In these cases, the sources are referred to as "orphan sources" because no known owner can be identified. They are one of the most frequently encountered sources of radioactivity in shipments received by scrap metal facilities.

Scrap yards and disposal sites attempt to detect orphan sources and other contaminated metals by screening incoming materials with sensitive radiation detectors before they can enter the processing stream and cause contamination. Housings that make the sources safe also make detection difficult. Further, if the source is buried in a load of steel, the steel acts as further shielding and thus these sources may elude detection. Consequently, there is always a potential for sources to become mixed within and impact scrap metal.

C. EXAMPLES OF COMMON RADIONUCLIDES

Table C.1 Examples of Common Radionuclides at Selected Types of Facilities

Facility Type	Common Radionuclides
Accelerator/Cyclotron	^{22}Na Activation products (e.g., ^{60}Co)
Aircraft Manufacturing and Maintenance Facility	^{3}H (dials and gauges) Magnesium-thorium alloys Nickel-thorium alloys ^{147}Pm (lighted dials and gauges) ^{226}Ra and progeny (radium dials) Depleted uranium
Cement Production Facility	Thorium series radionuclides Uranium series radionuclides
Ceramic Manufacturing Facility	Thorium series radionuclides Uranium series radionuclides
Fertilizer Plant	^{40}K Uranium series radionuclides
Fuel Fabrication Facility	^{99}Tc (reprocessing only) Enriched uranium Transuranics (e.g., ^{237}Np, ^{239}Pu) (reprocessing only)
Gaseous Diffusion Plant	^{99}Tc Enriched uranium Transuranics (e.g., ^{237}Np, ^{239}Pu)
Medical Imaging and Therapy Facility	^{60}Co ^{90}Sr ^{99m}Tc ^{131}I ^{137}Cs ^{192}Ir ^{201}Tl ^{226}Ra and progeny Depleted uranium collimators
Metal Foundry	^{40}K ^{60}Co ^{137}Cs Thorium series radionuclides Uranium series radionuclides
Munitions and Armament Manufacturing and Testing Facility	^{3}H (fire control devices) ^{226}Ra and progeny Depleted uranium

Table C.1 Examples of Common Radionuclides at Selected Types of Facilities (continued)

Facility Type	Common Radionuclides
Nuclear Medicine Laboratory or Pharmaceutical Laboratory	^{99m}Tc ^{131}I ^{137}Cs ^{192}Ir ^{201}Tl ^{226}Ra and progeny
Nuclear Power Reactor	Activation products (e.g., ^{55}Fe, ^{60}Co, ^{63}Ni) Fission products (e.g., ^{90}Sr, ^{137}Cs) Transuranics (e.g., ^{237}Np, ^{239}Pu)
Oil and Gas	^{226}Ra and progeny
Optical Glass Facility	Thorium series radionuclides Uranium series radionuclides
Paint and Pigment Manufacturing Facility	Thorium series radionuclides Uranium series radionuclides
Paper and Pulp Facility	Thorium series radionuclides Uranium series radionuclides
Radium Dial Painting	^{226}Ra and progeny
Rare Earth Facility	^{40}K Thorium series radionuclides Uranium series radionuclides
R&D Facility with Broad Scope License	^{3}H ^{14}C
Research Laboratory	^{3}H ^{14}C ^{22}Na ^{24}Na ^{32}P ^{57}Co ^{63}Ni ^{123}I ^{125}I
Scrap Metal Recycling Facility	^{60}Co ^{90}Sr ^{137}Cs ^{226}Ra and progeny Thorium series radionuclides Uranium series radionuclides

Table C.1 Examples of Common Radionuclides at Selected Types of Facilities (continued)

Facility Type	Common Radionuclides
Sealed Source Facility	^{60}Co ^{90}Sr ^{137}Cs ^{241}Am
Transuranic Facility	^{241}Am $^{238, 239, 240, 241}$Pu
Uranium Mill	^{238}U ^{230}Th ^{226}Ra and progeny Thorium series radionuclides Uranium series radionuclides
Waste Water Treatment Facility	Thorium series radionuclides Uranium series radionuclides
Widely Distributed General Commerce	^{3}H (exit signs) ^{40}K (naturally-occurring) ^{57}Co (lead paint analyzer) ^{60}Co (radiography source) ^{63}Ni (chemical agent detectors) ^{109}Cd (lead paint analyzer) ^{137}Cs (soil moisture density gauge, liquid level gauge) ^{192}Ir (radiography source) ^{226}Ra (watch dials) ^{241}Am (AmBe soil moisture density gauge, smoke detectors) Orphan sources

D. INSTRUMENTATION AND MEASUREMENT TECHNIQUES

D.1   Introduction

This appendix provides information on various field and laboratory equipment used to measure radiation levels and radioactive material concentrations. The descriptions provide information pertaining to the general types of available radiation detectors and the ways in which those detectors are utilized for various circumstances. Similar information may be referenced from MARSSIM Appendix H, "Description of Field Survey and Laboratory Analysis Equipment" (MARSSIM 2002), and NUREG-1761 Appendix B, "Advanced/Specialized Information" (NRC 2002). The information in this appendix is specifically designed to assist the user in selecting the appropriate radiological instrumentation and measurement technique during the implementation phase of the data life cycle (Chapter 5).

The following topics will be discussed for each instrumentation and measurement technique combination:

- **Instruments** – A description of the equipment and the typical detection instrumentation it employs;
- **Temporal Issues** – A synopsis of time constraints that may be encountered through use of the measurement technique;
- **Spatial Issues** – Limitations associated with the size and portability of the instrumentation as well as general difficulties that may arise pertaining to source-to-detector geometry;
- **Radiation Types** – Applicability of the measurement technique for different types of ionizing radiation;
- **Range** – The associated energy ranges for the applicable types of ionizing radiation;
- **Scale** – Typical sizes for the M&E applicable to the measurement technique; and
- **Ruggedness** – A summary of the durability of the instrumentation (note that this is frequently limited by the detector employed by the instrumentation; e.g., an instrument utilizing a plastic scintillator is inherently more durable than an instrument utilizing a sodium iodide crystal); suitable temperature ranges for proper operation of the instrumentation and measurement technique have been provided where applicable.

D.2   General Detection Instrumentation

This section summarizes the most common detector types used for the detection of ionizing radiation in the field. This will include many of the detector types incorporated into the measurement methods that are described in later sections of this chapter.

D.2.1   Gas-Filled Detectors

Gas-filled detectors are the most commonly used radiation detectors and include gas- ionization chamber detectors, gas-flow proportional detectors, and Geiger-Muller (GM) detectors. These detectors can be designed to detect alpha, beta, photon, and neutron radiation. They generally consist of a wire passing through the center of a gas-filled chamber with metal walls, which can be penetrated by photons and high-energy beta particles. Some chambers are fitted with Mylar windows to allow penetration by alpha and low-energy beta radiation. A voltage source is connected to the detector with the positive terminal connected to the wire and the negative terminal connected to the chamber casing to generate an electric field, with the wire serving as the anode, and the chamber casing serving as the cathode. Radiation ionizes the gas as it enters the chamber, creating free electrons and positively charged ions. The number of electrons and positively charged ions created is related to the properties of the incident radiation type (alpha particles produce many ion pairs in a short distance, beta particles produce fewer ion pairs due to their smaller size, and photons produce relatively few ion pairs as they are uncharged and interact with the gas significantly less than alpha and beta radiation). The anode attracts the free electrons while the cathode attracts the positively charged ions. The reactions among these ions and free electrons with either the anode or cathode produce disruptions in the electric field. The voltage applied to the chamber can be separated into different voltage ranges that distinguish the types of gas-filled detectors described below. The different types of gas-filled detectors are described in ascending order of applied voltage.

D.2.1.1 Ionization Chamber Detectors

Ionization chamber detectors consist of a gas-filled chamber operated at the lowest voltage range of all gas-filled detectors.[1] Ionization detectors utilize enough voltage to provide the ions with sufficient velocity to reach the anode or cathode. The signal pulse heights produced in ionization chamber detectors is small and can be discerned by the external circuit to differentiate among different types of radiation. These detectors provide true measurement data of energy deposited proportional to the charge produced in air, unlike gas-flow proportional and GM detectors which are detection devices. These detectors generally are designed to collect cumulative beta and photon radiation without amplification and many have a beta shield to help distinguish among these radiation types. These properties make ionization detectors excellent choices for measuring exposure rates from photon emission radiation in roentgens. These detectors can be deployed for an established period of time to collect data in a passive manner for disposition surveys. Ionization chamber detectors may assist in collecting measurements in inaccessible areas due to their availability in small sizes.

Another form of the ionization chamber detector is the pressurized ion chamber (PIC). As with other ionization chamber detectors, the PIC may be applied for M&E disposition surveys when a exposure-based action level is used. The added benefit of using PICs is that they can provide more accurate dose measurements because they compensate for the various levels of photon energies as opposed to other exposure rate meters (e.g., micro-rem meter), which are calibrated to a ^{137}Cs source. PICs can be used to cross-calibrate other exposure rate detectors applicable for surveying M&E, allowing the user to compensate for different energy levels and reduce or eliminate the uncertainty of underestimating or overestimating the exposure rate measurements.

D.2.1.2 Gas-Flow Proportional Detectors

The voltage applied in gas-flow proportional detectors is the next range higher than ionization chamber detectors, and is sufficient to create ions with enough kinetic energy to create new ion

---

[1] At voltages below the ionization chamber voltage range, ions will recombine before they can reach either the cathode or anode and do not produce a discernable disruption to the electric field.

pairs, called secondary ions. The quantity of secondary ions increases proportionally with the applied voltage, in what is known as the gas amplification factor. The signal pulse heights produced can be discerned by the external circuit to differentiate among different types of radiation. Gas-flow proportional detectors generally are used to detect alpha and beta radiation. Systems also detect photon radiation, but the detection efficiency for photon emissions is considerably lower than the relative efficiencies for alpha and beta activity. Physical probe areas for these types of detectors vary in size from approximately 100 cm^2 up to 600 cm^2. The detector cavity in these instruments is filled with P-10 gas which is an argon-methane mixture (90% argon and 10% methane). Ionizing radiation enters this gas-filled cavity through an aluminized Mylar window. Additional Mylar shielding may be used to block alpha radiation; a lower voltage setting may be used to detect pure alpha activity (NRC 1998b).

D.2.1.3 Geiger-Mueller Detectors

GM detectors operate in the voltage range above the proportional range and the limited proportional range.[2] This range is characterized by extensive gas amplification that results in what is referred to as an "avalanche" of ion and electron production. This mass production of electrons spreads throughout the entire chamber, which precludes the ability to distinguish among different kinds of radiation because all of the signals produced are the same size. GM detectors are most commonly used for the detection of beta activity, though they may also detect both alpha and photon radiation. GM detectors have relatively short response and dead times and are sensitive enough to broad detectable energy ranges for alpha, beta, photon, and neutron emissions (though they cannot distinguish which type of radiation produces input signals) to allow them to be used for surveying M&E with minimal process knowledge.[3]

GM detectors are commonly divided into three classes: "pancake", "end-window", and "side-wall" detectors. GM pancake detectors (commonly referred to as "friskers") have wide diameter, thin mica windows (approximately 15 cm^2 window area) that are large enough to allow them to be used to survey many types of M&E. Although GM pancake detectors are referenced beta and gamma detectors, the user should consider that their beta detection efficiency far exceeds their gamma detection efficiency. The end-window detector uses a smaller, thin mica window and is designed to allow beta and most alpha particles to enter the detector unimpeded for concurrent alpha and beta detection. The side-wall detector is designed to discriminate between beta and gamma radiation, and features a door that can be slid or rotated closed to shield the detector from beta emissions for the sole detection of photons. These detectors require calibration to detect for beta and gamma radiation separately. Energy-compensated GM detectors may also be cross-calibrated for assessment of exposure rates.

---

[2] The limited proportional range produces secondary ion pairs but does not produce reactions helpful for radiation detection, because the gas amplification factor is no longer constant.

[3] GM detectors may be designed and calibrated to detect alpha, beta, photon, and neutron radiation, though they are much better-suited for the detection of charged particles (i.e., alpha and beta particles) than neutral particles (i.e., photons and neutrons).

D.2.2   Scintillation Detectors

Scintillation detectors (sometimes referred to as "scintillators") consist of scintillation media that emits a light "output" called a scintillation pulse when it interacts with ionizing radiation. Scintillators emit low-energy photons (usually in the visible light range) when struck by high-energy charged particles; interactions with external photons cause scintillators to emit charged particles internally, which in turn interact with the crystal to emit low-energy photons. In either case, the visible light emitted (i.e., the low-energy photons) are converted into electrical signals by photomultiplier tubes and recorded by a digital readout device. The amount of light emitted is generally proportional to the amount of energy deposited, allowing for energy discrimination and quantification of source radionuclides in some applications.

D.2.2.1   Zinc Sulfide Scintillation Detectors

Zinc sulfide detector crystals are only available as a polycrystalline powder that are arranged in a thin layer of silver-activated zinc sulfide (ZnS(Ag)) as a coating or suspended within a layer of plastic scintillation material. The use of these thin layers makes them inherently dispositioned for the detection of high linear energy transfer (LET) radiation (radiation associated with alpha particles or other heavy ions). These detectors use an aluminized Mylar window to prevent ambient light from activating the photomultiplier tube (Knoll 1999). The light pulses produced by the scintillation crystals are amplified by a photomultiplier tube, converted to electrical signals, and counted on a digital scaler/ratemeter. Low LET radiations (particularly beta emissions) are detected at much lower detection efficiencies than alpha emissions and pulse characteristics may be used to discriminate beta detections from alpha detections.

D.2.2.2   Sodium Iodide Scintillation Detectors

Sodium iodide detectors are well-suited for detection of photon radiation. Energy-compensated sodium iodide detectors may also be cross-calibrated for assessment of exposure rates. Unlike ZnS(Ag), sodium iodide crystals can be grown relatively large and machined into varying shapes and sizes. Sodium iodide crystals are activated with trace amounts of thallium (hence the abbreviation NaI(Tl)), the key ingredient to the crystal's excellent light yield (Knoll, 1999). These instruments most often have upper- and lower-energy discriminator circuits and when used correctly as a single-channel analyzer, can provide information on the photon energy and identify the source radionuclides. Sodium iodide detectors can be used with handheld instruments or large stationary radiation monitors.

D.2.2.3   Cesium Iodide Scintillation Detectors

Cesium iodide detectors generally are similar to sodium iodide detectors. Like NaI(Tl), cesium iodide may be activated with thallium (CsI(Tl)) or sodium (CsI(Na)). Cesium iodide is more resistant to shock and vibration damage than NaI, and when cut into thin sheets it features malleable properties allowing it to be bent into various shapes. CsI(Tl) has variable decay times for various exciting particles, allowing it to help differentiate among different types of ionizing radiation. A disadvantage of CsI scintillation detectors is due to the fact that the scintillation emission wavelengths for CsI are longer than those produced by sodium iodide crystals; because almost all photomultiplier tubes are designed for NaI, there are optical incompatibilities that result in decreased intrinsic efficiencies for CsI detectors. Additionally, CsI scintillation detectors feature relatively long response and decay times for luminescent states in response to ionizing radiation (Knoll 1999).

D.2.2.4 Plastic Scintillation Detectors

Plastic scintillators are composed of organic scintillation material that is dissolved in a solvent and subsequently hardened into a solid plastic. Modifications to the material and specific packaging allow plastic scintillators to be used for detecting alpha, beta, photon, or neutron radiation. While plastic scintillators lack the energy resolution of sodium iodide and some other gamma scintillation detector types, their relatively low cost and ease of manufacturing into almost any desired shape and size enables them to offer versatile solutions to atypical radiation detection needs (Knoll 1999).

D.2.3   Solid State Detectors

Solid state detection is based on ionization reactions within detector crystals composed of an electron-rich (n-type or electron conductor) sector and an electron-deficient (p-type or hole conductor) sector. Reverse-bias voltage is applied to the detector crystal; forming a central region absent of free charge (this is termed the depleted region). When a particle enters this region, it interacts with the crystal structure to form hole-electron pairs. These holes and electrons are swept out of the depletion region to the positive and negative electrodes by the electric field, and the magnitude of the resultant pulse in the external circuit is directly proportional to the energy lost by the ionizing radiation in the depleted region.

Solid state detection systems typically employ silicon or germanium crystals[4] and utilize semiconductor technology (i.e., a substance whose electrical conductivity falls between that of a metal and that of an insulator, and whose conductivity increases with decreasing temperature and with the presence of impurities). Semiconductor detectors are cooled to extreme temperatures to utilize the crystal material's insulating properties to prevent thermal generation of noise. The use of semiconductor technology can achieve energy resolutions, spatial resolutions, and signal-to-noise ratios superior to those of scintillation detection systems.

D.3   Counting Electronics

Instrumentation requires a device to accumulate and record the input signals from the detector over a fixed period of time. These devices are usually electronic, and utilize scalers or rate-meters to display results representing the number of interaction events (between the detector and radionuclide emissions) within a period of time (e.g., counts per minute). A scaler represents the total number of interactions within a fixed period of time, while a rate-meter provides information that varies based on a short-term average of the rate of interactions.

---

[4] Solid state detection systems may also utilize crystals composed of sodium iodide, cesium iodide, or cadmium zinc telluride in non-semiconductor applications.

Scalers represent the simpler of these two counting approaches, because they record a single count each time an input signal is received from the detector. Scaling circuits typically are designed with scalers to allow the input signals to be cut by factors of 10, 100, or 1,000 to allow the input signals to be counted directly by electromechanical registers when counting areas with elevated radioactivity. Scalers generally are used when taking in situ measurements and are used to determine average activities.

Contemporary rate-meters utilize analog-to-digital converters to sample the pulse amplitude of the input signal received from the detector and convert it to a series of digital values. These digital values may then be manipulated using digital filters (or shapers) to average or "smooth" the data displayed. The counting-averaging technique used by rate-meters may be more helpful than scalers in identifying elevated activity. When using scalers in performing scanning surveys to locate areas of elevated activity, small areas of elevated activity may appear as very quick "blips" that are difficult to discern, while rate-meters continue to display heightened count rates once the detector has moved past the elevated activity, and display "ramped up" count rates immediately preceding the elevated activity as well. Rate-meters have the inherent limitation in that the use of their counting electronics varies the signals displayed by the meter because they represent a short-term average of the event rate. It is conceivable that very small areas of elevated activity (e.g., particle) might have their true activity concentrations "diluted" by the averaging of rate-meter counting electronics.

D.4   Hand-Held Instruments

This section discusses hand-held instruments, which may be used for in situ measurements or scanning surveys.

D.4.1   Instruments

In situ measurements with hand-held instruments typically are conducted using the detector types described in Section D.2. These typically are composed of a detection probe (utilizing a single detector) and an electronic instrument to provide power to the detector and to interpret data from the detector to provide a measurement display.

The most common types of hand-held detector probes are GM detectors, ZnS(Ag) alpha/beta scintillation detectors, and NaI(Tl) photon scintillation detectors. There are instances of gas-flow proportional detectors as hand-held instruments, though these are not as common because these detectors operate using a continuous flow of P-10 gas, and the accessories associated with the gas (e.g., compressed gas cylinders, gauges, tubing) make them less portable for use in the field.

D.4.2   Temporal Issues

Hand-held instruments generally have short, simple equipment set-ups requiring minimal time, often less than ten minutes. In situ measurement count times typically range from 30 seconds to two minutes. Longer count times may be utilized to increase resolution and provide lower minimum detectable limits. Typical scanning speeds are approximately 2.5 centimeters per second. Slower scanning speeds will aid in providing lower minimum detectable concentrations.

D.4.3   Spatial Issues

Detectors of hand-held instruments typically are small and portable, having little trouble fitting into and measuring most M&E. Spatial limitations are usually based on the physical size of the probe itself. The user must be wary of curved or irregular surfaces of M&E being surveyed. Detector probes generally have flat faces and incongruities between the face of the detector and the M&E being surveyed have an associated uncertainty. ZnS scintillation and gas-flow proportional detectors are known to have variations in efficiency of up to 10% across the face of the detector. Therefore, the calibration source used should have an area at least the size of the active probe area.

D.4.4   Radiation Types

Assortments of hand-held instruments are available for the detection of alpha, beta, photon, and neutron radiations. Table D.1 illustrates the potential applications for the most common types of hand-held instruments.

Table D.1 Potential Applications for Common Hand-Held Instruments

| | | | | | Detectable Energy Range | |
	Alpha	Beta	Photon	Neutron	Low End Boundary	High End Boundary
Ionization chamber detectors	NA	FAIR	GOOD	NA	40-60 keV	1.3-3 MeV
Gas-flow proportional detectors	GOOD	GOOD	POOR	POOR	5-50 keV	8-9 MeV
Geiger-Muller detectors	FAIR	GOOD	POOR	POOR	30-60 keV	1-2 MeV
ZnS(Ag) scintillation detectors	GOOD	POOR	NA	NA	30-50 keV	8-9 MeV
NaI(Tl) scintillation detectors	NA	POOR	GOOD	NA	40-60 keV	1.3-3 MeV
NaI(Tl) scintillation detectors (thin detector, thin window)	NA	FAIR	GOOD	NA	10 keV	60-200 keV
CsI(Tl) scintillation detectors	NA	POOR	GOOD	NA	40-60 keV	1.3-3 MeV
Plastic scintillation detectors	NA	FAIR	GOOD	NA	40-60 keV	1.3-3 MeV
BF$_3$ proportional detectors[5]	NA	NA	NA	GOOD	0.025 eV	100 MeV
^{3}He proportional detectors[5]	NA	NA	POOR	GOOD	0.025 eV	100 MeV

Notes:
GOOD   The instrument is well-suited for detecting this type of radiation.
FAIR   The instrument can adequately detect this type of radiation.
POOR   The instrument may be poorly suited for detecting this type of radiation.
NA   The instrument cannot detect this type of radiation.

---

[5] The use of moderators enables the detection of high-energy fast neutrons. Either BF$_3$ or ^{3}He gas proportional detectors may be used for the detection of fast neutrons, but ^{3}He are much more efficient in performing this function. BF$_3$ detectors discriminate against gamma radiation more effectively than ^{3}He detectors.

D.4.5   Range

The ranges of detectable energy using hand-held instruments are dependent upon the type of instrument selected and type of radiation. Some typical detectable energy ranges for common hand-held instruments are listed above in Table D.1. More detailed information pertaining to the ranges of detectable energy using hand-held instruments are available in the European Commission for Nuclear Safety and the Environment Report 17624 (EC 1998).

D.4.6   Scale

There is no definitive limit to the size of an object to be surveyed using hand-held instruments. Hand-held instruments may generally be used to survey M&E of any size; constraints are only placed by the practical sizing of M&E related to the sensitive area of the probe. Limitations may also be derived from the physical size of the detector probes used for surveying. The largest hand-held detector probes feature effective detection surface areas of approximately 175 to 200 cm^2. Detection probes larger than this may be of limited use with hand-held instruments.

D.4.7   Ruggedness

All varieties of hand-held instruments discussed here typically are calibrated for use in temperatures with lower ranges from -30 ° to -20 °C and upper ranges from 50 ° to 60 °C. The durability of a hand-held instrument depends largely upon the detection media (crystals, such as sodium iodide and germanium crystals are fragile and vulnerable to mechanical and thermal shock) and the presence of a Mylar (or similar material) window:

- **Ionization chamber detectors** – Ionization chamber detectors are susceptible to physical damage and may provide inaccurate data (including false positives) if exposed to mechanical shock.
- **Gas-flow proportional detectors** – Detection gas used with gas-flow proportional detectors may leak from seals such that these detectors are usually operated in the continuous gas flow mode; the use of flow meter gauges to continuously monitor the gas flow rate is recommended along with frequent quality control checks to ensure the detector still meets the required sensitivity; gas-flow proportional detectors may also use fragile Mylar windows to contain the detection gases, which renders the detectors vulnerable to puncturing and mechanical shock.
- **Geiger-Muller detectors** – GM tubes typically use fragile Mylar windows to contain the detection gases; the presence of a Mylar window renders the detector vulnerable to puncturing and mechanical shock.
- **ZnS(Ag) scintillation detectors** – Zinc sulfide is utilized as thin-layer polycrystalline powder in detectors and are noted for being vulnerable to mechanical shock; zinc sulfide detectors may use fragile Mylar windows, in which case the detector is vulnerable to puncturing and mechanical shock.
- **NaI(Tl) scintillation detectors** – Sodium iodide crystals are relatively fragile and can be damaged through mechanical shock; sodium iodide is also highly hydroscopic such that the crystals must remain environmentally sealed within the detector housing.

- **Plastic Scintillation Detectors** – Plastic scintillators typically are robust and resistant to damage from mechanical and thermal shock.

D.5   Volumetric Counters (Drum, Box, Barrel, Four-Pi Counters)

The term "box counter" is a generic description for a radiation measurement system that typically involves large area, four-pi (4-π) radiation detectors and includes the following industry nomenclature: tool counters, active waste monitors, surface activity measurement systems, and bag/barrel/drum monitors. Box counting systems are most frequently used for conducting in situ surveys of M&E that is utilized in radiologically controlled areas. These devices are best-suited for performing gross activity screening measurements on Class 2 and Class 3 M&E (NRC 2002). Typical items to be surveyed using box counters are hand tools, small pieces of debris, bags of trash, and waste barrels. Larger variations of box counting systems can count objects up to a few cubic meters in size. Because of potential problems with self-shielding, materials may need to be opened or partially disassembled prior to placing into a box counting system.

D.5.1   Instruments

Box counting systems typically consist of a counting chamber, an array of detectors configured to provide a 4-π counting geometry, and microprocessor-controlled electronics that allow programming of system parameters and data-logging. Systems typically survey materials for photon radiation and usually incorporate a shielded counting chamber and scintillation detectors (plastic scintillators or sodium iodide scintillation detectors). These systems most commonly utilize four or six detectors, which are situated on the top, bottom, and sides of the shielded counting chamber (Figure D.1). Some systems monitor M&E for beta activity, using a basic design similar to photon radiation detection systems, but utilizing gas-flow proportional counters. In rare cases, neutron detection has been used for criticality controls and counter-proliferation screening.

Box counting systems for alpha activity feature a substantial departure in design from beta/gamma detection systems. Alpha activity systems do not require heavy shielding to filter out ambient sources of radiation. These devices utilize air filters to remove dust and particulates from air introduced into counting chambers that incorporate airtight seals. Filtered air introduced into the counting chamber interacts with any surface alpha activity associated with the M&E.

Each alpha interaction with a surrounding air molecule produces an ion pair. These

Figure D-1 Example Volumetric Counter (Thermo 2005)

ion pairs are produced in proportion to the alpha activity per unit path length. This air (i.e., the ion pairs in the air) is then counted using an ion detector for quantification of the specific activity. The specific activity of the air in the counting chamber provides a total surface activity quantification for the M&E (BIL 2005).

D.5.2 Temporal Issues

Typically, box counting systems require approximately one to 100 seconds to conduct a measurement (Thermo 2005). The count times are dependent on a number of factors to include required measurement sensitivity and background count rates with accompanying subtraction algorithms. The count times for box counting typically are considered relatively short for most disposition surveys.

D.5.3 Spatial Issues

Because box counters typically average activity over the volume or mass of the M&E, the spatial distribution of radioactivity may be a significant limitation on the use of this measurement technique. The design of box counting systems is not suited to the identification of localized elevated areas, and therefore may not be the ideal choice when the disposition criteria is not based on average or total activity.

Some systems incorporate a turntable inside the counting chamber to improve measurement of difficult-to-measure areas or for heterogeneously distributed radioactivity. When practical, performing counts on objects in two different orientations (i.e., by rotating the M&E 90 or 180 degrees and performing a subsequent count) will yield more thorough and defensible data. Proper use of box counters includes segregating the M&E to be surveyed and promoting accurate measurements through uniform placement of items to be surveyed in the counting chamber. For example, a single wrench placed on its side in a box counter has different geometric implications from a tool of similar size standing up inside the counting chamber. Counting jigs for sources and M&E to be surveyed are frequently employed to facilitate consistent, ideal counting positions between the M&E and the counting chamber detector array.

D.5.4 Radiation Types

Box counting systems are intrinsically best-suited for the detection of moderate- to high- energy photon radiation. As described in Section D.5.1, specific systems may be designed for the detection of low-energy photon, beta, alpha, and in some cases neutron radiation. For proper calibration and utilization of box counters, it is often necessary to establish the radiation types and anticipated energy ranges prior to measurement.

D.5.5 Range

Photon radiation can typically be measured within a detectable energy range of 40 to 60 keV up to 1.3 to 3 MeV. For example, typical box counters positioned at radiological control area exit points are configured to alarm at a set point of 5,000 dpm total activity. The precise count time is adjusted automatically by setting the predetermined count rate to limit the error. Measurement times will range from 5 to 45 seconds in order to complete counts of this kind, depending on current background conditions (Thermo 2005). Lower detection capabilities are achievable by increasing count times or incorporating background reduction methodologies.

D.5.6 Scale

Size limitations pertaining to the M&E to be surveyed are inherently linked to the physical size of the counting chamber. Smaller box counting systems have a counting chamber of less than 0.028 cubic meters (approximately one cubic foot) and are often used for tools and other frequently used small items. The maximum size of box counters is typically driven by the logistics of managing the M&E to be measured, and this volume is commonly limited to a 55-gallon waste drum. Some box counting systems allow counts to be performed on oversized items protruding from the counting chamber with the door open.

D.5.7 Ruggedness

Many volumetric counter models feature stainless steel construction with plastic scintillation detectors and windowless designs, which translates to a rugged instrument that is resistant to mechanical shock.

D.6 Conveyorized Survey Monitoring Systems

Conveyorized survey monitoring systems automate the routine scanning of M&E. Conveyorized survey monitoring systems have been designed to measure materials such as soil, clothing (laundry monitors), copper chop (small pieces of copper), rubble, and debris. Systems range from small monitoring systems comprised of a single belt that passes materials through a detector array, to elaborate multi-belt systems capable of measuring and segregating material while removing extraneously large items. The latter type comprises systems that are known as segmented gate systems. These automated scanning systems segregate materials by activity by directing material that exceeds an established activity level onto a separate conveyor. Simpler conveyorized survey monitoring systems typically feature an alarm/shut-down feature that halts the conveyor motor and allows for manual removal of materials that have exceeded the established activity level.

D.6.1 Instruments

A typical conveyorized survey monitoring system consists of a motorized conveyor belt that passes materials through an array of detectors, supporting measurement electronics, and an automated data-logging system (Figure D.2). Systems typically survey materials for photon radiation and usually incorporate scintillation detectors (plastic scintillators or sodium iodide scintillation detectors) or high-purity germanium detectors. Scintillation detector arrays are often chosen for gross gamma activity screening. Conveyorized survey monitoring systems designed to detect radionuclide mixtures with a high degree of process knowledge work best using plastic scintillators, while systems categorizing material mixtures where the radionuclide concentrations are variable are better-suited to the use of sodium iodide scintillation detectors. Conveyorized survey monitoring systems designed for material mixtures where the radionuclide concentrations are unknown may be suitable for more expensive and maintenance-intensive high-purity germanium detector arrays, which will allow for quantitative measurement of complex photon energy spectra. An alternative method for screening materials for different photon energy regions of interest is to incorporate sodium iodide detector arrays with crystals of varying thickness to target multiple photon energies. Systems may also be fitted with gas flow proportional counters for the detection of alpha and beta emissions. Laundry conveyorized survey monitoring systems typically are designed for the detection of alpha and beta radiation, as the nature of clothes allows the survey media to be compressed, allowing the detector arrays to be close to or in contact with the survey media.

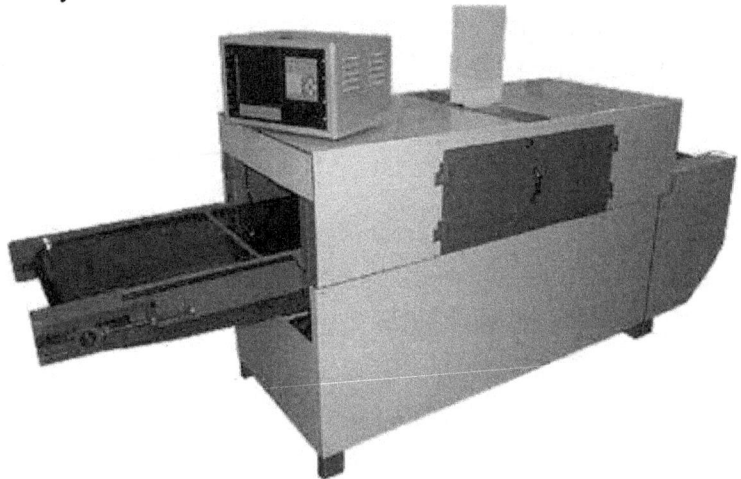

Figure D-2 Example Conveyorized Survey Monitoring System (Laurus 2001)

D.6.2 Temporal Issues

Typically, conveyorized survey monitoring systems require approximately one to six seconds to count a given field of detection (Novelec 2001a). Systems are designed to provide belt speeds ranging from 0.75 meters up to 10 meters (2.5 to 33 feet) per minute to accommodate the necessary response time for detection instrumentation (Thermo 2008; Eberline 2004). This yields processing times of 15 to 45 metric tons (16 to 50 tons) of material per hour for soil or construction demolition-type material conveyorized survey monitoring systems (NRC 2002).

D.6.3 Spatial Issues

The M&E that typically are surveyed by conveyorized survey monitoring systems may contain difficult-to-measure areas. Most systems employ the detector arrays in a staggered, off-set configuration, which allows the sensitive areas of the detectors to overlap with respect to the direction of movement. This off-set configuration helps to eliminate blind spots (i.e., locations where activity may be present but cannot be detected because the radiation cannot reach the detectors). Some systems are designed specifically for materials of relatively small particles of uniform size (e.g., soil), while others have been designed to accommodate heterogeneous materials like rubble and debris.

The data logging system accepts the signal pulses from the detector systems and stores the pulse data in counting scalers. The recorded values are continuously compared with pre-set alarm values corresponding to the selected action level(s). The detectors incorporate integral amplifiers which are routed to a PC containing multi-channel scaler hardware. The multi-channel scaler hardware allows data to be collected in a series of short, discrete scaler channels known as "time bins". The count time for each time bin is selected as a function of the speed of the conveyor belt. The time bin length is frequently set up to be half the length of "dwell time," which is the time the material aliquot to be surveyed spends within the detection field (Miller 2000).

The approach cited in the paragraph above ensures that activity present within the survey unit will be in full view of the detector for one complete time bin. Data collection is optimized by performing the measurement when the activity is concentrated (i.e., within an area of elevated activity) as well as when the activity is approximately homogenously distributed within a given material aliquot.

D.6.4 Radiation Types

Conveyorized survey monitoring systems generally are best-suited for the detection of photon radiation. Specific systems may be tailored for the detection of beta emissions of moderate energy and even alpha radiation by employing gas flow proportional counter detector arrays.

D.6.5 Range

Photon radiation can typically be measured with a detectable energy range from 50 keV up to 2 MeV. Conveyorized survey monitoring systems equipped to measure alpha and beta emissions can typically measure from 100 keV up to 6 MeV.

D.6.6 Scale

Most conveyorized survey monitoring systems are designed for soils or laundry, both of which are compressible media. Applicable sample/material heights range from 2 cm to 30 cm (Fuji 2008, Canberra 2008).

D.6.7 Ruggedness

Conveyorized survey monitoring systems have typical operating ranges from −20 °C to 50 °C. Conveyorized survey monitoring systems are often constructed from steel and with plastic scintillation detectors and windowless designs, which makes them generally resistant to damage from extraneous pieces of debris during scanning. Mechanical shock is not a typical concern for conveyorized survey monitoring systems because there is little need for moving these systems. For this reason conveyorized survey monitoring systems are seldom transported from one location to another.

D.7   In Situ Gamma Spectroscopy

In situ gamma spectroscopy (ISGS) systems combine the peak resolution capabilities of laboratory methods with instrumentation that is portable and rugged enough to be used in field conditions. These solid state systems can perform quantitative, multi-channel analysis of gamma-emitting isotopes in both solid and liquid media over areas as large as 100 m², enabling spectrographic analysis of M&E that assists the user in identifying constituent radionuclides and differentiating them from background radiation. ISGS system measurements can also provide thorough coverage within broad survey areas, minimizing the risk of failing to detect isolated areas of elevated radioactivity that could potentially be missed when collecting discrete samples.

D.7.1   Instruments

ISGS systems consist of a semiconductor detector, a cryostat, a multi-channel analyzer (MCA) electronics package that provides amplification and analysis of the energy pulse heights, and a computer system for data collection and analysis. Semiconductor detection systems typically employ a cryostat and a Dewar filled with liquid nitrogen (−196 °C). The cryostat transmits the cold temperature of the liquid nitrogen to the detector crystal, creating the extreme cold environment necessary for correct operation of the high-resolution semiconductor diode. ISGS systems may have electronic coolers as well.

ISGS systems use detectors referred to as N- and P-type detectors. N-type detectors contain small amounts of elements with five electrons in their outer electron shell (e.g., phosphorus, arsenic) within the germanium crystal (the inclusion of these elements within the germanium crystal is called "doping"). These result in free, unbonded electrons in the crystalline structure, providing a small negative current. P-type detectors utilize elements with less than four electrons in their outer electron shell (e.g., lithium, boron, gallium) are also used in doping to create electron holes, providing a small positive current. Use of these two varieties of doped germanium crystals provide different detection properties described below in Section D.7.5.

D.7.2   Temporal Issues

Setup for ISGS semiconductor systems may require one full day. The systems often require one hour to set up physically, six to eight hours for the semiconductor to reach the appropriate temperature operating range after the addition of liquid nitrogen, and quality control measurements may require another hour.[6] Count times using ISGS semiconductor systems tend to be longer than those associated with simpler detector systems for conducting static measurements, though this may be offset by enlarging the field-of-view. A measurement time of several minutes is common, depending on the intensity of the targeted gamma energies and the presence of attenuating materials.

Count times can be shortened by reducing the distance between the area being surveyed and the detector to improve the gamma incidence efficiency or by using a larger detector. Each option will ultimately help the detection system see more gamma radiation in a shorter time. Yet either

---

[6] It is important not to move the apparatus prematurely, as failure to allow the ISGS system to cool and equilibrate to its proper operating temperatures as may cause damage to the semiconductor detector.

approach creates greater uncertainty associated with the source-to-detector geometry. A slight placement error (e.g., a 0.5-cm placement error) will result in significantly higher quantification error at a distance of one centimeter than at a distance of 10 centimeters. Additionally, this technique for decreasing count times promotes an effect called cascade summing, a phenomena affecting detection of gamma radiation from radionuclides that emit multiple gamma photons in a single decay event (e.g., ^{60}Co, which yields gamma particles of 1.17 and 1.33 MeV). If both incident gammas deposit their energy in a relatively short period of time (i.e., when compared to the detector response time and/or the resolving time for the associated electronics), limitations of the detection system may prevent these individual photons from being distinguished (Knoll 1999).

D.7.3 Spatial Issues

ISGS semiconductor systems require calibration for their intended use. While ISGS semiconductor systems can be calibrated using traditional prepared radioactive sources, some ISGS systems have software that enables the user to calculate efficiencies by entering parameters such as elemental composition, density, stand-off distance, and physical dimensions. Supplied geometry templates assist in generating calibration curves that can be applied to multiple collected spectra. The high resolution of these systems coupled with advanced electronic controls for system parameters allows them to overcome issues related to source-to-detector geometry and produce quantitative concentrations of multiple radionuclides in a variety of media (e.g., soil, water, air filters). Because ISGS systems integrate all radioactivity within their field-of-view, lead shielding and collimation may be required to "focus" the field-of-view on a specified target for some applications.

D.7.4 Radiation Types

ISGS systems can accurately identify and quantify only photon-emitting radionuclides.

D.7.5 Range

ISGS systems can identify and quantify low-energy gamma emitters (50 keV with P-type detectors, 10 keV with N-type detectors) and high-energy gamma emitters (ISGS systems can be configured to detect gamma emissions upwards of 2.0 MeV). Specially designed germanium detectors that exhibit very little deterioration in resolution as a function of count rate use N-type detectors or planar crystals with a very thin beryllium window for the measurement of photons in the energy range 5 to 80 keV.

D.7.6 Scale

These systems therefore offer functional quantitative abilities to analyze small objects (e.g., samples) for radionuclides. They can also effectively detect radioactivity over areas as large as 100 m^2 or more (Canberra 2005a). With the use of an appropriate Dewar, the detector may be used in a vertical orientation to determine gamma isotope concentrations in the ground surface and shallow subsurface.

D.7.7 Ruggedness

ISGS semiconductor systems are fragile, because the extremely low temperatures utilized by the cryostat render portions of the system brittle and susceptible to damage if not handled with care. Some ISGS systems are constructed of more rugged materials and their durability is comparable to most hand-held instruments.

D.8 Hand-Held Radionuclide Identifiers

Hand-held radionuclide identifiers represent a relatively new addition to the radiation detection market, merging the portability of hand-held instruments with some of the analytical capabilities of ISGS systems. Hand-held radionuclide identifiers also feature data logging and storage capabilities (including user-definable radionuclide libraries) and the ability to transfer data to external devices. These devices are most commonly used for nuclear non-proliferation, where immediate isotope identification is more critical than low-activity detection sensitivity. Design parameters for hand-held radionuclide identifiers required by ANSI N42.34 (ANSI 2003) are user-friendly controls and intuitive menu structuring for routine modes of operation, enabling users without health physics backgrounds (e.g., emergency response personnel) to complete basic exposure rate or radionuclide identification surveys. These units also feature restricted "expert" survey modes of operation to collect activity concentration data for more advanced applications, including disposition surveys.

D.8.1 Instruments

Hand-held radionuclide identifiers consist of two general types: integrated systems and modular systems. The integrated systems have the detector and electronics contained in a single package; modular systems separate the detector from the electronics. These spectrometers employ small scintillators, typically NaI(Tl) or CsI(Tl), or room temperature solid semiconductors, such as cadmium zinc telluride (CZT), linked to multi-channel analyzers and internal radionuclide libraries to enable gamma-emitting radionuclide identification.

D.8.2 Temporal Issues

Hand-held radionuclide identifiers require minimal time to set up.[7] Depending upon the conditions in which data is being collected (i.e., climatic, environmental, the presence of sources of radiological interference), it may require seconds to several minutes for the unit to stabilize the input signals from the field of radiation and properly identify the radionuclides.

D.8.3 Spatial Issues

Detectors of hand-held radionuclide identifiers typically are small and portable. Spatial limitations are usually based on the physical size of the probe itself, and whether the probe is coupled internally within the casing or externally via an extension cord.

---

[7] The use of multi-point calibrations may add an estimated one to two hours to the time required for instrument set up.

D.8.4 Radiation Types

Hand-held radionuclide identifiers are most commonly used for the detection of photon radiation, although many devices have capabilities for detecting neutron and beta emissions (the detection of neutron radiation requires a different probe from the photon radiation probe).

D.8.5 Range

Photon radiation can typically be measured within a detectable energy range of 10 to 30 keV up to 2.5 to 3 MeV. Neutron radiation can typically be measured within a detectable energy range of 0.02 eV up to 100 MeV.

D.8.6 Scale

There is no definitive limit to the size of an object to be surveyed using hand-held radionuclide identifiers. Hand-held radionuclide identifiers may generally be used to survey M&E of any size; practical constraints are only imposed by the size of M&E related to the sensitive area of the probe.

D.8.7 Ruggedness

All varieties of hand-held radionuclide identifiers discussed here typically are calibrated for use in temperatures from −20 °C to 50 °C and feature seals or gaskets to prevent water ingress from rain, condensing moisture, or high humidity. Most hand-held radionuclide identifiers have a limited resistance to shock, though the durability of an instrument depends largely upon the detection media (e.g., NaI(Tl) crystals are fragile and vulnerable to mechanical and thermal shock).

D.9 Portal Monitors

Portal monitors screen access points to controlled areas, and are designed for detecting radioactivity above background. These systems are used for interdiction-type surveys, and generally do not provide radionuclide identification. Portal monitors are primarily designed to monitor activity on vehicles.

Historically, portal monitors have been used to detect radioactive materials at entrance points to scrap metal facilities and solid waste landfills, and radiological control area exit points within nuclear facilities to screen for the inadvertent disposal of radionuclides. The proximity of other items to be surveyed containing high concentrations of activity may influence the variability of the instrument background, because portal monitors survey activity by detecting small variations in ambient radiation (NRC 2002).

D.9.1 Instruments

Portal monitors can easily be arranged in various geometries that maximize their efficiencies. Most national and international standards, for example ANSI 42.35 (ANSI 2004) require both gamma- and neutron-detecting capabilities, but gamma-only versions are available. Portal monitors typically use large-area polyvinyl toluene scintillators (a form of plastic scintillators) to detect photon radiation and ^{3}He proportional tubes to detect neutrons.[8] Individual detectors may be cylindrical or flat. The detectors are usually arranged to form a detection field between two detectors, and items to be surveyed pass through the detection field (i.e., between the detectors) as shown in Figure D.3.

Figure D-3 Example Portal Monitor (Canberra 2005b)

The system usually consists of one or more detector array(s), an occupancy sensor, a control box, and a monitoring PC. The control box and monitoring PC store and analyze alarm and occupancy data, store and analyze all gamma and neutron survey data, and may even send data through an integrated internet connection. The monitoring PC also manages software that operates multiple arrangements of detector arrays as well as third party instruments. For example, security cameras can take high-resolution images of objects that exceed a radiation screening level (Novelec 2001b).

D.9.2 Temporal Issues

Count or integration times are very short, typically just a few seconds (NRC 2002). Set-up time in the field is variable, because temporary systems may require two hours to one half-day to set up, while permanent systems may require one week to install. For vehicular portal monitor systems, objects may typically pass through the field of detection at speeds of 8 to 9.5 kilometers per hour (Canberra 2005b). Most systems use speed correction algorithms to minimize the effects of variations in dwell time (i.e., the time a given area to be surveyed spends within the detection field).

D.9.3 Spatial Issues

There are a large number of factors that affect portal monitor performance. The isotopic content of a radioactive material can determine the ease of detection. For example, high-enriched uranium (HEU) is easier to detect in a gamma portal than low-enriched uranium (LEU) or natural uranium because of the larger gamma emission rate from ^{235}U.

---

[8] Neutron detectors use materials that detect thermal neutrons, which may be fast neutrons that are thermalized for detection through the use of moderators.

The chemical composition of a material is also important; background levels of radioactivity must also be considered. Neutron portals are an effective method for detecting plutonium in areas with large gamma backgrounds. The surface area and size of the detectors and distance between the detectors all affect the geometry and response of the system. In a large area system set-up, the closer together the detector arrays are, the better the geometric efficiencies are going to be. Finally, for each system there is a maximum passage speed through the portal that gives a counting time necessary to meet the required detection sensitivity.

D.9.4 Radiation Types

Portal monitors typically detect gamma radiation and can also be equipped to detect neutron radiation. Gamma portals often use integrated metal detectors to provide an indication of suspicious metal containers that could be used to shield radioactive materials. If the gamma radiation is not shielded adequately, the detector's alarm will sound. Portal monitors can detect radioactive material even if it is shielded with a material with a high atomic number, like lead.

D.9.5 Range

Photon radiation can typically be measured within a detectable energy range of 60 keV up to 2.6 MeV. Neutron radiation can typically be measured within a detectable energy range of 0.025 eV up to 100 MeV. Required detection sensitivities for gamma and neutron sources are described in ANSI 42.35, Table 3 (ANSI 2004). Portal monitors provide gross counts and cannot compute quantitative measurements (e.g., activity per unit mass).

D.9.6 Scale

Most systems are designed to monitor items ranging in size from bicycles and other small vehicles to tractor trailers, railroad cars, and even passenger airplanes (Canberra 2005b). The width of the detection field (i.e., space between the detector arrays) can usually be modified.

D.9.7 Ruggedness

Portal monitors have typical operating ranges from −20 ° to 55 °C, and some systems may be functional in temperatures as low as −40 °C according to ANSI 42.35 (ANSI 2004). Portal monitors are usually designed with weatherproofing to withstand prolonged outdoor use and exposure to the elements.

D.10 Sample with Laboratory Analysis

Laboratory analysis allows for more controlled conditions and more complex, less rugged instruments to provide lower detections limits and greater delineation among radionuclides than any measurement method that may be utilized in a field setting. For this reason, laboratory analyses are often applied as quality assurance measures to validate sample data collected using field equipment.

D.10.1   Instruments

This section provides a brief overview of instruments used for radiological analyses in a laboratory setting. For additional detail on these instruments, please refer to the accompanying section references in MARLAP.

D.10.1.1 Instruments for the Detection of Alpha Radiation

- **Alpha Spectroscopy with Multi-Channel Analyzer** – This system consists of an alpha detector housed in an evacuated counting chamber, a bias supply, amplifier, analog-to-digital converter, multi-channel analyzer, and computer. Samples are placed at a fixed distance from the solid state partially implanted silica for analysis, and the multi-channel analyzer yields an energy spectrum that can be used to both identify and quantify the radionuclides. The overall properties of the instrumentation allow for excellent peak resolution, although this technique often requires a complex chemical separation to obtain the best results.

- **Gas-Flow Proportional Counter** – The system consists of a gas-flow detector, supporting electronics, and an optional guard detector for reducing the background count rate. A thin window can be placed between the gas-flow detector and sample to protect the detector from contamination, or the sample can be placed directly into the detector. This system does not typically provide data useful for identifying radionuclides unless it is preceded by nuclide-specific chemical separations.

- **Liquid Scintillation Spectrometry** – Typically, samples will be subjected to chemical separations and the resulting materials placed in a vial with a scintillation cocktail. When the alpha particle energy is absorbed by the cocktail, light pulses are emitted, which are detected by photomultiplier tubes. One pulse of light is emitted for each alpha particle absorbed. The intensity of light emitted is related to the energy of the alpha. This system can provide data useful for identifying radionuclides if the system is coupled to a multi-channel analyzer.

- **Low-Resolution Alpha Spectrometry** – The system consists of a small sample chamber, mechanical pump, two-inch diameter silicon detector, multi-channel analyzer, readout module, and a computer. Unlike alpha spectroscopy with multi-channel analyzer, this method allows the technician to load samples for analysis without drying because the presence of moisture generally has negligible effects on the results. This method is therefore estimated to substantially reduce the time for analysis. However, the low resolution may limit the ability to identify individual radionuclides in a sample containing multiple radionuclides and thus may limit the applicability of this method (Meyer 1995).

- **Alpha Scintillation Detector** – This system is used primarily for the quantification of ^{226}Ra by the emanation and detection of ^{222}Rn gas. The system consists of a bubbler system with gas transfer apparatus, a vacuum flask lined with scintillating material called a Lucas Cell,[9] a photomultiplier tube, bias supply, and a scaler to record the count data.

---

[9] One end of a Lucas cell is covered with a transparent window for coupling to a photomultiplier tube and the remaining inside walls are coated with zinc sulfide.

D.10.1.2 Instruments for the Detection of Beta Radiation

- **Gas-Flow Proportional Counter** – The system consists of a gas-flow detector, supporting electronics, and an optional guard detector for reducing the background count rate. A thin window can be placed between the gas-flow detector and sample to protect the detector from non-fixed activity, or the sample can be placed directly into the detector. This technique does not provide data useful for identifying individual radionuclides unless it is preceded by nuclide-specific chemical separations.

- **Liquid Scintillation Spectrometry** – Typically, samples will be subjected to chemical separations and the resulting materials placed in a vial with a scintillation cocktail. When the beta particle energy is absorbed by the cocktail, light pulses are emitted, which are detected by photomultiplier tubes. One pulse of light is emitted for each beta particle absorbed. The intensity of light emitted is related to the energy of the beta. This system can provide data useful for identifying radionuclides if the system is coupled to a multi-channel analyzer. This system must be allowed to darken (i.e., equilibrate to a dark environment) prior to measurement.

D.10.1.3 Instruments for the Detection of Gamma or X-Radiation

- **High-Purity Germanium Detector with Multi-Channel Analyzer** – This system consists of a germanium detector connected to a cryostat (either mechanical or a Dewar of liquid nitrogen), high voltage power supply, spectroscopy grade amplifier, analog to digital converter, and a multi-channel analyzer. This system has high resolution for peak energies and is capable of identifying and quantifying individual gamma peaks in complex spectra. It is particularly useful when a sample may contain multiple gamma-emitting radionuclides and it is necessary to both identify and quantify all nuclides present.

- **Sodium Iodide Detector with Multi-Channel Analyzer** – This system consists of a sodium iodide detector, a high voltage power supply, an amplifier, an analog to digital converter, and a multi-channel analyzer. This system has relatively poor energy resolution and is not effective for identifying and quantifying individual gamma peaks in complex spectra. It is most useful when only a small number of gamma-emitting nuclides are present or when a gross-gamma measurement is adequate.

D.10.2   Temporal Issues

Laboratory analysis is usually controlled by the turnaround time involved in preparing and accurately measuring the collected samples. The sample matrix impacts the preparation time, because soils and bulk chemicals typically require more extensive preparation than liquids or smears. Table D.2 describes the typical preparation and counting times associated with the various analytical instruments and methods described in Section D.10.1. Additional issues that may result in extended time for sample preparation and analysis are described in MARLAP.

Table D.2 Typical Preparation and Counting Times

	Typical Preparation Time	Typical Counting Time
Alpha Spectroscopy with Multi-Channel Analyzer	1 to 7 days	100 to 1,000 minutes
Gas-Flow Proportional Counter	Hours to days	10 to 1,000 minutes
Liquid Scintillation Spectrometer	Minutes,[10] hours to 2 days[11]	>60 to 300 minutes
Low-Resolution Alpha Spectroscopy	Minutes (DOE, 1995)	10 to 1,000 minutes
High-Purity Germanium (HPGe) Detector with Multi-Channel Analyzer	Minutes to 1 day	10 to 1,000 minutes
Sodium Iodide (NaI) Detector with Multi-Channel Analyzer	Minutes to 1 day	1 to 1,000 minutes
Alpha Scintillation Detector	1 to 4 days; 4 to 28 days[12]	10 to 200 minutes

D.10.3   Spatial Issues

This section addresses issues related to detector-M&E geometry and provides information on the range of impacts resulting from dissenting geometries between the calibration source and the measured sample. Other topics may include detector dimensions and problems positioning instruments.

D.10.3.1 Alpha Spectroscopy with Multi-Channel Analyzer

Sample geometry (lateral positioning on a detector shelf) in some detectors may be a small source of additional uncertainty. Uncertainty in the preparation of the actual calibration standards as well as the applicability of the calibration standards to the sample analysis should also be considered.

D.10.3.2 Gas-Flow Proportional Counter

Even deposition of sample material on the planchette is critical to the analytical process. In some analyses, ringed planchettes may aid in the even deposition of sample material. An uneven deposition may result in an incorrect mass-attenuation correction as well as introducing a position-dependent bias to the analysis. The latter situation arises from the fact that gas-flow proportional counters are not radially symmetric, so rotation of an unevenly deposited sample by 45° may drastically change the instrument response.

---

[10] Minimal preparation times are possible if the sample does not require concentration prior to being added to the liquid scintillation cocktail vial.

[11] Longer preparation times are necessary for speciation of low-energy beta emitters.

[12] Longer count times represent the necessary time for in-growth of ^{222}Rn for ^{226}Ra analyses.

D.10.3.3 Liquid Scintillation Spectrometer

For gross counting, samples (e.g., smears and filters) can be placed directly into a liquid scintillation counter (LSC) vial with liquid scintillation cocktail, and counted with no preparation. There are samples with more complicated matrices that require chemical separation prior to being placed and counted in LSC vials. Calibration sources are also kept and counted in these vials, so the geometry of the source and the sample compared to the detector generally are similar.

D.10.3.4 Low-Resolution Alpha Spectroscopy

Sample geometry (lateral positioning on a detector shelf) in some detectors may be a small source of additional uncertainty. Uncertainty in the preparation of the actual calibration standards as well as the applicability of the calibration standards to the sample analysis should be considered.

D.10.3.5 High-Purity Germanium Detector with Multi-Channel Analyzer

Geometry considerations are most important for spectroscopic gamma analyses. Sample positioning on the detector may significantly affect the analytical results, depending on the size and shape of the germanium crystal. Moreover, the instrument is calibrated with a source that should be the same physical size, shape, and weight as the samples to be analyzed.[13] Discrepancies between the volume or density of the sample and the source introduce additional uncertainty to the analytical results.

Sample homogeneity is a critical factor in gamma spectroscopy analyses, particularly with relatively large samples. For example, sediment settling during the course of analysis of a turbid aqueous sample will result in a high bias from any activity contained in the solid fraction. Likewise, the positioning of areas containing elevated activity in a solid sample will create a bias in the overall sample activity (the activity will be disproportionately high if the particle is located at the bottom of the sample, and the activity will be disproportionately low if it is located at the top of the sample).

D.10.3.6 Sodium Iodide Detector with Multi-Channel Analyzer

The spatial considerations for NaI detectors are the same as those listed above for high-purity germanium detectors.

D.10.3.7 Alpha Scintillation Detectors

Accurate sample analysis depends heavily on the complete dissolution of the ^{226}Ra or other radionuclides of interest in the bubbler solution. Adequate sample preparation will help ensure that spatial issues do not influence results, as the apparatus itself minimizes any other potential geometry-related sources of error or uncertainty.

---

[13] Some software packages allow a single calibration geometry to be modeled to assimilate the properties of other geometries.

D.10.4 Radiation Types

Table D.3 describes the types of radiation that each laboratory instrument and method can measure.

Table D.3 Radiation Applications for Laboratory Instruments and Methods

	Alpha	Beta	Photon	Neutron	Differentiate Radiation Types	Identify Specific Radionuclides
Alpha Spectrometry with a Multi-Channel Analyzer	GOOD	NA	NA	NA	NA	GOOD
Gas-Flow Proportional Counter	GOOD	GOOD	POOR	NA	FAIR	POOR
Liquid Scintillation Spectrometer	POOR	GOOD[14]	POOR	NA	FAIR	FAIR
Low-Resolution Alpha Spectroscopy	GOOD	NA	NA	NA	NA	FAIR[15]
High-Purity Germanium Detector with Multi-Channel Analyzer	NA	NA	GOOD	NA	NA	GOOD
Sodium Iodide Detector with Multi-Channel Analyzer	NA	NA	GOOD	NA	NA	FAIR
Alpha Scintillation Detector	GOOD	NA	NA	NA	NA	FAIR

Notes:
GOOD   The instrumentation and measurement technique is well-suited for this application
FAIR   The instrumentation and measurement technique can adequately perform this application
POOR   The instrumentation and measurement technique may be poorly suited for this application
NA   The instrumentation and measurement technique cannot perform this application

D.10.5 Range

All of the instrumentation discussed here has physical limitations as to the amount of activity that can be analyzed. This limitation arises primarily from the ability of the detector to recover after an ionizing event, and the speed with which the component electronics can process the data. Typically, a count rate on the order of 10^6 counts per second taxes the physical limitations of most detectors. Other practical considerations, (such as the potential to impact the detector with non-fixed activity) often override the physical limitations of the counting system.

There are energy range limitations as well. For example: window proportional counters are poor choices for very low energy beta emitters; some gamma spectrometers have poor efficiencies at low energies; and some systems are not calibrated for high-energy gammas. Table D.4 describes the energy range that each instrument and method can be used to determine, and the maximum activity per sample that the method can be used to count.[16]

---

[14] This system is designed for the detection of low-energy beta particles.

[15] The low resolution may limit the ability to identify individual radionuclides in a sample containing multiple radionuclides.

[16] David Burns, Paragon Analytics, Inc., private communication with Nick Berliner, Cabrera Services, Inc., March 2005.

Table D.4 Typical Energy Ranges and Maximum Activities

	Energy Range	Maximum Activity
Alpha Spectrometry with Multi-Channel Analyzer	3 to 8 MeV	<10 Bq (<270 pCi)
Gas-Flow Proportional Counter	3 to 8 MeV (α) 100 to 2,000 keV (β)	35 Bq (946 pCi)
Liquid Scintillation Spectrometer	>3 MeV 15 to 2,500 keV (β); >1.5 MeV (β)[17]	100,000 Bq (2.7 μCi)
Low-Resolution Alpha Spectrometry	3 to 8 MeV (α)	<10 Bq (<270 pCi)
High-Purity Germanium (HPGe) Detector with Multi-Channel Analyzer	50 to >2,000 keV (P-type detector); 5 to 80 keV (N-type detector)	370 Bq (10,000 pCi)
Sodium Iodide (NaI) Detector with Multi-Channel Analyzer	>80 to 2,000 keV	370 Bq (10,000 pCi)
Alpha Scintillation Detector	All α emission energies	<10 Bq (<270 pCi)

D.10.6 Scale

There is no minimum sample size required for a given analysis. Smaller sample sizes will necessarily result in elevated detection limits. Minimum sample sizes (e.g., 0.1 gram) may be specified in order to ensure that the sample is reasonably representative given the degree of homogenization achieved in the laboratory. Typical liquid and solid sample sizes are noted in Table D.5.

Table D.5 Typical Liquid and Solid Sample Sizes

	Typical Liquid Sample Size	Typical Solid Sample Size
Alpha Spectrometry with Multi-Channel Analyzer	1 liter	2 grams; 50 grams[18]
Gas-Flow Proportional Counter	1 liter	2 grams
Liquid Scintillation Spectrometer	<10 milliliters; 1 liter[19]	<0.5 grams; 500 grams
Low-Resolution Alpha Spectrometry	1 liter	2 grams; 50 grams[17]
High-Purity Germanium (HPGe) Detector with Multi-Channel Analyzer	4 liters	1 kilogram
Sodium Iodide (NaI) Detector with Multi-Channel Analyzer	4 liters	1 kilogram
Alpha Scintillation Detector	1 liter	2 grams

---

[17] Very high-energy beta emitters may be counted using liquid scintillation equipment without liquid scintillation cocktails by the use of the Cerenkov light pulse emitted as high energy charged particles move through water or similar substances.

[18] The use of sample digestion processes allows the processing of larger sample masses.

[19] Direct depositing of sample material into the scintillation cocktail limits the sample size to the smaller sample sizes noted; prepared analyses may use substantially larger sample quantities as noted (this applies to both liquid and solid sample matrices).

D.10.7 Ruggedness

Ruggedness does not hold relevance to laboratory analyses, because they are performed in a controlled environment that precludes the instrumentation from being exposed to conditions requiring durability.

E. DISPOSITION CRITERIA

E.1 Department of Energy

Disposition criteria specified by DOE regulations and orders are found in the Code of Federal Regulations, Title 10 (especially 10 CFR 835, Occupational Radiation Protection) and in applicable DOE Orders (especially DOE Order 5400.5, Radiation Protection of the Public and the Environment). The DOE regulations and orders govern the conduct of DOE employees and contractors in the operation of DOE facilities and in the disposition of real property (e.g., buildings and land) and non-real property ("personal property" such as materials, equipment, materials in containers, clothing, etc.). The DOE Order requirements are applicable to DOE activities only and are enforceable as contractual provisions in most DOE contracts. DOE rules are enforceable under 10 CFR Part 820. The following list of DOE requirements is not exhaustive. In addition, a listing of some non-mandatory guidance documents is also provided.

E.1.1 10 CFR 835 (Non-Exhaustive Excerpts)

E.1.1.1 § 835.405 Receipt of Packages Containing Radioactive Material

(a) If packages containing quantities of radioactive material in excess of a Type A quantity (as defined at 10 CFR 71.4) are expected to be received from radioactive material transportation, arrangements shall be made to either:

> (1) Take possession of the package when the carrier offers it for delivery; or
> (2) Receive notification as soon as practicable after arrival of the package at the carrier's terminal and to take possession of the package expeditiously after receiving such notification.

(b) Upon receipt from radioactive material transportation, external surfaces of packages known to contain radioactive material shall be monitored if the package:

> (1) Is labeled with a Radioactive White I, Yellow II, or Yellow III label (as specified at 49 CFR 172.403 and 172.436–440); or
> (2) Has been transported as low specific activity material (as defined at 10 CFR 71.4) on an exclusive use vehicle (as defined at 10 CFR 71.4); or
> (3) Has evidence of degradation, such as packages that are crushed, wet, or damaged.

(c) The monitoring required by paragraph (b) of this section shall include:

> (1) Measurements of removable contamination levels, unless the package contains only special form (as defined at 10 CFR 71.4) or gaseous radioactive material; and
> (2) Measurements of the radiation levels, unless the package contains less than a Type A quantity (as defined at 10 CFR 71.4) of radioactive material.

(d) The monitoring required by paragraph (b) of this section shall be completed as soon as practicable following receipt of the package, but not later than 8 hours after the beginning of the working day following receipt of the package.

E.1.1.2   § 835.605 Labeling Items and Containers

Except as provided at § 835.606, each item or container of radioactive material shall bear a durable, clearly visible label bearing the standard radiation warning trefoil and the words "Caution, Radioactive Material" or "Danger, Radioactive Material." The label shall also provide sufficient information to permit individuals handling, using, or working in the vicinity of the items or containers to take precautions to avoid or control exposures.

E.1.1.3   § 835.606 Exceptions to Labeling Requirements

(a) Items and containers may be excepted from the radioactive material labeling requirements of § 835.605 when:

> (1) Used, handled, or stored in areas posted and controlled in accordance with this subpart and sufficient information is provided to permit individuals to take precautions to avoid or control exposures; or
> (2) The quantity of radioactive material is less than one tenth of the values specified in appendix E of this part; or
> (3) Packaged, labeled, and marked in accordance with the regulations of the Department of Transportation or DOE Orders governing radioactive material transportation; or
> (4) Inaccessible, or accessible only to individuals authorized to handle or use them, or to work in the vicinity; or
> (5) Installed in manufacturing, process, or other equipment, such as reactor components, piping, and tanks; or
> (6) The radioactive material consists solely of nuclear weapons or their components.

(b) Radioactive material labels applied to sealed radioactive sources may be excepted from the color specifications of § 835.601(a).

E.1.1.4   § 835.1101 Control of Material and Equipment

(a) Except as provided in paragraphs (b) and (c) of this section, material and equipment in contamination areas, high contamination areas, and airborne radioactivity areas shall not be released to a controlled area if:

> (1) Removable surface contamination levels on accessible surfaces exceed the removable surface contamination values specified in appendix D of this part; or
> (2) Prior use suggests that the removable surface contamination levels on inaccessible surfaces are likely to exceed the removable surface contamination values specified in Appendix D of this part.

(b) Material and equipment exceeding the removable surface contamination values specified in Appendix D of this part may be conditionally released for movement on-site from one radiological area for immediate placement in another radiological area only if appropriate monitoring is performed and appropriate controls for the movement are established and exercised.

(c) Material and equipment with fixed contamination levels that exceed the total contamination values specified in Appendix D of this part may be released for use in controlled areas outside of radiological areas only under the following conditions:

(1) Removable surface contamination levels are below the removable surface contamination values specified in Appendix D of this part; and (2) The material or equipment is routinely monitored and clearly marked or labeled to alert personnel of the contaminated status.

E.1.1.5   § 835.1102 Control of Areas

(a) Appropriate controls shall be maintained and verified which prevent the inadvertent transfer of removable contamination to locations outside of radiological areas under normal operating conditions.

(b) Any area in which contamination levels exceed the values specified in appendix D of this part shall be controlled in a manner commensurate with the physical and chemical characteristics of the contaminant, the radionuclides present, and the fixed and removable surface contamination levels.

(c) Areas accessible to individuals where the measured total surface contamination levels exceed, but the removable surface contamination levels are less than, corresponding surface contamination values specified in Appendix D of this part, shall be controlled as follows when located outside of radiological areas:

(1) The area shall be routinely monitored to ensure the removable surface contamination level remains below the removable surface contamination values specified in Appendix D of this part; and

(2) The area shall be conspicuously marked to warn individuals of the contaminated status.

(d) Individuals exiting contamination, high contamination, or airborne radioactivity areas shall be monitored, as appropriate, for the presence of surface contamination.

(e) Protective clothing shall be required for entry to areas in which removable contamination exists at levels exceeding the removable surface contamination values specified in Appendix D of this part.

E.1.2   Appendix D to 10 CFR 835 – Surface Contamination Values

The data presented in Appendix D are to be used in identifying the need for posting of contamination and high contamination areas in accordance with § 835.603(e) and (f) and identifying the need for surface contamination monitoring and control in accordance with §§ 835.1101 and 835.1102.

Table E.1 Surface Contamination Values[1] in dpm/100 cm[2] as Reported in Appendix D to 10 CFR 835

Radionuclide	Removable[2,4]	Total (Fixed+Removable)[2,3]
U-nat, U-235, U-238, and associated decay products	1,000[7]	5,000[7]
Transuranics, Ra-226, Ra-228, Th-230, Th-228, Pa-231, Ac-227, I-125, I-129	20	500
Th-nat, Th-232, Sr-90, Ra-223, Ra-224, U-232, I-126, I-131, I-133	200	1,000
Beta-gamma emitters (nuclides with decay modes other than alpha emission or spontaneous fission) except Sr-90 and others noted above[5]	1,000	5,000
Tritium and tritiated compounds[6]	10,000	N/A

[1] The values in this appendix, with the exception noted in footnote 5, apply to radioactive contamination deposited on, but not incorporated into the interior or matrix of, the contaminated item. Where surface contamination by both alpha-and beta-gamma-emitting nuclides exists, the limits established for alpha-and beta-gamma-emitting nuclides apply independently.

[2] As used in this table, dpm (disintegrations per minute) means the rate of emission by radioactive material as determined by correcting the counts per minute observed by an appropriate detector for background, efficiency, and geometric factors associated with the instrumentation.

[3] The levels may be averaged over one square meter provided the maximum surface activity in any area of 100 cm[2] is less than three times the value specified. For purposes of averaging, any square meter of surface shall be considered to be above the surface contamination value if: (1) From measurements of a representative number of sections it is determined that the average contamination level exceeds the applicable value; or (2) it is determined that the sum of the activity of all isolated spots or particles in any 100 cm[2] area exceeds three times the applicable value.

[4] The amount of removable radioactive material per 100 cm[2] of surface area should be determined by swiping the area with dry filter or soft absorbent paper, applying moderate pressure, and then assessing the amount of radioactive material on the swipe with an appropriate instrument of known efficiency. (Note: The use of dry material may not be appropriate for tritium.) When removable contamination on objects of surface area less than 100 cm[2] is determined, the activity per unit area shall be based on the actual area and the entire surface shall be wiped. It is not necessary to use swiping techniques to measure removable contamination levels if direct scan surveys indicate that the total residual surface contamination levels are within the limits for removable contamination.

[5] This category of radionuclides includes mixed fission products, including the Sr-90 which is present in them. It does not apply to Sr-90 which has been separated from the other fission products or mixtures where the Sr-90 has been enriched.

[6] Tritium contamination may diffuse into the volume or matrix of materials. Evaluation of surface contamination shall consider the extent to which such contamination may migrate to the surface in order to ensure the surface contamination value provided in this appendix is not exceeded. Once this contamination migrates to the surface, it may be removable, not fixed; therefore, a "Total" value does not apply.

[7] (alpha)

DOE Order 5400.5 (Non-exhaustive Excerpts) from Chapter II

5. Release of Property Having Residual Radioactive Material

(a) Release of Real Property. Release of real property (land and structures) shall be in accordance with the guidelines and requirements for residual radioactive material presented in Chapter IV. These guidelines and requirements apply to both DOE-owned facilities and to private properties that are being prepared by DOE for release. Real properties owned by DOE that are being sold to the public are subject to the requirements of Section 120(h) of the Comprehensive Environmental Response Compensation and Liability Act (CERCLA), as amended, concerning hazardous substances, and to any other applicable Federal, State, and local requirements. The requirements of 40 CFR Part 192 are applicable to properties remediated by DOE under Title I of the Uranium Mill Tailings Radiation Control Act (UMTRCA).

(b) Release of Personal Property. Personal property, which potentially could be contaminated, may be released for unrestricted use if the results of a survey with appropriate instruments indicate that the property is less than the contamination limits presented in Figure IV-1.

(c) Release of Materials and Equipment.

(1) Surface Contamination Levels. Prior to being released, property shall be surveyed to determine whether both removable and total surface contamination (Including contamination present on and under any coating) are in compliance with the levels given in Figure IV-1 and that the contamination has been subjected to the ALARA process.

(2) Potential for Contamination. Property shall be considered to be potentially contaminated if it has been used or stored in radiation areas that could contain unconfined radioactive material or that are exposed to beams of particles capable of causing activation (neutrons, protons, etc.).

(3) Surveys. Surfaces of potentially contaminated property shall be surveyed using instruments and techniques appropriate for detecting the limits stated in Figure IV-1.

(4) Inaccessible Areas. Where potentially contaminated surfaces are not accessible for measurement (as in some pipes, drains, and ductwork), such property may be released after case-by-case evaluation and documentation based on both the history of its use and available measurements demonstrate that the unsurveyable surfaces are likely to be within the limits given in Figure IV-1.

(5) Records. The records of released property shall include:

(a) A description or identification of the property;

(b) The date of the last radiation survey;

(c) The identity of the organization and the individual who performed the monitoring operation;

(d) The type and identification number of monitoring instruments;

(e) The results of the monitoring operation; and

(f) The identity of the recipient of the released material.

(6) Volume Contamination. No guidance is currently available for release of material that has been contaminated in depth, such as activated material or smelted contaminated metals (e.g., radioactivity per unit volume or per unit mass). Such materials may be released if criteria and survey techniques are approved by EH-1.

E.1.3 DOE Guidance and Similar Documents

The following discussion summarizes DOE policy, practice, and guidance for the disposition of personal property, including materials and equipment from several DOE guidance documents.

"Application of DOE 5400.5 requirements for release and control of property containing residual radioactive material," a guidance memorandum dated November 17, 1995. This guidance memorandum explains the procedures through which authorized limits can be approved for the disposition of waste materials to sanitary waste landfills. It also discusses the disposition criteria for certain radionuclides. Finally, it delegates some responsibilities for the approval of release of volumetrically contaminated materials to DOE field office managers when specified conditions are met.

Table E.2 Figure IV-1, from DOE Order 5400.5, as Supplemented in November, 1995 Memorandum: Surface Activity Guidelines – Allowable Total Residual Surface Activity (dpm/100 cm^2)[1]

Radionuclides[2]	Average[3,4]	Maximum[4,5]	Removable[4,6]
Group 1 - Transuranics, I-125, I-129, Ac-227, Ra-226, Ra-228, Th-228, Th-230, Pa-231	100	300	20
Group 2 - Th-natural, Sr-90, I-126, I-131, I-133, Ra-223, Ra-224, U-232, Th-232	1,000	3,000	200
Group 3 - U-natural, U-235, U-238, and associated decay products, alpha emitters	5,000	15,000	1,000
Group 4 - Beta-gamma emitters (radionuclides with decay modes other than alpha emission or spontaneous fission) except Sr-90 and others noted above[7]	5,000	15,000	1,000
Tritium (applicable to surface and subsurface)[8]	N/A	N/A	10,000

[1] As used in this table, dpm (disintegrations per minute) means the rate of emission by radioactive material as determined by correcting the counts per minute measured by an appropriate detector for background, efficiency, and geometric factors associated with the instrumentation.

[2] Where surface contamination by both alpha- and beta-gamma-emitting radionuclides exists, the limits established for alpha- and beta-gamma-emitting radionuclides should apply independently.

[3] Measurements of average contamination should not be averaged over an area of more than 1 m^2. For objects of less surface area, the average should be derived for each such object.

[4] The average and maximum dose rates associated with surface contamination resulting from beta-gamma emitters should not exceed 0.2 mrad/h and 1.0 mrad/h, respectively, at 1 cm.

[5] The maximum contamination level applies to an area of not more than 100 cm^2.

[6] The amount of removable material per 100 cm^2 of surface area should be determined by wiping an area of that size with dry filter or soft absorbent paper, applying moderate pressure, and measuring the amount of radioactive material on the wiping with an appropriate instrument of known efficiency. When removable contamination on objects of surface area less than 100 cm^2 is determined, the activity per unit area should be based on the actual area and the entire surface should be wiped. It is not necessary to use wiping techniques to measure removable contamination levels if direct scan surveys indicate that the total residual surface contamination levels are within the limits for removable contamination.

[7] This category of radionuclides includes mixed fission products, including the Sr-90 which is present in them. It does not apply to Sr-90 which has been separated from the other fission products or mixtures where the Sr-90 has been enriched.

[8] Property recently exposed or decontaminated, [sic] should have measurements (smears) at regular time intervals to ensure that there is not a build-up of contamination over time. Because tritium typically penetrates material it contacts, the surface guidelines in group 4 are not applicable to tritium. The Department has reviewed the analysis conducted by the DOE Tritium Surface Contamination Limits Committee ("Recommended Tritium Surface Contamination Release Guides," February 1991), and has assessed potential doses associated with the release of property containing residual tritium. The Department recommends the use of the stated guideline as an interim value for removable tritium. Measurements demonstrating compliance of the removable fraction of tritium on surfaces with this guideline are acceptable to ensure that non-removable fractions and residual tritium in mass will not cause exposures that exceed DOE dose limits and constraints.

"Control and Release of Property with Residual Radioactive Material for use with DOE Order 5400.5, Radiation Protection of the Public and the Environment," DOE G 441.1-XX, a draft guidance document approved for interim use and issued on May 1, 2002. This guidance document contains detailed discussions of the disposition approaches for real and personal property, as well as summaries of DOE's policies regarding the disposition or release of property.

"Cross-Cut Guidance on Environmental Requirements for DOE Real Property Transfers (Update)," DOE/EH-413/97-12, originally issued October, 1997, revised March, 2005. This guidance document contains a summary of various environmental requirements for the release or transfer of real property.

"Managing the Release of Surplus and Scrap Materials," January 19, 2001, from DOE Secretary Richardson to all DOE elements. This memorandum provides direction as well as guidance regarding the release of property from DOE radiological control. It also restricts the release of metal from radiological areas for recycle until certain steps are taken by DOE.

E.2   International Organizations

In general, each country establishes its own disposition criteria for materials and equipment. These national disposition criteria may be consistent with guidance promulgated by multi-national organizations, such as the International Atomic Energy Agency (IAEA) or the European Commission (EC). One example of widely accepted regulations is the "Advisory Material for the IAEA Regulations for the Safe Transport of Radioactive Material SAFETY GUIDE No. TS-G-1.1 (ST-2)." The references listed below provide the detailed information on guidance from the IAEA and the EC. URLs are provided for internet access of this information. Disposition criteria from specific nations should be obtained from those nations.

E.2.1   International Atomic Energy Agency (IAEA)

Advisory Material for the IAEA Regulations for the Safe Transport of Radioactive Material SAFETY GUIDE No. TS-G-1.1 (ST-2):
http://www-pub.iaea.org/MTCD/publications/PDF/Pub1109_scr.pdf.

Planning and Preparing for Emergency Response to Transport Accidents Involving Radioactive Material, SAFETY GUIDE No. TS-G-1.2 (ST-3)
http://www-pub.iaea.org/MTCD/publications/PDF/Pub1119_scr.pdf.

Application of the Concepts of Exclusion, Exemption and Clearance SAFETY GUIDE No. RS-G-1.7: http://www-pub.iaea.org/MTCD/publications/PDF/Pub1202_web.pdf.

E.2.2   European Commission

The publication list for radiation protection may be found on the EC website at:
http://europa.eu.int/comm/energy/nuclear/radioprotection/publication_en.htm. Contact information for most of the authorities in the European Union may be found in Annex 3, in the last pages of publication 139, "A review of consumer products containing radioactive substances in the European Union," which can be found at:
http://europa.eu.int/comm/energy/nuclear/radioprotection/publication/doc/139_en.pdf.

Radiation protection publications pertaining to disposition criteria for materials and equipment include:

> 134: Evaluation of the application of the concepts of exemption and clearance for practices according to title III of Council Directive 96/29/Euratom of 13 May 1996 in EU Member States, Volume 1, Main Report:
> http://europa.eu.int/comm/energy/nuclear/radioprotection/publication/doc/134_en.pdf.

122: Practical Use of the Concepts of Clearance and Exemption Part I: Guidance on General Clearance Levels for Practices:
http://europa.eu.int/comm/energy/nuclear/radioprotection/publication/doc/122_part1_en.pdf.

122: Practical Use of the Concepts of Clearance and Exemption Part II: Application of the Concepts of Exemption and Clearance to Natural Radiation Sources:
http://europa.eu.int/comm/energy/nuclear/radioprotection/publication/doc/122_part2_en.pdf.

114: Definition of Clearance Levels for the Release of Radioactively Contaminated Buildings and Building Rubble:
http://europa.eu.int/comm/energy/nuclear/radioprotection/publication/doc/114_en.pdf.

European legislation related to the transport of radioactive materials (database):
http://europa.eu.int/comm/energy/nuclear/transport/legislation_en.htm.

E.3   Nuclear Regulatory Commission

Disposition criteria specified by NRC regulations are found in the Code of Federal Regulations, Title 10 (10 CFR). NRC regulations in 10 CFR are structured in Parts, which apply to respective areas of applicability. For example, 10 CFR Part 20 addresses "Standards for Protection against Radiation." The regulatory citations below indicate the specific Part by the number to the left of the decimal point, for example, §20.2003 is in 10 CFR Part 20, and 2003 indicates a specific portion. In this appendix only the radiological component of those criteria pertaining to quantitative measurement attributes are listed; there are almost always additional regulatory requirements. "Disposition criteria" refers to the quantitative radiological portion of the complete criteria. In some circumstances, disposition criteria are not addressed in the regulations, and these cases are handled by existing policy and practices. A list of NRC disposition criteria, which is not necessarily exhaustive, follows.

E.3.1   § 20.2003 Disposal by Release into Sanitary Sewerage.

(2) The quantity of licensed or other radioactive material that the licensee releases into the sewer in 1 month divided by the average monthly volume of water released into the sewer by the licensee does not exceed the concentration listed in table 3 of appendix B to part 20; and

(4) The total quantity of licensed and other radioactive material that the licensee releases into the sanitary sewerage system in a year does not exceed 5 curies (185 GBq) of hydrogen-3, 1 curie (37 GBq) of carbon-14, and 1 curie (37 GBq) of all other radioactive materials combined.

E.3.2   § 20.2005 Disposal of Specific Wastes

(a) A licensee may dispose of the following licensed material as if it were not radioactive
    (1) 0.05 microcurie (1.85 kBq), or less, of hydrogen-3 or carbon-14 per gram of medium used for liquid scintillation counting; and
    (2) 0.05 microcurie (1.85 kBq), or less, of hydrogen-3 or carbon-14 per gram of animal tissue, averaged over the weight of the entire animal.

E.3.3   § 35.92 Decay-in-Storage

(a) A licensee may hold byproduct material with a physical half-life of less than 120 days for decay-in-storage before disposal without regard to its radioactivity if it–

    (1) Monitors byproduct material at the surface before disposal and determines that its radioactivity cannot be distinguished from the background radiation level with an appropriate radiation detection survey meter set on its most sensitive scale and with no interposed shielding.

E.3.4   § 35.315 Safety Precautions

(4) Either monitor material and items removed from the patient's or the human research subject's room to determine that their radioactivity cannot be distinguished from the natural background radiation level with a radiation detection survey instrument set on its most sensitive scale and with no interposed shielding, or handle the material and items as radioactive waste.

E.3.5   § 36.57 Radiation Surveys

(e) Before releasing resins for unrestricted use, they must be monitored before release in an area with a background level less than 0.5 microsievert (0.05 millirem) per hour. The resins may be released only if the survey does not detect radiation levels above background radiation levels. The survey meter used must be capable of detecting radiation levels of 0.5 microsievert (0.05 millirem) per hour.

E.3.6   Appendix A to Part 40–Criteria Relating to the Operation of Uranium Mills and the Disposition of Tailings or Wastes Produced by the Extraction or Concentration of Source Material from Ores Processed Primarily for Their Source Material Content

(6) The design requirements in this criterion for longevity and control of radon releases apply to any portion of a licensed and/or disposal site unless such portion contains a concentration of radium in land, averaged over areas of 100 square meters, which, as a result of byproduct material, does not exceed the background level by more than: (i) 5 picocuries per gram (pCi/g) of radium-226, or, in the case of thorium byproduct material, radium-228, averaged over the first 15 centimeters (cm) below the surface, and (ii) 15 pCi/g of radium-226, or, in the case of thorium byproduct material, radium-228, averaged over 15-cm thick layers more than 15 cm below the surface.

E.3.7   § 71.4 Definitions

The following terms are as defined here for the purpose of this part. To ensure compatibility with international transportation standards, all limits in this part are given in terms of dual units: The International System of Units (SI) followed or preceded by U.S. standard or customary units. The U.S. customary units are not exact equivalents but are rounded to a convenient value, providing a functionally equivalent unit. For the purpose of this part, either unit may be used.

A_1 means the maximum activity of special form radioactive material permitted in a Type A package. This value is either listed in Appendix A, Table A-1, of this part, or may be derived in accordance with the procedures prescribed in Appendix A of this part.

A_2 means the maximum activity of radioactive material, other than special form material, LSA, and SCO material, permitted in a Type A package. This value is either listed in Appendix A, Table A-1, of this part, or may be derived in accordance with the procedures prescribed in Appendix A of this part.

Low Specific Activity (LSA) material means radioactive material with limited specific activity which is nonfissile or is excepted under §71.15, and which satisfies the descriptions and limits set forth below. Shielding materials surrounding the LSA material may not be considered in determining the estimated average specific activity of the package contents. LSA material must be in one of three groups:

(1) LSA−I
- (i) Uranium and thorium ores, concentrates of uranium and thorium ores, and other ores containing naturally occurring radioactive radionuclides which are not intended to be processed for the use of these radionuclides;
- (ii) Solid unirradiated natural uranium or depleted uranium or natural thorium or their solid or liquid compounds or mixtures;
- (iii) Radioactive material for which the A_2 value is unlimited; or
- (iv) Other radioactive material in which the activity is distributed throughout and the estimated average specific activity does not exceed 30 times the value for exempt material activity concentration determined in accordance with Appendix A.

(2) LSA−II
- (i) Water with tritium concentration up to 0.8 TBq/L (20.0 Ci/L); or
- (ii) Other material in which the activity is distributed throughout and the average specific activity does not exceed $10^{-4}A_2$/g for solids and gases, and $10^{-5}A_2$/g for liquids.

(3) LSA−III. Solids (e.g., consolidated wastes, activated materials), excluding powders, that satisfy the requirements of § 71.77, in which:
- (i) The radioactive material is distributed throughout a solid or a collection of solid objects, or is essentially uniformly distributed in a solid compact binding agent (such as concrete, bitumen, ceramic, etc.);
- (ii) The radioactive material is relatively insoluble, or it is intrinsically contained in a relatively insoluble material, so that even under loss of packaging, the loss of radioactive material per package by leaching, when placed in water for 7 days, would not exceed 0.1 A_2; and
- (iii) The estimated average specific activity of the solid does not exceed $2\times10^{-3}A_2$/g.

Low toxicity alpha emitters means natural uranium, depleted uranium, natural thorium; uranium-235, uranium-238, thorium-232, thorium-228 or thorium-230 when contained in ores or physical or chemical concentrates or tailings; or alpha emitters with a half-life of less than 10 days.

Surface Contaminated Object (SCO) means a solid object that is not itself classed as radioactive material, but which has radioactive material distributed on any of its surfaces. SCO must be in one of two groups with surface activity not exceeding the following limit:

(1) SCO-I: A solid object on which:

    (i) The non-fixed contamination on the accessible surface averaged over 300 cm^2 (or the area of the surface if less than 300 cm^2) does not exceed 4 Bq/cm^2 (10^4 microcurie/cm^2) for beta and gamma and low toxicity alpha emitters, or 0.4 Bq/cm^2 (10^{-5} microcurie/cm^2) for all other alpha emitters;

    (ii) The fixed contamination on the accessible surface averaged over 300 cm^2 (or the area of the surface if less than 300 cm^2) does not exceed 4×10^4 Bq/cm^2 (1.0 microcurie/cm^2) for beta and gamma and low toxicity alpha emitters, or 4×10^3 Bq/cm^2 (0.1 microcurie/cm^2) for all other alpha emitters; and

    (iii) The non-fixed contamination plus the fixed contamination on the inaccessible surface averaged over 300 cm^2 (or the area of the surface if less than 300 cm^2) does not exceed 4×10^4 Bq/cm^2 (1 microcurie/cm^2) for beta and gamma and low toxicity alpha emitters, or 4×10^3 Bq/cm^2 (0.1 microcurie/cm^2) for all other alpha emitters.

(2) SCO-II: A solid object on which the limits for SCO-I are exceeded and on which:

    (i) The nonfixed contamination on the accessible surface averaged over 300 cm^2 (or the area of the surface if less than 300 cm^2) does not exceed 400 Bq/cm^2 (10^2 microcurie/cm^2) for beta and gamma and low toxicity alpha emitters or 40 Bq/cm^2 (10^3 microcurie/cm^2) for all other alpha emitters;

    (ii) The fixed contamination on the accessible surface averaged over 300 cm^2 (or the area of the surface if less than 300 cm^2) does not exceed 8×10^5 Bq/cm^2 (20 microcuries/cm^2) for beta and gamma and low toxicity alpha emitters, or 8×10^4 Bq/cm^2 (2 microcuries/cm^2) for all other alpha emitters; and

    (iii) The non-fixed contamination plus the fixed contamination on the inaccessible surface averaged over 300 cm^2 (or the area of the surface if less than 300 cm^2) does not exceed 8×10^5 Bq/cm^2 (20 microcuries/cm^2) for beta and gamma and low toxicity alpha emitters, or 8×10^4 Bq/cm^2 (2 microcuries/cm^2) for all other alpha emitters.

E.3.8 § 71.14 Exemption for Low-Level Materials

(a) A licensee is exempt from all the requirements of this part with respect to shipment or carriage of the following low-level materials:

    (1) Natural material and ores containing naturally occurring radionuclides that are not intended to be processed for use of these radionuclides, provided the activity concentration of the material does not exceed 10 times the values specified in Appendix A, Table A-2, of this part.

    (2) Materials for which the activity concentration is not greater than the activity concentration values specified in Appendix A, Table A-2 of this part, or for which the consignment activity is not greater than the limit for an exempt consignment found in Appendix A, Table A-2, of this part.

(b) A licensee is exempt from all the requirements of this part, other than §§ 71.5 and 71.88, with respect to shipment or carriage of the following packages, provided the packages do not contain any fissile material, or the material is exempt from classification as fissile material under § 71.15:

(1) A package that contains no more than a Type A quantity of radioactive material;

(2) A package transported within the United States that contains no more than 0.74 TBq (20 Ci) of special form plutonium-244; or

(3) The package contains only LSA or SCO radioactive material, provided--

(i) That the LSA or SCO material has an external radiation dose of less than or equal to 10 mSv/h (1 rem/h), at a distance of 3 m from the unshielded material; or

(ii) That the package contains only LSA-I or SCO-I material.

E.3.9 § 110.22 General License for the Export of Source Material

(3) Th-227, Th-228, U-230, and U-232 when contained in a device, or a source for use in a device, in quantities of less than 100 millicuries of alpha activity (3.12 micrograms Th-227, 122 micrograms Th-228, 3.7 micrograms U-230, 4.7 milligrams U-232) per device or source.

E.3.10 § 110.23 General License for the Export of Byproduct Material

(2) Actinium-225 and -227, americium-241 and -242m, californium-248, -249, -250, -251, -252, -253, and -254, curium-240, -241, -242, -243, -244, -245, -246 and -247, einsteinium-252, -253, -254 and -255, fermium-257, gadolinium-148, mendelevium-258, neptunium-235 and -237, polonium-210, and radium-223 must be contained in a device, or a source for use in a device, in quantities of less than 100 millicurie of alpha activity (see Sec. 110.2 for specific activity) per device or source, unless the export is to a country listed in Sec. 110.30. Exports of americium and neptunium are subject to the reporting requirements listed in paragraph (b) of this section.

(3) For americium-241, exports must not exceed one curie (308 milligrams) per shipment or 100 curies (30.8 grams) per year to any country listed in Sec. 110.29, and must be contained in industrial process control equipment or petroleum exploration equipment in quantities not to exceed 20 curies (6.16 grams) per device or 200 curies (61.6 grams) per year to any one country.

(5) For polonium-210, the material must be contained in static eliminators and may not exceed 100 curies (22 grams) per individual shipment.

(6) For tritium in any dispersed form, except for recovery or recycle purposes (e.g., luminescent light sources and paint, accelerator targets, calibration standards, labeled compounds), exports must not exceed the quantity of 10 curies (1.03 milligrams) or less per item, not to exceed 1,000 curies (103 milligrams) per shipment or 10,000 curies (1.03 grams) per year to any one country. Exports of tritium to the countries listed in Sec. 110.30 must not exceed the quantity of 40 curies (4.12 milligrams) or less per item, not to exceed 1,000 curies (103 milligrams) per shipment or 10,000 curies (1.03 grams) per year to any one country, and exports of tritium in luminescent safety devices installed in aircraft must not exceed a quantity of 40 curies (4.12 milligrams) or less per light source.

E.3.11 Policies and Practices

Disposition criteria for the release of materials and equipment that are not specified in NRC regulations are determined by the current policies and practices. NRC's current approaches for making decisions on disposition of solid materials is different for materials licensees, i.e., industrial, research, and medical facilities, and for reactors, which include power, test, and research reactors. These are summarized in Table E-3, and discussed in more detail below.

For non-reactor licensees—materials licensees—licensee requests for release of solid material will continue to be evaluated using the nuclide concentration tables in Regulatory Guide 1.86 and its equivalent, Fuel Cycle Policy and Guidance Directive FC 83-23. Many materials licensees obtain approval, as a license condition, to routinely use these guidelines. For residual radioactivity within the volume of solid materials (for example, within a concrete or soil matrix), non-reactor licensee requests for release of solid material may continue to be approved under a disposal request (10 CFR 20.2002); a license termination plan; decommissioning plan review; or other specific license amendment. In verifying that the dose from such release is maintained ALARA and below the limits of our regulations in 10 Part 20, approval of a release is possible. The disposition of materials with volumetrically distributed radioactivity from materials licensees is considered on a case-by-case basis with a reference of an annual individual dose criterion of a "few mrem per year (a few 0.01 mSv/a)."

Non-reactor licensees, that is, materials licensees, and reactor licensees have essentially the same detection level criteria for surface activity. But for materials licensees, radioactivity below these detection level criteria is allowed—detectable radioactivity is not allowed at any level for reactor licensees.

For reactor licensees, licensees may release of solid material using the "no detectable" policy of NRC's Inspection and Enforcement Circular 81-07 and Information Notices 85-92 and 88-22. For reactors, the policy is that released material can have no detectable licensed radioactivity. The levels of detection are specified by each reactor licensee's procedures and are frequently consistent with a now discontinued Regulatory Guide issued in 1974. In practice, these detection levels for radioactivity on surfaces are: 5/6 Bq /cm^2 (5,000 dpm/100 cm^2) total β-γ and $^1/_6$ Bq/cm^2 (1,000 dpm/100 cm^2) removable β-γ. Non-detection at these levels of detectability was considered to result in potential doses to an individual significantly less than 5 mrem/y (<<0.05 mSv/a) from any non-detectable radioactivity that could remain on surfaces.

Detection levels for α-emitting radioactivity are specified as 1/60 Bq/cm^2 (100 dpm/100 cm^2) total and 1/300 Bq/cm^2 (20 dpm/100 cm^2) for removable α-emitting radioactivity. For volumetric radioactivity from reactors, the detection levels are from guidance written in the late 1970s and specifies β-γ concentrations in the general range of 3–4 Bq/kg (81–108 pCi/kg).

Table E.3 Summary of NRC Disposition Criteria from Current Practices for the Release of Materials and Equipment

	Surficial Radioactivity	Volumetric Radioactivity
Reactor Licenses	β-γ: Non-detectable [MDC 5/6 Bq/cm^2; 1/6 Bq/cm^2 removable]	β-γ: Non-detectable [MDC in General range of ≈ 3-4 Bq/kg]
	α: Non-detectable [MDC 1/60 Bq/cm^2; 1/300 Bq/cm^2 removable]	α: Non-detectable [MDC not indicated]
Materials Licenses	β-γ: 5/6 Bq/cm^2; 1/6 Bq/cm^2 removable[1]	β-γ: Case-by-case [Reference to a few 0.01 mSv in a year]
	α: 1/60 Bq/cm^2; 1/300 Bq/cm^2 removable[2]	α: Case-by-case [Reference to a few 0.01 mSv in a year]

[1] Except Sr-90, I-126, I-131, and I-133, where 1/6 Bq/cm^2 and 1/30 Bq/ cm^2 removable applies; and except I-125, and I-129 where 1/60 Bq/cm^2 and 1/300 Bq/cm^2 removable applies.

[2] Except natural U, U-235, U-238, and associated decay products where 5/6 Bq/cm^2 and 1/6 Bq/cm^2 removable applies; and except transuranics, Ra-226, Ra-228, Th-230, Th-228, Pa-231, and Ac-227, where 1/60 Bq/cm^2 and 1/300 Bq/cm^2 removable applies.

E.3.12 Issues Related to International Trade

With regard to issues relating to international trade of solid materials released from facilities, NRC's regulations contain requirements for export and import of material and could be considered in handling materials that meet established international clearance criteria and, at the same time, do not meet the guidelines for NRC licensees. Among other things, these regulations require that "the proposed import does not constitute an unreasonable risk to the public health and safety."

REFERENCES

10 CFR 20.1003. *Standards for Protection Against Radiation, Definitions*. Code of Federal Regulations (CFR) Title 10, Energy, Part 20.1003, December 19, 2002. http://www.nrc.gov/reading-rm/doc-collections/cfr/part020/ (accessed September 1, 2008).

10 CFR 36.57. *Licenses and Radiation Safety Requirements for Irradiators, Radiation Surveys*. Code of Federal Regulations (CFR) Title 10, Energy, Part 36.57, January 1, 2005. http://www.nrc.gov/reading-rm/doc-collections/cfr/part036/part036-0057.html (accessed September 1, 2008).

10 CFR 50.2. *Domestic Licensing of Production and Utilization Facilities, Definitions*. Code of Federal Regulations (CFR) Title 10, Energy, Part 50.2, January 13, 1998. http://www.nrc.gov/reading-rm/doc-collections/cfr/part050/ (accessed September 1, 2008).

10 CFR 71.4. *Packaging And Transportation of Radioactive Material*. Code of Federal Regulations (CFR) Title 10 (Energy), Part 71.4, September 28, 1995. http://www.nrc.gov/reading-rm/doc-collections/cfr/part071/full-text.html (accessed October 24, 2008).

10 CFR 820. *Procedural Rules for DOE Nuclear Activity*. Code of Federal Regulations (CFR) Title 10, Energy, Part 820, August 17, 1993. http://www.hss.energy.gov/enforce/rands/10CFR820.pdf (accessed October 24, 2008).

10 CFR 835. *Occupational Radiation Protection*. Code of Federal Regulations (CFR) Title 10, Energy, Part 835, Subpart E, "Monitoring of Individuals and Areas," 401, "General Requirements" November 4, 1998. http://www.eh.doe.gov/radiation/10cfr835/835gpo.pdf (accessed September 1, 2008).

10 CFR 172. *Hazardous Materials Table, Special Provisions, Hazardous Materials Communications, Emergency Response Information, Training Requirements, and Security Plans*. Code of Federal Regulations (CFR) Title 49 (Transportation), Subpart E, "Labeling," 403, "Class 7 (radioactive) material" April 15, 1976, 436, "Radioactive White-I Label," 438, "Radioactive Yellow-II Label," 440 "Radioactive Yellow-III Label" December 20, 1991. http://ecfr.gpoaccess.gov/cgi/t/text/text-idx?c=ecfr;sid=f5b01bccf82655893c893d14885a307f;rgn=div5;view=text;node=49%3A2.1.1.3.8;idno=49;cc=ecfr#49:2.1.1.3.8.9.25.22 (accessed October 24, 2008).

49 CFR 173.433. *Requirements for Determining Basic Radionuclide Values and for the Listing of Radionuclides on Shipping Papers and Labels*. Code of Federal Regulations (CFR) Title 49, Transportation, Part 173.433, September 23, 2005. http://ecfr.gpoaccess.gov/cgi/t/text/text-idx?c=ecfr;sid=f5b01bccf82655893c893d14885a307f;rgn=div5;view=text;node=49%3A2.1.1.3.8;idno=49;cc=ecfr#49:2.1.1.3.8.9.25.22 (accessed October 24, 2008).

Abelquist, E. 2001. *Decommissioning Health Physics: A Handbook for MARSSIM Users*. Institute of Physics Publishing, Philadelphia, PA.

American National Standards Institute (ANSI) 1989. *Performance Specifications for Health Physics Instrumentation-Portable Instrumentation for Use in Extreme Environmental Conditions*. ANSI N42.17C.

American National Standards Institute (ANSI) 1994. *Calibration and Usage of Thallium-Activated Sodium Iodide Detector Systems for Assay of Radionuclides*. ANSI N42.12.

American National Standards Institute (ANSI) 1997. *Radiation Protection Instrumentation Test and Calibration, Portable Survey Instruments*. ANSI N323A.

American National Standards Institute (ANSI) 1999. *Surface and Volume Radioactivity Standards for Clearance*. ANSI N13.12.

American National Standards Institute (ANSI) 2003a. *Performance Criteria for Hand-held Instruments for the Detection and Identification of Radionuclides*. ANSI N42.34.

American National Standards Institute (ANSI) 2003b. *Performance Specifications for Health Physics Instrumentation-Portable Instrumentation for Use in Normal Environmental Conditions*. ANSI N42.17A.

American National Standards Institute (ANSI) 2004. *American National Standard for Evaluation and Performance of Radiation Protection Portal Monitors for Use in Homeland Security*. ANSI N42.35.

Beck H.L., Gogolak C.V., Miller K.M., and Lowder W.M. 1980. "Perturbations on the Natural Radiation Environment Due to the Utilization of Coal as an Energy Source," in *Natural Radiation Environment III*, U.S. Department of Energy CONF-780422.

BIL 2005. "IonSens® 208 Large Item Monitor." BIL Solutions. http://www.bilsolutions.co.uk/pdf/datasheets_new/ionsens208largeitemsmonitor.pdf (accessed September 1, 2008).

Canberra 2005a. "Considerations for Environmental Gamma Spectroscopy Systems." Canberra, Inc. http://www.canberra.com/literature/972.asp (accessed September 1, 2008).

Canberra 2005b. "RadSentry™ Security Portals for SNM and Other Radionuclides." Canberra, Inc. http://www.canberra.com/products/1211.asp (accessed September 1, 2008).

Canberra 2008. "Automatic Conveyor Monitor for Soil and Debris." Canberra, Inc. http://www.canberra.com/products/795.asp (accessed September 1, 2008).

Currie, L.A. 1968. "Limits for Qualitative Detection and Quantitative Determination: Application to Radiochemistry." *Analytical Chemistry* 40(3): 586–593.

DOE 1987. *The Environmental Survey Manual, Appendix A – Criteria for Data Evaluation*. DOE/EH-0053, DOE, Office of Environmental Audit, Washington, D.C. (DE88-000254), August.

DOE 1993. *Radiation Protection of the Public and the Environment*. DOE Order 5400.5, Change 2, U.S. Department of Energy, Washington, DC, January. http://homer.ornl.gov/oepa/guidance/risk/54005.pdf (accessed September 1, 2008).

DOE 2005. *RESRAD Recycle Version 3.10*. U.S. Department of Energy, Argonne National Laboratory. http://web.ead.anl.gov/resrad/home2/index.cfm (accessed September 1, 2008).

Eberline 2004. "Segmented Gate System." Eberline Services. http://www.eberlineservices.com/documents/SGSBrochure_000.pdf (accessed September 1, 2008).

Eckerman, K.F., Westfall, R.J., Ryman, J.C., and Cristy, M. 1993. *Nuclear Decay Data Files of the Dosimetry Research Group*, ORNL-TM-12350.

Eicholz G.G., Clarke F.J., and Kahn, B. 1980. "Radiation Exposure From Building Materials," in *Natural Radiation Environment III*, U.S. Department of Energy CONF-780422.

Electric Power Research Institute (EPRI) 2003. *Operational Changes and Impacts on LLW Scaling Factors*. D. James, EPRI Report 1008017, Electric Power Research Institute, Palo Alto, CA, December.

U.S. Environmental Protection Agency (EPA) 1980. *Upgrading Environmental Radiation Data, Health Physics Society Committee Report HPSR-1*. EPA 520/1-80-012, EPA, Office of Radiation Programs, Washington, D.C. (PB81-100364), August.

U.S. Environmental Protection Agency (EPA) 1992a. *Guidance for Data Usability in Risk Assessment, Part A*. Office of Solid Waste and Emergency Response (OSWER) Directive 9285.7-09A, Environmental Protection Agency, Office of Emergency and Remedial Response, Washington, D.C. (PB92-963356), April.

U.S. Environmental Protection Agency (EPA) 1992b. *Guidance for Data Usability in Risk Assessment, Part B*. Office of Solid Waste and Emergency Response (OSWER) Directive 9285.7-09B, Environmental Protection Agency, Office of Emergency and Remedial Response, Washington, D.C. (PB92-963362), May.

U.S. Environmental Protection Agency (EPA) 2000. *Evaluation of EPA's Guidelines for Technologically Enhanced Naturally Occurring Radioactive Materials (TENORM)*. Report to Congress, EPA 402-R-00-01, U.S. Environmental Protection Agency.

U.S. Environmental Protection Agency (EPA) 2001. *Guidance for Preparing Standard Operating Procedures*, EPA QA/G-6. EPA/240/B-01/004, U.S. Environmental Protection Agency, Office of Environmental Information, Washington, DC, March.

U.S. Environmental Protection Agency (EPA) 2002a. *Guidance for Quality Assurance Project Plans*, EPA QA/G-5. EPA/240/R-02/009, U.S. Environmental Protection Agency, Office of Environmental Information, Washington, DC, December.

U.S. Environmental Protection Agency (EPA) 2002b. *Calculating Upper Confidence Limits For Exposure Point Concentrations At Hazardous Waste Sites*. Office of Solid Waste and Emergency Response (OSWER) Directive 9285.6-10, Environmental Protection Agency, Office of Emergency and Remedial Response, Washington, DC.

U.S. Environmental Protection Agency (EPA) 2002c. *Guidance for Developing Quality Systems for Environmental Programs*, EPA QA/G-1. EPA/240/R-02/008, U.S. Environmental Protection Agency, Office of Environmental Information, Washington, DC, November.

U.S. Environmental Protection Agency (EPA) 2006a. *Guidance on Systematic Planning Using the Data Quality Objectives Process*, EPA QA/G-4. EPA/240/B-06/001, U.S. Environmental Protection Agency, Office of Environmental Information, Washington, DC, February.

U.S. Environmental Protection Agency (EPA) 2006b. *Data Quality Assessment: Statistical Tools for Practitioners*, EPA QA/G-9R. EPA/240/B-06/002, U.S. Environmental Protection Agency, Office of Environmental Information, Washington, DC, February.

U.S. Environmental Protection Agency (EPA) 2006c. *Data Quality Assessment: A Reviewer's Guide*, EPA QA/G-9S. EPA/240/ B-06/003, U.S. Environmental Protection Agency, Office of Environmental Information, Washington, DC, February.

U.S. Environmental Protection Agency (EPA) 2006d. *Technical Support Center for Monitoring and Site Characterization: Software for Calculating Upper Confidence Limits (UCLs), ProUCL Version 4.00.02.* http://www.epa.gov/esd/tsc/software.htm (accessed September 1, 2008).

European Commission for Nuclear Safety and the Environment (EC) 1998. *Handbook on Measurement Methods and Strategies at Very Low Levels and Activities.* Report 17624, ISBN 92-828-3163-9.

Fuji 2008. Radioactive Contamination Monitors. Hasegawa, T., Hashimoto, T. Hashimoto, M. Fuji Electronics, http://www.fujielectric.co.jp/eng/company/tech/pdf/r50-4/05.pdf (accessed September 1, 2008).

Gilbert, R.O. and Simpson, J.C. 1992. *Statistical Methods for Evaluating the Attainment of Cleanup Standards.* Pacific Northwest Laboratory Report PNL-7409, Rev. 1, http://www.osti.gov/energycitations/product.biblio.jsp?osti_id=6637176 (accessed November 10, 2008).

GUM Workbench 2006. GUM Workbench Version 1.2 Demo Installer and User Manual. http://www.gum.dk/download/download.html (accessed September 1, 2008).

Hatch, L. L., Rentos, P. G., Godbey, F. W., and Schrems, E. L. 1978. *Self-Evaluation of Occupational Safety and Health Programs.* DHEW (NIOSH) 78-187," U.S. Department of Health, Education, and Welfare.

Hobbs T.G. 2000. *Radioactivity Measurements on Glazed Ceramic Surfaces.* J. Res. Natl. Inst. Stand. Technol., 105, 275-283.

International Commission on Radiological Protection (ICRP) 1989. *Optimization and Decision-Making in Radiological Protection: Annals of the ICRP Volume 20/1.* ICRP Publication 55.

International Organization for Standardization (ISO) 1988. *Evaluation of Surface Contamination – Part 1: Beta Emitters and Alpha Emitters*. ISO-7503-1 (1st Edition), Geneva, Switzerland.

International Organization for Standardization (ISO) 1995. *Guide to the Expression of Uncertainty in Measurement*. ISO Guide 98, GUM, Geneva, Switzerland.

International Organization for Standardization (ISO) 1996. *International Vocabulary of Basic and General Terms in Metrology*. ISO Guide 99, VIM, Geneva, Switzerland, 1996.

International Organization for Standardization (ISO) 1997. *Capability of Detection – Part 1: Terms and Definitions*. ISO 11843-1, Geneva, Switzerland.

Knoll, G.F. 1999. *Radiation Detection and Measurement, 3rd Edition*. John Wiley & Sons, Inc., New York, NY.

Kragten 1994. J. Analyst, 119, 2161-2165.

Laurus 2001. "Gamma Solid Waste Monitor WM-295." Laurus Systems, Inc.. http://www.laurussystems.com/products/gamma_solid_waste_monitor.htm (accessed September 1, 2008).

Lewis V., Woods M., Burgess P., Green S., Simpson J., and Wardle J. 2000. *The Measurement Good Practice Guide No. 49: The Assessment of Uncertainty in Radiological Calibration and Testing*. National Physical Laboratory, Teddington, Middlesex, UK, TW11 0LW, February. http://eig.unige.ch/nucleaire/articles/gpg49.pdf, (accessed September 1, 2008).

MARLAP 2004. *Multi-Agency Radiological Laboratory Analytical Protocols Manual*. Nuclear Regulatory Commission NUREG-1576, Environmental Protection Agency EPA 402-B-04-001A, National Technical Information Service NTIS PB2004-105421, July. http://www.epa.gov/radiation/marlap/links.html (accessed November 10, 2008).

MARSSIM 2002. *Multi-Agency Radiation Survey and Site Investigation Manual* (Revision 1). Nuclear Regulatory Commission NUREG-1575 Rev. 1, Environmental Protection Agency EPA 402-R-97-016 Rev. 1, Department of Energy DOE EH-0624 Rev. 1, August. http://www.epa.gov/radiation/marssim/obtain.html (accessed November 10, 2008).

McCroan 2006. GUMCalc freeware. http://www.mccroan.com/GumCalc.htm (accessed September 1, 2008).

Meyer, K. and Lucas, A. 1995. "Assays of Thick Soil Samples Using Low-Resolution Alpha Spectroscopy." 41st Bioassay Analytical and Environmental Radiochemistry Conference, 1995. http://www.lanl.gov/BAER-Conference/BAERCon-41p038.pdf (accessed September 1, 2008).

Miller E,. Peters J., Nichols D. 2000. "Release Surveying of Scrap Metals with the IonSens™ Conveyor." Waste Management 2000 Conference, Tucson, AZ, February. http://www.bilsolutions.co.uk/file_download.php?file_id=4 (accessed September 1, 2008).

National Academy of Sciences (NAS) 1999. *Evaluation of Guidelines for Technologically Enhanced Naturally Occurring Radioactive Materials (TENORM)*. Committee on Evaluation of EPA Guidelines for Exposure to Naturally Occurring Radioactive Materials, National Research Council, National Academy of Sciences, National Academy Press, p. 72.

National Bureau of Standards (NBS) 1963. *Experimental Statistics*. NBS Handbook 91, National Bureau of Standards, Gaithersburg, MD.

National Council on Radiation Protection and Measurements (NCRP) 1988a. *Exposure of the Population in the United States and Canada from Natural Background Radiation*. NCRP Report No. 94, Bethesda, MD.

National Council on Radiation Protection and Measurements (NCRP) 1988b. *Exposure of the U.S. Population from Consumer Products and Miscellaneous Sources*. NCRP Report No. 95, Bethesda, MD.

National Council on Radiation Protection and Measurements (NCRP) 1993. *Limitation of Exposure to Ionizing Radiation*. NCRP Report No. 116, MD.

National Council on Radiation Protection and Measurements (NCRP) 2002. *Managing Potentially Radioactive Scrap Metal*. NCRP Report No. 141, Bethesda, MD.

National Institute of Standards and Technology (NIST) 1994. *Guidelines for Evaluating and Expressing the Uncertainty of NIST Measurement Results*. NIST Technical Note 1297, Gaithersburg, MD. http://physics.nist.gov/Document/tn1297.pdf (accessed September 1, 2008).

National Institute of Standards and Technology (NIST) 1996. *Tables of X-Ray Mass Attenuation Coefficients and Mass Energy-Absorption Coefficients*. Hubbell, J.H. and Seltzer, S.M., Gaithersburg, MD, April. http://physics.nist.gov/PhysRefData/XrayMassCoef/cover.html (accessed September 1, 2008).

National Institute of Standards and Technology (NIST) 1998. *XCOM: Photon Cross Sections Database*. http://physics.nist.gov/PhysRefData/Xcom/Text/XCOM.html (accessed September 1, 2008).

National Institute of Standards and Technology (NIST) 2006. "NIST/SEMATECH e-Handbook of Statistical Methods." http://www.itl.nist.gov/div898/handbook/ (accessed September 1, 2008).

Novelec 2001a. "Technical Information Systems and Machines." Novelec Nuclear Instrumentation, 2001. http://www.novelec.fr/descriptionUK-CRL01.htm#T2C%20and%20T2C-AB (accessed September 1, 2008).

Novelec 2001b. "Technical Information Radiation Survey." Novelec Nuclear Instrumentation, 2001. http://www.novelec.fr/descriptionUK-MSA.htm#MSA-P and MSA-C (accessed September 1, 2008).

U.S. Nuclear Regulatory Commission (NRC) 1977. *Operating Philosophy for Maintaining Occupational Radiation Exposures As Low As Is Reasonably Achievable*. Regulatory Guide 8.10. Washington, DC.

U.S. Nuclear Regulatory Commission (NRC) 1982. *Information Relevant to Ensuring that Occupational Radiation Exposures at Medical Institutions Will Be As Low As Reasonably Achievable*. Regulatory Guide 8.18. Washington, DC.

U.S. Nuclear Regulatory Commission (NRC) 1984. *Lower Limit of Detection: Definition and Elaboration of a Proposed Position for Radiological Effluent and Environmental Measurements*. NUREG/CR-4007. Washington, DC.

U.S. Nuclear Regulatory Commission (NRC) and Oak Ridge Associated Universities 1992. *Manual for Conducting Radiological Surveys in Support of License Termination, Draft Report for Comment*. NUREG/CR-5849. Washington, DC.

U.S. Nuclear Regulatory Commission (NRC) 1993. *ALARA Levels for Effluents from Materials Facilities*. Regulatory Guide 8.37. Washington, DC.

U.S. Nuclear Regulatory Commission (NRC) 1994. *Background as a Residual Radioactivity Criterion for Decommissioning, Draft Report*. NUREG-1501. Office of Nuclear Regulatory Research, Washington, DC.

U.S. Nuclear Regulatory Commission (NRC) 1995. *Proposed Methodologies for Measuring Low Levels of Residual Radioactivity for Decommissioning*. NUREG-1506, Draft Report for Comment. Washington, DC.

U.S. Nuclear Regulatory Commission (NRC) 1998a. *A Nonparametric Statistical Methodology for the Design and Analysis of Final Status Decommissioning Surveys* (Revision 1). NUREG-1505 Rev. 1. Office of Nuclear Regulatory Research, Washington, DC.

U.S. Nuclear Regulatory Commission (NRC) 1998b. *Minimum Detectable Concentrations with Typical Radiation Survey Instruments for Various Contaminants and Field Conditions*. NUREG-1507. Office of Nuclear Regulatory Research, Washington, DC.

U.S. Nuclear Regulatory Commission (NRC) 2000. *Low-Level Waste Classification, Characterization, and Assessment: Waste Streams and Neutron-Activated Metals*. NUREG/CR-6567, D.E. Robertson, C.W. Thomas, S.L. Pratt, E.A. Lepel, and V.W. Thomas, Pacific Northwest Laboratory.

U.S. Nuclear Regulatory Commission (NRC) 2002a. *Radiological Surveys for Controlling Release of Solid Materials*. NUREG-1761. Office of Nuclear Regulatory Research, Washington, DC.

U.S. Nuclear Regulatory Commission (NRC) 2002b. *Information Relevant to Ensuring that Occupational Radiation Exposures at Uranium Mills Will Be As Low As Is Reasonably Achievable*. Regulatory Guide 8.31. Washington, DC.

U.S. Nuclear Regulatory Commission (NRC) 2003a. *Radiological Assessments for Clearance of Materials from Nuclear Facilities*. NUREG-1640. Office of Nuclear Regulatory Research, Washington, DC.

U.S. Nuclear Regulatory Commission (NRC) 2003b. *Radiological Toolbox Version 1.0.0.* Eckerman, K.F. and Sjoreen, A.L. http://www.nrc.gov/about-nrc/regulatory/research/radiological-toolbox.html (accessed September 1, 2008).

Occupational Safety and Health Administration (OSHA) 2002. *Job Hazard Analysis.* OSHA Publication 3071. http://www.osha.gov/Publications/osha3071.pdf (accessed September 1, 2008).

Pacific Northwest Laboratory (PNL) 1988. *Health Physics Manual of Good Practices for Reducing Radiation Exposure to Levels that are As Low As Reasonably Achievable (ALARA).* Munson, L.H., et al. PNL-6577, UC-610, , Richland, Washington. http://www.pnl.gov/bayesian/refs/ALARAMAN88.PDF (accessed September 1, 2008).

Srom, D.J. *Counting Statistics Utility for Comparing Eight Decision Rules.* http://www.pnl.gov/bayesian/strom/strompub.htm (accessed September 1, 2008).

Thermo 2005. "SAM11 Small articles and tools monitor for low level gamma measurement" Thermo Fisher Scientific, Inc. http://www.thermo.com/eThermo/CMA/PDFs/Product/productPDF_23747.pdf (accessed September 1, 2008).

Thermo 2008. "FHT 3031 CCM Contamination Monitor." Thermo Fisher Scientific, Inc.. http://www.thermo.com/com/cda/product/detail/0,1055,15815,00.html (accessed September 1, 2008).

United Nations Scientific Committee on the Effects of Atomic Radiation (UNSCEAR) 2000. *Sources and Effects of Ionizing Radiation.* Report to the General Assembly, ISBN 92-1-142238-8. http://www.unscear.org/unscear/en/publications.html (accessed September 1, 2008).

Vetter, T.W. 2006. *Quantifying Measurement Uncertainty in Analytical Chemistry – A Simplified Practical Approach.* http://www.cstl.nist.gov/acd/839.03/Uncertainty.pdf (accessed September 1, 2008).

GLOSSARY

Accessible Area is an area that can be easily reached or obtained. In many cases an area must be physically accessible to perform a measurement. However, radioactivity may be measurable even if an area is not physically accessible. See in this glossary *measurable radioactivity*.

Action Level is the numerical value that causes a decision-maker to choose one of the alternative actions. In the context of MARSAME, the numerical value is the radionuclide concentration or level of radioactivity corresponding to the disposition criterion, and the alternative actions are determined by the selection of a disposition option.

Alternative Action is the choice between two mutually exclusive possibilities. See in this glossary *decision rule*.

Ambient Radiation is radiation that is currently present in the surrounding area. Ambient radiation may include natural background, intrinsic radiation from surrounding materials, intrinsic radiation from the item(s) being measured, contamination, or radiation from nearby machines (e.g., x-ray machines when operating) depending on the local conditions. Ambient radiation changes with season, time, location, weather, and other environmental conditions.

Background Radiation (as defined in Nuclear Regulatory Commission regulations) is radiation from cosmic sources; naturally occurring radioactive material including radon (except as a decay product of source or special nuclear material); and global fallout as it exists in the environment from the testing of nuclear explosive devices or from past nuclear accidents such as Chernobyl that contribute to background radiation and are not under the control of the licensee. "Background radiation" does not include radiation from source, byproduct or special nuclear materials regulated by the Nuclear Regulatory Commission (10 CFR 20.1003). See in this glossary *distinguishable from background*.

Biased Measurements are measurements performed at locations selected using professional judgment based on unusual appearance, location relative to known contamination areas, high potential for residual radioactivity, and general supplemental information. Biased measurements are not included in the statistical evaluation of survey unit data because they violate the assumption of randomly selected, independent measurements. Instead, biased measurement results are individually compared to the action levels. Biased measurements are also called judgment measurements (MARSSIM 2002).

Calibration Function is the function that relates the net instrument signal to activity (e.g., relates counts to disintegrations or radiations).

Categorization is the act of determining whether M&E are impacted or non-impacted. This is a departure from MARSSIM where this decision was included in the definition of *classification*.

Class 1 M&E are impacted M&E that have, or had, the following: (1) highest potential for, or known, radionuclide concentration(s) or radioactivity above the action level(s); (2) highest potential for small areas of elevated radionuclide concentration(s) or radioactivity; and (3) insufficient evidence to support reclassification as Class 2 or Class 3. Such potential may be based on historical information and process knowledge, while known radionuclide concentration(s) or radioactivity may be based on preliminary surveys. See in this glossary *Class 2, Class 3, classification,* and *impacted.*

Class 2 M&E are impacted M&E that have, or had, the following: (1) low potential for radionuclide concentration(s) or radioactivity above the action level(s); and (2) little or no potential for small areas of elevated radionuclide concentration(s) or radioactivity. Such potential may be based on historical information, process knowledge, and preliminary surveys. See in this glossary *Class 1, Class 3, classification,* and *impacted.*

Class 3 M&E are impacted M&E that have, or had, the following: (1) little or no potential for radionuclide concentrations(s) or radioactivity above background; and (2) insufficient evidence to support categorization as non-impacted. See in this glossary *Class 1, Class 2, classification, impacted,* and *non-impacted.*

Classification is the act or result of separating impacted M&E or survey units into one of three designated classes: Class 1, Class 2, or Class 3. Classification is the process of determining the appropriate level of survey effort based on estimates of activity levels and comparison to action levels, where the activity estimates are provided by historical information, process knowledge, and preliminary surveys. See in this glossary *Class 1, Class 2, Class 3,* and *impacted.*

Clearance is the removal of radiological regulatory controls from materials and equipment. Clearance is a subset of release. See in this glossary *release, restricted release,* and *unrestricted release.*

Combined Standard Uncertainty is the standard uncertainty of an output estimate calculated by combining the standard uncertainties of the input estimates. The combined standard uncertainty of y is denoted by $u_c(y)$. See also in this glossary *expanded uncertainty, input estimate, measurement method uncertainty, output estimate,* and *standard uncertainty.*

Combined Variance is the square of the combined standard uncertainty. The combined variance of y is denoted by $[u_c(y)]^2$. See in this glossary *combined standard uncertainty.*

Concentration is activity per unit mass or volume (e.g., Bq/kg, pCi/g, or Bq/m^3) or activity per unit area (e.g., Bq/m^2 or dpm/100 cm^2).

Conceptual Model is an idealized model or map of a component or area to be surveyed and the associated radionuclides or radioactivity expected to be present, and is intended to aid in describing or designing the survey. The initial conceptual model is based on the results of the initial assessment. Additional data is used to update the conceptual model iteratively throughout the development, implementation, and assessment of the disposition survey. See in this glossary *initial assessment.*

Coverage Factor (k) is the value multiplied by the combined standard uncertainty $u_c(y)$ to give the expanded uncertainty, U. See in this glossary *combined standard uncertainty* and *expanded uncertainty.*

Coverage Probability is the approximate probability that the reported uncertainty interval will contain the value of the measurand. See in this glossary *measurand.*

Critical Value in the context of radiation detection is the minimum measured value (e.g., of the instrument signal or the radionuclide concentration) required to give a specified probability that a positive (non-zero) amount of radioactivity is present in the material being measured. The critical value is the same as the critical level or decision level in publications by Currie (Currie 1968 and NRC 1984).

Data Life Cycle is the process of planning the survey, implementing the survey plan, and assessing the survey results prior to making a decision (MARSSIM 2002).

Data Quality Objectives (DQOs) are qualitative and quantitative statements derived from the DQO process that clarify [the survey] technical and quality objectives, define the appropriate type of data, and specify tolerable levels of potential decision errors that will be used as the basis for establishing the quality and quantity of data needed to support decisions (MARSSIM 2002).

Data Quality Objectives Process is a systematic strategic planning tool based on the scientific method that identifies and defines the type, quality, and quantity of data needed to satisfy a specific use (MARSSIM 2002). See also in this glossary *data quality objectives*.

Data Quality Assessment (DQA) is a scientific and statistical evaluation that determines whether data are the right type, quality and quantity to support their intended use (EPA 2006b).

Decision Rule is an "if...then" statement consisting of three parts: action level(s), parameter of interest, and alternative actions. A theoretical decision rule is developed early in the planning process assuming ideal data are available to support a disposition decision (see Chapter 3). An operational decision rule is developed based on the measurements that will be performed as part of the final disposition survey (see Chapter 4).

Detection Capability is a generic term describing the capability of a measurement process to distinguish small amounts of radioactivity from zero. It may be expressed in terms of the minimum detectable concentration. See in this glossary *minimum detectable concentration*.

Difficult-to-Measure Radioactivity is radioactivity that is not measurable until the M&E to be surveyed is prepared. Preparation of M&E may be relatively simple (e.g., cleaning) or more complicated (e.g., disassembly or complete destruction). Given sufficient resources, all radioactivity can be made measurable; however, it is recognized that increased survey costs can outweigh the benefit of some dispositions.

Discrimination Limit is the level of radioactivity selected by the members of the planning team that can be reliably distinguished from the action level. The lower bound of the gray region (LBGR) for Scenario A and the upper bound of the gray region (UBGR) for Scenario B are examples of discrimination limits. See also in this glossary *lower bound of the gray region*, *upper bound of the gray region*, *Scenario A*, and *Scenario B*.

Disposition is the future use, fate, or final location for something (e.g., recycle, reuse, disposal).

Disposition Decision is the selection among alternative actions to determine acceptable future use, fate, or final location for something (e.g., recycle, reuse, disposal)..

Disposition Survey is a radiological survey designed to collect information to support a disposition decision.

Distinguishable from Background is the radionuclide concentration or radioactivity that is statistically different from the background level of that radionuclide concentration or radioactivity in similar M&E. See in this glossary *background radiation, measurable radioactivity, minimum detectable concentration, measurement quality objectives.*

Energy Resolution is the quantifiable ability of a measurement method to distinguish between radiations with different energies.

Environmental Radioactivity is radioactivity from the environment where the M&E are located. Environmental radioactivity includes background radiation as well as inherent radioactivity and radioactivity from nearby sources.

Evaluation Function is a mathematical expression that allows the user to compare options and draw a conclusion or calculate a result.

Expanded Uncertainty is the product, U, of the combined standard uncertainty of a measured value, y, and a coverage factor, k, chosen so that the interval from $y - U$ to $y + U$ has a desired high probability of containing the value of the measurand. See in this glossary *combined standard uncertainty, coverage factor,* and *measurand.*

Fluence is the number of photons or particles passing through a cross-sectional area. The international standard (SI) unit for fluence is m^{-2}.

Frequency Plot is a chart plotting the number of data points against their measured values.

Graded Approach is the process of basing the level of application of managerial controls applied to an item or work according to the intended use of the results and the degree of confidence needed in the quality of the results. See in this glossary *data quality objectives process.*

Gray Region is the range of radionuclide concentrations or quantities between the discrimination limit and the action level, where the consequence of making a decision error is relatively minor. See in this glossary *action level, discrimination limit, lower bound of the gray region,* and *upper bound of the gray region.*

Impacted is a term applied to M&E that are not classified as non-impacted. M&E with a reasonable potential to contain radionuclide concentration(s) or radioactivity above background are considered impacted (10 CFR 50.2). See in this glossary *background radiation* and *non-impacted.*

Inherent Radioactivity is radioactivity resulting from radionuclides that are an essential constituent of the material being measured (e.g., ^{40}K in fertilizer containing potassium).

Initial Assessment (IA) is an investigation to collect existing information describing materials and equipment and is similar to the historical site assessment (HSA) described in MARSSIM.

Input Quantity is any of the quantities in a mathematical measurement model whose values are measured and used to calculate the value of another quantity, called the output variable.

Instrument Efficiency is the ratio between the instrument net reading and the surface emission rate of a source under given geometrical conditions (ISO 1988). For a given instrument, the instrument efficiency depends on the energy of the radiations emitted by the source. See in this glossary *source efficiency* and *total efficiency*.

Interdiction is an increase in the level of radiological control or a decision not to accept control from another party. Examples of interdiction include identification of radioactive material that results in the initiation of radiological controls or identification of unauthorized movement of radioactive material.

Interdiction Survey is the collection of data to support an interdiction decision regarding M&E. In general, interdiction surveys are used to accept or refuse to accept control of M&E that are potentially radioactive. The goal of an interdiction survey often is to detect radioactive M&E that should be controlled. In some cases, an interdiction survey may result in the impoundment of radioactive M&E that represent an unacceptable risk to human health or the environment.

Interference is the presence of other radiation or radioactivity, chemicals, background noise, instrument noise, or other factors that hinders the ability to analyze for the radiation or radioactivity of interest.

Intrinsic Radioactivity See in this glossary *inherent radioactivity.*

Lower Bound of the Gray Region (LBGR) is the radionuclide concentration or level of radioactivity that corresponds with the lowest value in the range where the consequence of decision errors is relatively minor. For Scenario A, the LBGR corresponds to the discrimination limit. For Scenario B, the LBGR corresponds to the action level. See in this glossary *action level, discrimination limit, gray region, Scenario A*, and *Scenario B*.

Mathematical Model is the general characterization of a process, object, or concept in terms of mathematics, which enables the relatively simple manipulation of variables to be accomplished in order to determine how the process, object, or concept would behave in different situations.

Materials and Equipment (M&E) are items considered for disposition that include metals, concrete, dispersible bulk materials, tools, equipment, piping, conduit, furniture, solids, liquids, and gases in containers, etc. M&E are considered non-real property distinguishable from buildings and land, which are considered real property. See in this glossary *disposition* and *non-real property*.

Measurand is a particular quantity subject to measurement (ISO 1996).

Measurement Method Uncertainty See in this glossary *method uncertainty.*

Measurement Quality Objectives (MQOs) are a statement of a performance objective or requirement for a particular method performance characteristic (MARLAP 2004).

Measurement Standard Deviation See in this glossary *standard deviation (of measurement).*

Measurement Uncertainty See in this glossary *uncertainty (of measurement).*

Measurable Radioactivity is radioactivity that can be quantified using known or predicted relationships developed from historical information, process knowledge, or preliminary measurements as long as the relationships are developed, verified, and validated as specified in the *data quality objectives* (DQOs) and *measurement quality objectives* (MQOs).

Median is the middle value of the data set when the number of data points is odd, or the average of the two middle values when the number of data points is even.

Method Uncertainty, u_M, is the predicted uncertainty of the measured value that would be calculated if the method were applied to a hypothetical sample with specified concentration.

Minimum Detectable Activity (MDA) is the minimum detectable value of activity for a measurement. See in this glossary *minimum detectable value*.

Minimum Detectable Concentration (MDC) is the minimum detectable value of the radionuclide or radioactivity concentration for a measurement. See in this glossary *minimum detectable value*.

Minimum Detectable Value is an estimate of the smallest true value of the measurand that ensures a specified high probability, $1 - \beta$, of detection. This definition presupposes that an appropriate detection criterion has been specified (e.g., critical value). See in this glossary *measurand* and *critical value*.

Minimum Quantifiable Concentration (MQC) is the smallest value of the concentration that ensures the relative standard deviation of a measurement of M&E with that concentration does not exceed a specified value, usually 10%.

Non-impacted is a term applied to M&E where there is no reasonable potential to contain radionuclide concentration(s) or radioactivity above background (10 CFR 50.2). See in this glossary *background radiation* and *impacted*.

Non-Real Property is property that is not real property. See in this glossary *real property* and *materials and equipment (M&E)*.

Null Hypothesis, or baseline condition, is a tentative assumption about the true, but unknown, radionuclide concentration or level of radioactivity that can be retained or rejected based on the available evidence. When hypothesis testing is applied to disposition decisions, the data are used to select between a presumed baseline condition (the null hypothesis) and an alternate condition (the alternative hypothesis). The null hypothesis is retained until evidence demonstrates with a previously specified probability that the baseline condition is false.

Output Quantity is the quantity in a mathematical measurement model whose value is calculated from the measured values of other quantities in the model. See in this glossary *input quantity*.

Planning Team is the group of people who perform the DQO process. Members may include the decision-maker (senior manager), site manager, representatives of other data users, senior program and technical staff, someone with statistical expertise, and a quality assurance and quality control advisor (such as a QA manager) (EPA 2000a).

Posting Plot is a map of the survey unit with the data values entered at the measurement locations. This type of plot potentially reveals heterogeneities in the data, especially possible patches of elevated radioactivity.

Preliminary Survey is any survey performed prior to the disposition survey in MARSAME, and is generally performed to provide information required to support the design of the disposition survey. See also in this glossary *disposition survey*.

Process Knowledge is information concerning the characteristics, history of prior use, and inherent radioactivity of the materials and equipment being considered for release. Process knowledge is obtained through a review of the operations conducted in facilities or areas where materials and equipment may have been located and the processes where the materials and equipment were involved.

Radioactive Materials consist of any material, equipment or system component determined or suspected to contain radionuclides in excess of inherent radioactivity. Radioactive material includes activated material, sealed and unsealed sources, and substances that emit radiation. See in this glossary *inherent radioactivity*.

Radiological Controls are any means, method or activity (including engineered or administrative) designed to protect personnel or the environment from exposure to a radiological risk.

Radionuclides or Radiations of Concern are radionuclides or radiations that are present at a concentration or activity that may pose an unacceptable risk to human health or the environment. In MARSAME, the term radionuclides or radiations of concern is used to describe the radionuclides or radiations that are actually measured during the disposition survey. See also in this glossary *radionuclides or radiations of potential concern* and *disposition survey*.

Radionuclides or Radiations of Potential Concern are radionuclides or radiations that are identified during the initial assessment as potentially being associated with the M&E being investigated. See also in this glossary *initial assessment* and *radionuclides or radiations of concern*.

Ratemeter is an instrument that indicates the counting rate of an electronic counter. In the context of radiological measurements, a ratemeter displays the counting rate from a radiation detector. The averaging time for calculating the rate is determined by the time constant of the meter. See in this glossary *scaler*.

Real Property, in the MARSAME context, means developed or undeveloped land, fixed buildings and structures, or surface and subsurface soil remaining in place. Real property is outside the scope of MARSAME. See in this glossary *materials and equipment (M&E)* and *non-real property*.

Recycle is beneficial reuse of constituent materials incorporated within the M&E. A hammer that is melted down as scrap metal so the component metals can be reused is an example of recycle.

Reference Material is material of similar physical, radiological, and chemical characteristics as the M&E considered for disposition. Reference material provides information on the level of radioactivity that would be present if the M&E being investigated had not been radiologically impacted. See in this glossary *impacted*.

Relative Standard Uncertainty is the ratio of the standard uncertainty of a measured result to the result itself. The relative standard uncertainty of x may be denoted by $u_r(x)$. See in this glossary *standard uncertainty*.

Release is a reduction in the level of radiological control, or a transfer of control to another party. Release includes clearance. Examples of release (other than clearance) include recycle, reuse, disposal as waste, or transfer of control of radioactive M&E from one authorized user to another. See also in this glossary *reuse*, *recycle*, *restricted release*, and *clearance*.

Release Survey is a type of disposition survey designed to collect information to support a release decision. See also in this glossary *disposition survey* and *release*.

Restricted Release is a reduction in the level of radiological control, or transfer of control to another party, where restrictions are placed on how the released items will be used or transferred. Maintaining a tool crib in a radiologically controlled area restricts reuse of those tools to that radiologically controlled area, and tools returned to the tool crib represent a restricted release of those tools. See also in this glossary *reuse*, *recycle*, *release*, and *clearance*.

Reuse is the continued use of M&E for their original purpose(s). An example of reuse is a hammer that continues to be used as a hammer.

Ruggedness is the relative stability of a measurement technique's performance when small variations in method parameter values are made.

Sampling Standard Deviation, σ_S, is the theoretical true value of the variability of radionuclide concentration or radioactivity in space and time (i.e., the variation of the true but unknown concentrations from place to place and from time to time). The extent of the survey unit, the physical sizes of the measured material, and the choice of measurement locations affects the sampling standard deviation.

Scaler is an electronic counter that displays the aggregate of a number of signals, which usually occur too rapidly to be recorded individually. In the context of radiological measurements, a scaler records the number of counts from a radiation detector over a specified time interval. See in this glossary *ratemeter*.

Scenario A uses a null hypothesis that assumes the level of radioactivity associated with the M&E exceeds the action level. Scenario A is sometimes referred to as "presumed not to comply" or "presumed not clean."

Scenario B uses a null hypothesis that assumes the level of radioactivity associated with the M&E is less than or equal to the action level. Scenario B is sometimes referred to as "indistinguishable from background" or "presumed clean."

Secular Equilibrium is the condition in which the initial member of the decay series has a longer half-life than any subsequent members of the series. Secular equilibrium is achieved when the activities for all members of the decay series are equal to the activity of the precursor radionuclide.

Segregation is the process of separating or isolating from a main body or group. In the context of disposition surveys, segregation is based on the physical and radiological attributes of the M&E being investigated and is used to help control measurement method uncertainty.

Sensitivity Coefficient for an input estimate, x_i, used to calculate an output estimate, $y = f(x_1, x_2, ..., x_N)$, is the value of the partial derivative, $\partial f/\partial x_i$, evaluated at $i = x_1, x_2, ..., x_N$. The sensitivity coefficient represents the ratio of the change in y to a small change in x_i.

Sentinel Measurement is a biased measurement performed at a key location to provide information specific to the objectives of the *initial assessment* (IA).

Significance Level is, in the context of a hypothesis test, a specified upper limit for the probability of a *Type I decision error*.

Sign Test is a non-parametric statistical test used to evaluate radionuclide-specific disposition survey results if the radionuclide being measured is not present in background, or is present at such a small fraction of the action level as to be considered insignificant.

Smear is a non-quantitative test for the presence of removable radioactive materials in which the suspected surface or area is wiped with a filter paper or other substance, which is then tested for the presence of radioactivity. The surface area tested may be related to the release criterion. Smear is also referred to as a smear test, swipe, or wipe.

Source Efficiency is the ratio between the number of particles of a given type above a given energy emerging from the front face of a source or its window per unit time and the number of particles of the same type created or released within the source (for a thin source) or its saturation layer thickness (for a thick source) per unit time (ISO 1988). See also in this glossary *instrument efficiency* and *total efficiency*.

Specific Activity is the radioactivity per unit mass for a specified radionuclide.

Specificity is the ability of the measurement method to measure the radionuclide of concern in the presence of interferences.

Spectrometry is a measurement across a range of energies. The measurement of alpha particles by energy is called alpha spectrometry.

Spectroscopy is the measurement and analysis of electromagnetic spectra produced as the result of the emission or absorption of energy by various substances. The measurement of gamma-ray emissions from a substance is called gamma spectroscopy.

Standard Operating Procedure (SOP) is a written document that details the method for an operation, analysis, or action with thoroughly prescribed techniques and steps, and that is officially approved as the method for performing certain routine or repetitive tasks (MARSSIM 2002).

Standard Deviation (of Measurement), σ_M, is a theoretical parameter describing the variability in the distribution of the measurement. See also in this glossary *uncertainty (of measurement)*.

Standard Uncertainty is the uncertainty of a measured value expressed as an estimated standard deviation, often called a "1-sigma" (1σ) uncertainty (MARLAP 2004). The standard uncertainty of a value x is denoted by $u(x)$. See also in this glossary *uncertainty (of measurement)*.

Standardized Initial Assessment is a set of instructions or questions that are used to perform the initial assessment, usually documented in a standard operating procedure. See also in this glossary *initial assessment* and *standard operating procedure*.

Surficial Radioactive Material is radioactive material distributed on any of the surfaces of a solid object. Surficial radioactive material may be either removable by non-destructive means (such as casual contact, wiping, brushing, or washing) or fixed to the surface.

Surrogate Measurement is a measurement where one radionuclide is quantified and used to demonstrate compliance with the release criterion for additional radionuclide(s) based on known or accepted relationships between the measured radionuclide and unmeasured radionuclide(s).

Survey Unit for M&E is the specific lot, amount, or piece of M&E on which measurements are made to support a disposition decision concerning the same specific lot, amount, or piece of M&E.

Total Efficiency is the product of the instrument efficiency and the source efficiency. See in this glossary *instrument efficiency* and *source efficiency*.

Traceability is the "property of the result of a measurement or the value of a standard whereby it can be related to stated references, usually national or international standards, through an unbroken chain of comparisons all having stated uncertainties" (ISO 1996).

Type I Decision Error occurs when the null hypothesis is rejected when it is actually true. The Type I decision error rate, or significance level, is represented by α. See in this glossary *null hypothesis* and *significance level*.

Type II Decision Error occurs when the null hypothesis is not rejected when it is actually false. The Type II decision error rate is denoted by β. See in this glossary *null hypothesis*.

Uncertainty (of Measurement), $u(x)$, is a parameter, associated with the result of a measurement, x, that characterizes the dispersion of the values that could reasonably be attributed to the measurement of x. It is the estimated value of $\sigma(x)$ obtained from the propagation of uncertainty. See also in this glossary See also in this glossary *standard deviation (of measurement)*.

Unrestricted Release is the removal of radiological regulatory controls from materials and equipment. See in this glossary *release* and *clearance*.

Upper Bound of the Gray Region (UBGR) is the radionuclide concentration or level of radioactivity that corresponds with the highest value in the range where the consequence of decision errors is relatively minor. For Scenario A, the UBGR corresponds to the action level. For Scenario B, the UBGR corresponds to the discrimination limit. See in this glossary *action level*, *discrimination limit*, *gray region*, *Scenario A*, and *Scenario B*.

Volumetric Radioactive Material is radioactive material that is distributed throughout or within the materials or equipment being measured, as opposed to a surficial distribution. Volumetric radioactive material may be homogeneously (e.g., uniformly activated metal) or heterogeneously (e.g., activated reinforced concrete) distributed throughout the M&E.

Wilcoxon Rank Sum (WRS) Test is a non-parametric statistical test used to evaluate disposition survey results when the radionuclide being measured is present in background by comparing the results to measurements performed using an appropriately chosen reference material.